Soccer Analytics

Sports analytics is on the rise, with top soccer clubs, bookmakers, and broadcasters all employing statisticians and data scientists to gain an edge over their competitors.

Many popular books have been written exploring the mathematics of soccer. However, few provide details on how soccer data can be analysed in real life. This book addresses this issue via a practical 'route one' approach designed to show readers how to successfully tackle a range of soccer-related problems using the easy-to-learn computer language R. Through a series of easy-to-follow examples, this book explains how R can be used to:

- Download and edit soccer data
- Produce graphics and statistics
- Predict match outcomes and final league positions
- Formulate betting strategies
- Rank teams
- Construct passing networks
- Assess match play

Soccer Analytics: An Introduction Using R is a comprehensive introduction to soccer analytics aimed at all those interested in analysing soccer data, be they fans, gamblers, coaches, sports scientists, or data scientists and statisticians wishing to pursue a career in professional soccer. It aims to equip the reader with the knowledge and skills required to confidently analyse soccer data using R, all in a few easy lessons.

Clive Beggs is Emeritus Professor of Applied Physiology in the Carnegie School of Sport at Leeds Beckett University in the UK. He is both a physiologist and a bio-engineer who has worked for many years with leading research teams around the world on a wide variety of medical and sport-related projects – publishing many scientific papers in both fields. With a background in mathematical modelling of clinical and biological systems, he also has expertise in data analysis and machine learning, which he regularly uses in his sport performance work. Clive is both an amateur runner and a soccer fan, and it is his lifelong interest in sport and mathematics that has prompted him to write this book.

Chapman & Hall/CRC Data Science Series

Reflecting the interdisciplinary nature of the field, this book series brings together researchers, practitioners, and instructors from statistics, computer science, machine learning, and analytics. The series will publish cutting-edge research, industry applications, and textbooks in data science.

The inclusion of concrete examples, applications, and methods is highly encouraged. The scope of the series includes titles in the areas of machine learning, pattern recognition, predictive analytics, business analytics, Big Data, visualization, programming, software, learning analytics, data wrangling, interactive graphics, and reproducible research.

Recently Published Titles

Practitioner's Guide to Data Science
Hui Lin and Ming Li

Natural Language Processing in the Real World: Text Processing, Analytics, and Classification
Jyotika Singh

Telling Stories with Data: With Applications in R
Rohan Alexander

Big Data Analytics: A Guide to Data Science Practitioners Making the Transition to Big Data
Ulrich Matter

Data Science for Sensory and Consumer Scientists
Thierry Worch, Julien Delarue, Vanessa Rios De Souza, and John Ennis

Data Science in Practice
Tom Alby

Introduction to NFL Analytics with R
Bradley J. Congelio

Soccer Analytics: An Introduction Using R
Clive Beggs

For more information about this series, please visit: https://www.routledge.com/Chapman--HallCRC-Data-Science-Series/book-series/CHDSS

Soccer Analytics

An Introduction Using R

Clive Beggs

CRC Press is an imprint of the
Taylor & Francis Group, an **informa** business
A CHAPMAN & HALL BOOK

First edition published 2024
by CRC Press
2385 NW Executive Center Drive, Suite 320, Boca Raton FL 33431

and by CRC Press
4 Park Square, Milton Park, Abingdon, Oxon, OX14 4RN

CRC Press is an imprint of Taylor & Francis Group, LLC

Library of Congress Cataloging-in-Publication Data
Names: Beggs, Clive, author.
Title: Soccer analytics : an introduction using R / Clive Beggs.
Description: Boca Raton, FL : CRC Press, 2024. | Series: Data science
series | Includes index. | Summary: “Sports analytics is on the rise, with top soccer clubs, bookmakers, and broadcasters all employing statisticians and data scientists to gain an edge over their competitors. Many popular books have been written exploring the mathematics of soccer. However, few supply details on how soccer data can be analysed in real-life. The book addresses this issue via a practical route one approach designed to show readers how to successfully tackle a range of soccer related problems using the easy-to-learn computer language R. Through a series of easy-to-follow examples, the book explains how R can be used to: Download and edit soccer data Produce graphics and statistics Predict match outcomes and final league positions Formulate betting strategies Rank teams Construct passing networks Assess match play Soccer Analytics: An Introduction Using R is a comprehensive introduction to soccer analytics aimed at all those interested in analysing soccer data, be they fans, gamblers, coaches, sports scientists, or data scientists and statisticians wishing to pursue a career in professional soccer. It aims to equip the reader with the knowledge and skills required to confidently analyse soccer data using R, all in a few easy lessons”— Provided by publisher.
Identifiers: LCCN 2023035054 (print) | LCCN 2023035055 (ebook) |
ISBN 9781032357836 (hardcover) | ISBN 9781032357584 (paperback) | ISBN 9781003328568 (ebook)
Subjects: LCSH: Soccer—Statistical methods. | R (Computer program
language)—Statistical methods. | Sports science.
Classification: LCC GV943 .B4174 2024 (print) | LCC GV943 (ebook) |
DDC 796.3340072/7—dc23/20231027
LC record available at https://lccn.loc.gov/2023035054
LC ebook record available at https://lccn.loc.gov/2023035055

ISBN: 9781032357836 (hbk)
ISBN: 9781032357584 (pbk)
ISBN: 9781003328568 (ebk)

DOI: 10.1201/9781003328568

Typeset in Plantino
by codeMantra

Disclaimer

Although the publisher and the author have made every effort to ensure that the information in this book was correct at press time and while this publication is designed to provide accurate information in regard to the subject matter covered, the publisher and the author assume no responsibility for errors, inaccuracies, omissions, or any other inconsistencies herein and hereby disclaim any liability to any party for any loss, damage, or disruption caused by errors or omissions, whether such errors or omissions result from negligence, accident, or any other cause.

This book is purely for educational purposes. Users of the R code contained within this book do so at their own risk, with the author bearing no responsibility for the use to which the code is applied.

The accompanying files contained in the SoccerAnalytics GitHub repository are covered by the MIT licence, which permits the user to copy, modify, merge, and utilise the software free of charge. Under the MIT licence, "the software is provided 'as is', without warranty of any kind, express or implied, including but not limited to the warranties of merchantability, fitness for a particular purpose, and noninfringement. In no event shall the authors or copyright holders be liable for any claim, damages, or other liability, whether in an action of contract, tort, or otherwise, arising from, out of, or in connection with the software or the use or other dealings in the software."

Contents

Examples Index

5. Predicting End-of-Season League Position

6. Predicting Soccer Match Outcomes

7. Betting Strategies

8. Who Are the Key Players? Using Passing Networks to Analyse Match Play

9. Which Is the Best Team? Ranking Systems in Soccer

Preface

Soccer fans love numbers and statistics, and they will spend hours discussing how many goals various players have scored or where a particular team is likely to finish in the league. Perhaps this is not too surprising because soccer leagues are essentially mathematical constructs in which numbers represent the difference between success and failure. After all, in the English Premier League at the end of the 2018–2019 season, Liverpool finished with a record total of 97 points, which normally would have been cause for great celebration – except that on this occasion, Manchester City won the league with 98 points! So, the margin between success and failure can be very fine, and it is this that makes soccer exciting. Uncertainty plays a major role in the sport, and therefore managers, players, owners, fans, bookmakers, and broadcasters are all, to a greater or lesser extent, interested in soccer stats.

Most bookshops bear fulsome testimony to the unlikely love affair between *association football* (more commonly known as soccer or *football* in Europe) and mathematics, with their shelves containing many popular books on the role of statistics and data analysis in soccer. For the most part, these books discuss quirky aspects of the game and explain to the reader, in a non-numerical way, how mathematics and data science can help explain many of the phenomena observed on and off the pitch. These books, however, while entertaining, give few practical details on how such data can be analysed in real-life situations, which means that those wanting to learn more about soccer analytics have to resort to surfing the Internet and reading dry textbooks on data science. For most people, this can be a particularly challenging experience, which is why I have written this book, a practical and easy-to-follow introduction to soccer analytics with plenty of example codes in R (an easy-to-learn statistical programming language), which can be copied and adapted by the reader for their own purposes.

As a university academic who has taught thousands of students, I know only too well how difficult learning a new subject can be, and so in writing this text, my aim has been to produce a practical 'how-to' book that will enable the reader to competently analyse soccer data within a short period of time. To this end, the chapters are built around a series of structured examples, each of which contains fully functioning R code designed to teach the reader new skills (e.g., downloading and manipulating soccer data from the Internet, predicting match outcomes, formulating betting strategies, performing statistical tests). In short, this book is intended to get the reader up and running with R as quickly as possible so that they can start to analyse soccer data for themselves.

This book is aimed at all those with an interest in analysing soccer data, be they fans interested in soccer stats; those interested in sports betting;

sports scientists (students and professionals); coaches wanting to develop their analytical skills; or data scientists and statisticians simply wishing to pursue a career in professional soccer. It is an introductory text that takes a practical 'route one' approach and is designed to appeal to a wide readership. With this in mind, the book is focused on soccer, rather than on statistics, with the examples primarily aimed at solving practical problems commonly encountered in soccer. As such, the statistical and machine learning techniques introduced in the various chapters are primarily a means to an end – practical solutions designed to solve real-life problems. For consistency and ease of reference, the examples in this book are mostly drawn from the English Premier League, which is widely followed around the world, with millions of Liverpool, Chelsea, Arsenal, Manchester United, and Manchester City fans to be found on every continent.

As someone who has taught university students from widely different disciplines, I am acutely aware that many readers may not have a background in mathematics or statistics. If this applies to you, don't worry – this book requires no prior knowledge of statistics or coding. Chapter 2 contains a full tutorial, which will help you learn R and enable you to become competent at coding in just a few easy lessons. Also, for those readers unfamiliar with statistics and machine learning, 'key concept boxes' have been included in the chapters to explain, in simple terms, the technical concepts introduced in this book. My experience is that once people feel confident with some basic coding (which can be mastered in just a few hours), they want to experiment and are happy to play with code copied from the Internet.

This book arises from my experience teaching students in a leading university sports science department. What amazed me was just how receptive my students were to R, despite the fact that most had no background in mathematics or computer programming. Within a few days, most were up and running, and eager to use R to analyse their own data. They instantly saw the potential of R and how it could help them with their work. What's more, they began to teach their friends and colleagues how to code in R, which over time changed the culture of the department and revolutionised its work on sport performance analysis.

In writing this book, I have tried to bring to the table not only the lessons that I have learned as a university lecturer but also my experience as a scientist and data analyst, having worked in various disciplines (medicine, sport, bio-engineering, etc.) with a number of leading research teams around the world. While this may sound grandiose, in reality it means trying to make sense of complex, messy data; identifying what is important and what is not important; trying to keep things simple and avoid drowning in data; and communicating complex ideas to audiences who are not necessarily numerically literate – all of which are common challenges for data analysts, whether they work in sport or not.

As I explain in Chapter 11, becoming a successful analyst is far more than simply using advanced statistical and machine learning techniques.

Rather, it involves softer skills such as knowing what the data means and how to interpret it; knowing which techniques to use and which to avoid; and knowing how to communicate results in an appropriate manner. Sadly, many students and would-be analysts are not always aware of this. Therefore, this book contains lots of practical tips and strategies to help you avoid the common pitfalls that novice analysts frequently fall into when tackling soccer data.

Anyway, I hope that you enjoy reading this book and that it helps equip you with the skills necessary to analyse soccer (or football, as I often call it) data. Once you become confident and have mastered a little coding in R, you will quickly become aware that there is a whole new world out there, just waiting to be explored. I have never ceased to be amazed by the journeys that I've embarked upon with a few lines of well-drafted code. I hope very much that as you read this book, you too will become intrigued and curious enough to set off on your own analytical adventure – I can assure you that wherever you end up (and sometimes you may be surprised!), your achievement will be well worth the effort. Good luck – and bon voyage!

Clive Beggs

1

Soccer Analytics: The Way Ahead

Over the years, many popular books have been written exploring the mathematics of association football (soccer). For the most part, these books discuss quirky aspects of the game and explain to the reader in a non-numerical way how mathematics and statistics can explain many of the phenomena observed on and off the pitch. Some, such as the excellent *The Numbers Game: Why Everything You Know About Soccer Is Wrong* [1], also explain how elite soccer clubs employ data scientists and use sophisticated analytical techniques borrowed from the financial sector to acquire insights that give them an advantage over their competitors.

However, while these books are entertaining, few give any practical details on how football data can be analysed in real life, which means that those wanting to learn more about soccer analytics have to resort to surfing the Internet and reading dry textbooks on data science, all of which can be a challenging experience. Consequently, there is a gap in the market – a gap that is addressed by this book, which is a practical and easy-to-understand introduction to soccer analytics. As such, this book has been specifically designed to show the reader how to analyse soccer data and enable them to be up and running in a few easy lessons.

If you have opened this book and are reading these words, then chances are that you are interested in sport and also, to a greater or lesser extent, in data analytics. If so, then you are not alone because sports analytics is a discipline that is on the rise, with top soccer clubs, bookmakers, and broadcasters all employing statisticians and data scientists to give them an edge over their competitors. Indeed, such is the demand for data analytics in sport that a number of specialist companies such as Opta Sports, StatDNA, and Prozone have emerged in recent years, specifically to provide soccer clubs, broadcasters, bookmakers, etc. with the technical data and analytical expertise necessary to compete at the highest level. Collectively, this means that data analysis in sport, and in soccer analytics in particular, is now a big business, with the global market thought to be worth about 1.79 billion US dollars in 2021 [2].

This book is aimed at all those with an interest in soccer 'stats' who want to develop their analytical skills, be they fans wanting to learn how to predict match results; people interested in sports betting; coaches and sports scientists wanting to analyse training data; or indeed, data scientists interested in pursuing a career in professional football. The analysis vehicle that we shall use in this book is the statistical programming language R, which is easy to

DOI: 10.1201/9781003328568-1

learn and, perhaps more importantly, is a free open-source software package and thus widely used. As such, this book is an accessible introduction to the use of R for analysing soccer data.

Rather than being a formal textbook, this book is intended to be a practical 'how to' text aimed at enabling the reader to competently analyse soccer data within a short period of time. To this end, the chapters are built around a series of structured examples, each of which contains fully functioning R code designed to teach the reader new skills (e.g., downloading soccer data from the Internet, predicting match outcomes, and performing statistical tests). To make things easy, all the R scripts used in this book, together with the data files, have been deposited on the GitHub repository (which can be found at: https://github.com/cbbeggs/SoccerAnalytics), and readers are encouraged to copy and adapt these to suit their own needs.

1.1 Learning R

Don't worry if you are not familiar with R or computer programming in general, because no prior knowledge of coding is required to utilise and learn from this book. A full tutorial is provided in Chapter 2, which will help you learn R. This will enable you to become competent at coding in a few easy lessons. Also, for those readers unfamiliar with statistics and machine learning, *key concept boxes* are included in the text, which explain in simple terms many of the more technical concepts discussed in the book.

So why bother learning R? The simple answer is that R is the leading open-source statistical programming language in the world [3]. Because it is free and user-friendly, it is widely used both in universities and in industry for statistical and data analysis work. It is therefore an ideal tool for modelling and analysing sports data. Indeed, it is now regularly used both in professional sport and in sports science as a tool for analysing the performance of athletes and teams [4–6]. In addition, R is easy to learn and can be mastered to a reasonable level by individuals with a non-mathematical background relatively quickly. Having taught R to students in a university sports department, the author's own experience is that after attending his two-day course, most students became proficient at coding, despite many coming from a non-mathematical background and almost all having no previous experience of computer programming. So on a personal level, because it is easy to learn, many people find R to be an excellent vehicle with which to analyse football data.

1.2 How to Use This Book

The philosophy underpinning this book is one of learning by copying. When children learn a language, they do so by copying and experimenting in an iterative process that involves making lots of mistakes. At first, they simply copy the words that they hear without any real understanding of how these words should be used, but as they learn and become more confident, they soon adapt and synthesise the words to form sentences. Although this might seem like a rather random process, it is in fact a very successful strategy that enables young children to master complex languages relatively quickly. Similarly, the best way to learn a computer language like R is to copy code that other more experienced programmers have written and experiment with it. Don't start your learning process by trying to write completely new computer code from scratch; that is the slow way to learn! Instead, find some existing R code that is both interesting and accessible, and simply copy and run it (see Chapter 2 for more details) to see what happens. As you copy, play, and experiment, you will quickly develop new skills and be amazed by how much you have learned in a very short period of time.

This book assumes that the best way to learn to program in R is to copy code that others have written and adapt it to suit your own needs. To this end, the example codes presented in this book have been structured in such a way as to teach you how to program in R as well as enable you to analyse soccer data. You are therefore encouraged to copy the various R code examples in this book from the GitHub repository (https://github.com/cbbeggs/SoccerAnalytics) and to load these into RStudio to see how they run (see Chapter 2). You are also encouraged to adapt and improve the example codes in this book, as this will help you develop your programming skills.

In order to get a feel for what R can do, in this first chapter, a few snippets of code are presented as a kind of 'taster session'. These are designed to demonstrate: (i) how useful R can be and (ii) how an awful lot can be achieved with just a few lines of code. In other words, the aim of this first chapter is simply to whet the reader's appetite for R. If you can already programme in R, then go ahead and run the code examples in this chapter. However, if you are a beginner and new to R, don't worry; just sit back and go with the flow! The important thing for now is to get a feel of what R looks like. You can learn to code later using the R tutorial in Chapter 2. The code snippets in this first chapter are purely intended to demonstrate the wide range of tasks that can be accomplished using R and to show its versatility. If you are a beginner, a good strategy is to revisit and run the code snippets in this chapter after you have completed the tutorial in Chapter 2.

1.3 Data Analytics and Football

Before looking in detail at what R can do and how it can be used, it is perhaps worth taking a few minutes to consider why data analysis has become so ubiquitous in elite soccer in recent years. Those of us with long memories know that professional football has traditionally been the sport of the 'working man'. It is a passionate sport built on 'blood and guts' mingled with 'wizardry and skill', that means a great deal to millions of fans – a sentiment brilliantly captured by Bill Shankly, Liverpool's long-time manager, when he said: *"Some people believe football is a matter of life and death, I am very disappointed with that attitude. I can assure you it is much, much more important than that"* [7].

Sentiments like this are a far cry from the world of high finance and economics, the traditional home of data analytics and machine learning. Therefore, it will come as no surprise to hear that, in the past, professional soccer has been somewhat reluctant to embrace the newfangled analytical techniques offered by statisticians and data scientists. Managers are creatures of habit and are generally much happier to trust their own gut instincts and the evidence of their eyes. However, things are changing, and this change is being driven as much by events off the pitch as anything on it.

The truth is that football is much more than what happens on the pitch. Each week, millions of passionate supporters enjoy watching their teams play, either live or on television (TV), with many buying replica kits, memorabilia, etc. Given this, it is not difficult to see that soccer is more than a mere source of entertainment; rather, it is an issue of identity for many people, and as such, it is of great importance. Consequently, football is not something that stays on the pitch; rather, it travels to the home, the workplace, the pub, or wherever fans spend time together discussing opinions, facts, and figures related to soccer. TV companies know this and spend huge sums of money securing the broadcasting rights to competitions in an attempt to grow audiences and increase advertising revenue. For example, the various deals for the broadcasting rights to the English Premier League (EPL) for the period of 2022–2025 have been estimated to be worth in excess of £10 billion [8]. Broadcasters are prepared to pay such vast sums of money because they know that huge numbers of fans will watch the matches on TV and that this will boost not only subscriptions but also advertising revenue. Furthermore, the broadcasters know that fans will also spend hours watching TV pundits (expert commentators) express opinions in pre-match build-up and after-match post-mortem sessions, all of which can be used to raise prime advertising revenue. The newspapers follow a similar strategy because they know that high-quality soccer-related content boosts sales and thus increases advertising revenue.

One problem that the media companies have is providing enough interesting content to keep their viewers and readers engaged, and this has led to the rise of specialist sports analytics companies like Opta Sports, which provide broadcasters and newspapers with soccer data and analysis. Opta Sports was founded in 1996 to analyse the EPL and was first contracted by Sky Sports for their TV broadcasts during the 1996–1997 season. Since then, they have gone on to become one of the leading sport analytics companies in the world, supplying many prominent broadcasters and newspapers with football data and specialist analytical expertise. With the advent of Opta and other specialist sports analytics companies like Prozone, StatDNA, etc., the market for soccer analytics has steadily grown, with broadcasters, newspapers, and websites all wanting up-to-date data and analysis. In fact, the total global sports analytics market was valued at 1.79 billion US dollars in 2021 [2]. Collectively, this explosion in sports analytics has helped to promote an interest in football stats amongst fans and also led to soccer (and sports in general) analytics becoming a respected sub-discipline within the much larger field of data science.

One area where soccer analytics have become ubiquitous is in the sports betting industry. Indeed, it would not be an exaggeration to say that it is the lifeblood of the sports betting industry. Sports betting is big business, with the global market thought to be approximately 1.5 trillion euros in 2020 [9]. Of this, the global amount wagered on the EPL alone in season 2019–2020, was over 68 billion euros worldwide, an amount that dwarfs both the income generated by the EPL clubs and the sums paid to the EPL by broadcasters.

Bookmakers make money by accurately calculating the outcome probabilities of football matches and other sporting events and then setting betting odds that will ensure that they make a profit. To do this, they need to be able to accurately predict the likely outcome of matches. Bookmakers therefore need to employ an army of statisticians and data scientists whose job it is solely to make predictions and set odds that will ensure that the betting companies make money. In addition, they often use data collected by third-party sports analytics companies like Opta Sports. Having access to such data gives the bookmakers the edge and allows them to develop statistical and machine learning models that produce better predictions.

It is within the context of these off-the-pitch matters that data analytics in professional soccer needs to be viewed. Elite professional soccer is dominated by high finance, with huge sums of money invested by owners in playing squads in order to compete at the highest level. Naturally, with so much money at stake, owners of clubs want to ensure that sound investment choices are made. With transfer fees in the tens of millions of pounds, owners want to know the likely investment return that will be achieved when purchasing target players. Not only is it important to know how many goals a new signing is likely to score, but also how much the player will boost ticket and merchandise sales. Also, it is important to have an estimate of the potential value of the player when they are eventually sold on to another club in a few years. Naturally, to answer these questions, soccer clubs, like any other company in the corporate sector, have learned to embrace data analytics.

With respect to on-the-pitch matters, professional soccer has been much slower to embrace state-of-the-art data analytics, probably because most managers and coaches come from an era before the benefits of data science became widely accepted. However, from a somewhat sceptical start, professional soccer clubs have come to realise that data analytics can give them an edge over their competitors, with most top football clubs now employing data analysts [10]. For example, data scientists can run computer simulations to assess how opposing teams might react during matches to tactical changes or substitutions. In so doing, coaches can gain insights that may prove helpful when making strategic changes during matches. Rather than relying purely on gut feelings, such simulations provide coaches and managers with hard evidence to support any decisions that they might make. Consequently, data analysis has the potential to improve the quality of on-the-pitch decision-making and thus give teams an edge over their opponents.

With the decision to embrace data analytics, professional soccer has unwittingly unleashed an arms race, with the top football clubs needing to continually upgrade their analytical expertise in order to stay ahead of the competition. As a result, many clubs have had to turn to the likes of Opta and Prozone for specialist support. In fact, in 2012, Arsenal spent £2.165 million to acquire one such company, StatDNA [11], such was their need for advanced data analytics. When one considers that it is now possible to track individual players during matches and training sessions using computer-linked video and Global Positioning System technologies, it is not difficult to appreciate why specialist analytical expertise is required. Such systems can collect positional data from players at a rate of 5–10 times a second [12], which means that one single match can potentially generate more than 54,000 data points per player. This is a huge amount of data, far more than can be managed using a spreadsheet. Therefore, in order to perform any meaningful analysis, it is necessary to employ tools like R together with sophisticated data science techniques. Given this, it is easy to appreciate why advanced data analytics are becoming more popular in professional football.

1.4 Communicating Results in Tables and Plots

Now that we have a feel for the depth and breadth of soccer analytics as a subject, we can turn our attention to some of the ways in which R can be used to help analyse football data. In the following sections, a few example code snippets are presented that have been created to illustrate how R can be used to address a variety of problems commonly encountered in soccer analytics. These are primarily designed to give the reader a taste of what R can do. As stated above, if you are new to R, the best strategy here is probably to go with the flow and concentrate on what R can do rather than on the code itself.

Once you have learned how to do some coding in Chapter 2, you can always return to the examples presented here and run them for yourself. Remember the mantra, *'copy to learn'*! So for now, all you really need to know is that in R, the symbol '<-' is equivalent to '=', and that the '#' symbol is the 'remark' symbol, which enables you to insert comments and remarks into your script without R thinking that your remarks are part of the computing code. In the code snippets in this first chapter, the lines of executable R code are denoted with a shaded background, and the output produced is denoted in the text with a '##' symbol at the start of the line or table.

So where do we start? Well, something that every data analyst has to do is be able to present data and results in a way that can be easily understood. Soccer analysts are no exception to this because they frequently have to present results to members of the public, managers, etc. who may not be particularly numerate. So, being able to communicate results in an appropriate manner is an important skill that soccer analysts must learn and one with which R can help, as we will see in the following example.

Often, when writing reports or preparing presentations, it is necessary to produce tables and plots that convey important information. These can be produced quickly and easily in R, which contains many built-in functions specifically designed to assist with the production of graphics. To illustrate this, we will utilise the following code and frequency data for the various teams that have won the EPL (1992–2021) to produce a simple bar chart and pie chart.

In this first snippet of R code, we will simply enter the names of all the teams that have won the EPL, together with the number of times each team has been champion. We do this by creating two vectors (i.e., lists of data), one called 'Clubs,' which contains the club names, and the other called 'Titles', which contains the number of times each team has won the EPL. After this, we use the 'cbind.data.frame' function to combine these into a data object called a data frame (see Chapter 2 for more details) to form a table, which we display using the 'print' command. However, don't worry about understanding each line of code, as all the commands will be explained in Chapter 2. For now, all that is important is that you appreciate what the code does.

```
Clubs <- c("Arsenal","Blackburn Rovers","Chelsea","Leicester City", "Liverpool","Man City",
        "Man United")
Titles <- c(3,1,5,1,1,6,13)
epl.dat <- cbind.data.frame(Clubs,Titles)
print(epl.dat)
```

This produces the following table showing how many times each club has won the EPL.

```
##
             Clubs  Titles
1          Arsenal       3
2 Blackburn Rovers       1
3          Chelsea       5
4   Leicester City       1
5        Liverpool       1
6         Man City       6
7       Man United      13
```

Now that the data is in R, we can produce a bar chart using the following code. Again, don't worry about the specific commands used, as these will be explained in Chapter 2.

```
# Bar plot
barplot(epl.dat$Titles, names.arg=epl.dat$Clubs, ylim=c(0,14))
title("Bar chart of EPL champions 1992-2021")
```

This produces the simple bar chart shown in Figure 1.1.

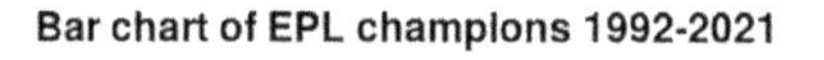

FIGURE 1.1
Bar chart showing the number of times each team has won the English Premier League from 1992 to 2021.

Similarly, a simple grey-scale pie chart can be produced as follows:

```
# Pie chart
colours = gray.colors(length(epl.dat$Clubs))
pie(Titles, labels = Clubs , col=colours,main="Pie Chart of EPL champtions 1992-2021")
```

This produces the pie chart shown in Figure 1.2.

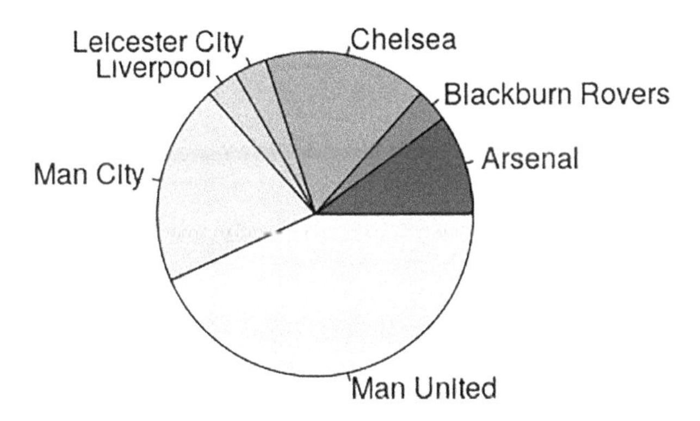

FIGURE 1.2
Pie chart showing the breakdown of the teams that have won the English Premier League from 1992 to 2021.

Although the example above is purely for illustrative purposes, we can clearly see that it is possible to produce pretty reasonable-looking plots with just a few lines of code in R. However, R is capable of producing much more sophisticated plots than the ones illustrated above. For example, Figure 1.3 shows the network of passes that occurred between the various players in the Germany team during the 2014 FIFA World Cup final against Argentina. The graph essentially shows who passed to whom during the match, with the thickness of the arrows indicating the strength of the relationships between the respective players. From this, it can be seen that during the match, the interactions between Lahm and Boateng were particularly strong, with Lahm passing the ball 23 times to Boateng, who made 22 passes to Lahm in return. By comparison, those between Lahm and Klose were much weaker, with Lahm making only six passes to Klose and receiving two in return. Passing networks are discussed in detail in Chapter 8, which shows how graphs like the one presented in Figure 1.3 can be constructed in R.

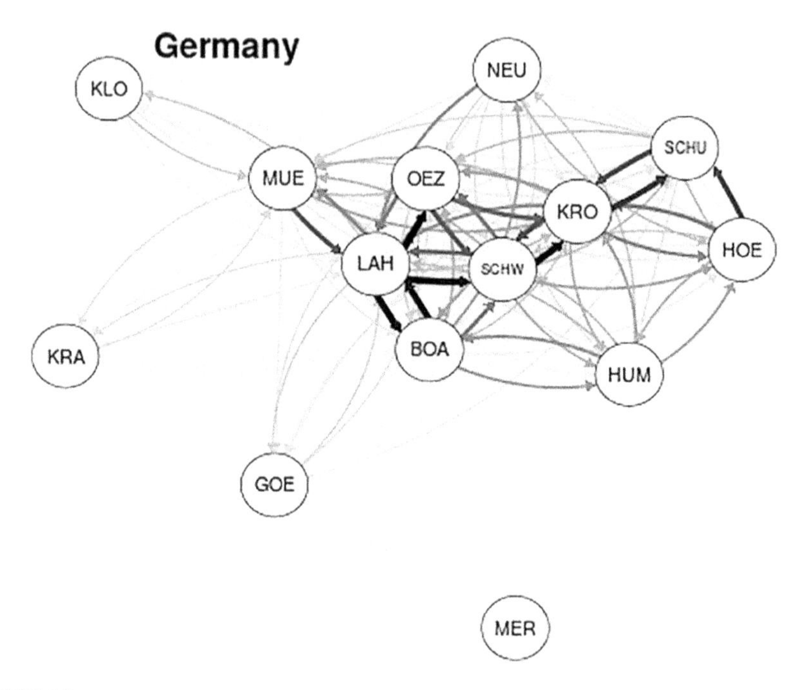

FIGURE 1.3
Passing network for the Germany team during the 2014 FIFA World Cup final against Argentina. The graph shows the network of who passed to whom during the match.

1.5 Is There Any Difference between Managers?

The issue of whether or not managers make a difference to the performance of soccer teams is a hotly debated one on which most football fans have an opinion. Here we will illustrate how R can be used to investigate this question. In so doing, we will also introduce the concept of statistical testing, which is something that data analysts frequently use to evaluate whether or not an observed effect is real and not just something that occurred by chance (see Chapter 2 for more details). R is an excellent tool for undertaking statistical testing.

Top football clubs sack and replace team managers with remarkable regularity. For example, since Alex Ferguson retired in May 2013, Manchester United have had six managers (at the time of writing in summer 2022), excluding two very short caretaker appointments (Ryan Giggs in 2014 and Michael Carrick in 2021). So why so many changes of manager? Well, the simple answer is that managers are sacked when there is a perception amongst club owners that things are not going well on the pitch. Of course, the decision to replace a manager is not necessarily always the correct one to make; many managers have been sacked because of temporary dips in form that would otherwise have corrected themselves in a process called 'regression to the mean' [13–15]. Furthermore, it is often the case that replacing the manager makes little difference to the long-term performance of teams [16]. As such, this suggests that in football, managerial changes are often made based on gut feelings rather than on any objective evidence.

To illustrate how R can be used to assess managerial performance, let's look at the managerial changes that have occurred at Manchester United since Alex Ferguson retired. Excluding Giggs and Carrick's short interregnums, the records of the various managers are presented in Table 1.1. From which we see that although Alex Ferguson (Figure 1.4) had the highest win ratio; he won 59.7% of the matches in which he was in charge. Jose Mourinho got close to this, achieving a win ratio of 58.3%. By comparison, Ralf Rangnick only managed a win ratio of 37.9%, which was the lowest exhibited by any of the managers.

On the face of it, it would appear that Alex Ferguson was clearly the best manager, with the others all performing less well. But how can we be sure that this was down to superior skill on the part of Alex Ferguson rather than just good luck (i.e., pure chance)? After all, Jose Mourinho's record was pretty close to that of Alex Ferguson. So how can we say with certainty that one manager is better than another? Well, sports scientists often use statistical hypothesis testing to settle the matter. The subject of statistical significance and hypothesis testing is covered in Chapter 2, and so we will not go into any more detail here. But suffice to say, scientists and statisticians generally become confident that an observed effect is real and not due to chance when the significance is $p<0.05$, in which case they say that the result or observation is statistically significant.

TABLE 1.1

Record of the Various Manchester United Managers from Alex Ferguson to Ralf Rangnick

Manager	Duration	Wins	Draws	Losses	Wins:Losses	Win (%)
Ferguson	1986–2013	895	338	267	3.352:1	59.7%
Moyes	2013–2014	27	9	15	1.800:1	52.9%
Van Gaal	2014–2016	54	25	24	2.250:1	52.4%
Mourinho	2016–2018	84	32	28	3.000:1	58.3%
Solskjaer	2018–2021	91	37	40	2.275:1	54.2%
Rangnick	2021–2022	11	10	8	1.375:1	37.9%

FIGURE 1.4
Alex Ferguson, Manchester United's legendary manager who led them to thirteen EPL championships, five FA Cups, four Football League Cups, and two UEFA Champions Leagues.

Source: Shutterstock/Paolo Bona. With permission.

So let's ask ourselves: Was Alex Ferguson's performance better than that of Ralf Rangnick? Well, on the face of it, the answer is 'yes', because the former had a win ratio of 59.7%, while the latter's was only 37.9%. But how can we be sure the observed difference (what statisticians call the effect) between the two managers is not due to natural variance in the data (i.e., chance)? Is there any statistical evidence that one manager is better than the other? Well, one way to do this in R is to use a chi-square test of independence and the frequency data in Table 1.1, as follows: The chi-square test is a popular non-parametric statistical test that can be used to compare the frequencies with which observed events occur in two groups (see Key Concept Box 1.1).

In order to perform the chi-square test in R, we first need to construct a frequency (contingency) table of the win, draw, and lose counts. Here, the data are inputted as vectors (i.e., lists of data), which are then combined together using the 'cbind.data.frame' function to produce a data frame called 'mu. record'. As before, the data frame is displayed using the 'print' command.

```
Manager <- c("Ferguson","Moyes","van Gaal","Mourinho","Solskjaer","Rangnick")
Wins <- c(895,27,54,84,91,11)
Draws <- c(338,9,25,32,37,10)
Losses <- c(267,15,24,28,40,8)

mu.record <- cbind.data.frame(Manager,Wins,Draws,Losses)
print(mu.record)
```

Which produces:

```
##
    Manager Wins Draws Losses
1  Ferguson  895   338    267
2     Moyes   27     9     15
3  van Gaal   54    25     24
4  Mourinho   84    32     28
5 Solskjaer   91    37     40
6  Rangnick   11    10      8
```

Now we can use the following code to select the two managers that we want to compare, which in this case are Alex Ferguson and Ralf Rangnick.

```
man1 <- "Ferguson"
man2 <- "Rangnick"

Manager1 <- mu.record[mu.record$Manager == man1,]
Manager2 <- mu.record[mu.record$Manager == man2,]
```

We then use the 'as.table' function in R to construct a contingency table (here called 'ContTab') of the frequency results for Ferguson and Rangnick, as follows:

```
# Create a table called ContTab
temp = as.matrix(rbind(Manager1[,c(2:4)], Manager2[,c(2:4)]))
ContTab <- as.table(temp)
dimnames(ContTab) = list(Manager = c("Manager 1", "Manager 2"),
            Outcome = c("Wins","Draws", "Loses"))
print(ContTab)
```

Which produces:

```
##
             Outcome
Manager    Wins Draws Loses
Manager 1   895   338   267
Manager 2    11    10     8
```

Finally, we perform the chi-square test, as follows:

```
chsqRes = chisq.test(ContTab) # This displays the results summary
print(chsqRes)
```

Which produces:

```
##      Pearson's Chi-squared test

data:  ContTab
X-squared = 5.5681, df = 2, p-value = 0.06179
```

From this, we see, rather surprisingly, that the observed difference between Alex Ferguson and Ralf Rangnick did not reach statistical significance. This is because $p=0.062$ is greater than the threshold of $p=0.05$. Strictly speaking, statisticians would actually say here that team performance (i.e., win, draw, or lose) is independent of the variable 'manager'. In other words, we cannot conclusively demonstrate that team performance is related to the choice of manager, and therefore we cannot rule out the possibility that the observed difference between the two managers might purely be down to chance. After all, Alex Ferguson might have had a lot of good luck and Ralf Rangnick a lot of bad luck.

While the above example illustrates how a chi-square test can be performed using R, it perhaps teaches us a much more important lesson, namely, that when it comes to assessing performance in sport, statistical significance is probably the wrong tool to use. Often in elite sport the margins are so fine that there is no significant difference between who comes first and who is second or third, which in statistical terms means that there is no difference between the competitors. Yet in reality, one competitor came first, one second, and another third – and this matters very much in sport! So, while the chi-square test might not show a significant statistical difference between the performances of Alex Ferguson and Ralf Rangnick, to all sane observers, it is abundantly clear that Ferguson's record at Manchester United was much superior to that of Rangnick.

KEY CONCEPT BOX 1.1: CHI-SQUARE TEST

If we toss a fair coin 100 times, we would expect it to land 50 times on heads and 50 times on tails. However, this is an ideal expectation, and in reality, some natural variation will occur around this ratio. So, if we toss a coin 100 times, we might easily obtain, say, 52 heads and 48 tails or 47 heads and 53 tails, as both results are highly likely to occur. However, if the coin landed 90 times on heads and only 10 times on tails, then we might get suspicious that the coin was not fair and that it was biased in some way. But how can we objectively test this? Well, one way in which we can test if the coin is biased or not is by using a chi-square test, which is a test that can determine whether a statistically significant difference exists between the expected and observed frequency counts in one or more categories in a contingency table.

The chi-square test, first developed by Karl Pearson at the end of the 19th century, is a widely used technique for determining whether or not a statistically significant difference exists between the expected and observed frequency counts. It is underpinned by a very simple idea, namely that if there is no difference between two classes (say, Class A and Class B) in a population, then if we take a sample from that population, we would expect to see both classes equally represented in the sample. So if, for example, we toss a coin 100 times, we would expect that it would land approximately 50 times on heads and approximately 50 times on tails. However, if there is a great imbalance in the observed head/tails ratio, then the chi-square test will tell us that the result is highly unlikely to have occurred by chance and that therefore the coin might be biased.

The chi-square test is well suited to testing hypotheses that involve categorical (e.g., win, lose, draw) or dichotomous (e.g., yes/no) data. This type of categorical data is generally analysed by frequency of occurrence, with observations assigned to mutually exclusive categories. With respect to this, the chi-square test can be used to compare the relative frequency of occurrence over two or more categories. So, for example, the chi-square test can be used to assess the win/lose/draw performance of football managers, as illustrated in Section 1.5.

To illustrate how the chi-square test works, imagine that we toss a coin 160 times and record the outcome. If the coin is fair, then we would expect it to land on heads for 50% of the tosses and on tails for 50% of the tosses. However, in our experiment, we found that the coin landed heads on 88 occasions and tails on 72 occasions. So is this a fair result (within the bounds of reasonable possibility), or is the coin biased (loaded)?

In order to answer this, we first construct the following contingency table containing the observed and expected frequency counts.

Results	Heads	Tails
Observed	88	72
Expected	80	80

From this, we can see that the number of degrees of freedom is one, because once we have entered the number of observed heads into the table, the number of observed tails is calculated automatically. So we only have the freedom to enter one observed value.

From the contingency table, we can see that there are certainly more heads and fewer tails than we would normally expect, but is this all that unusual? So the next thing that we can do is compute the chi-square statistic, as follows:

$$\text{Chi-square statistic: } \chi^2 = \frac{(88-80)^2}{80} + \frac{(72-80)^2}{80} = 1.6 \quad (1.1)$$

Having done this, we can use R or statistical tables to find the probability (p-value) that the chi-square statistic is greater or equal to 1.6, assuming one degree of freedom, which in this case turns out to be $p=0.206$. In other words, what the chi-square test tells us here is that 20.6% of observations will have a chi-square statistic that is greater than or equal to 1.6. So although the observed coin toss result is reasonably uncommon, it is not that unusual and is likely to occur in about one in five coin toss experiments. Therefore, we cannot assume that the coin is biased.

The version of the chi-square test described above is called the goodness of fit test, and this is used to assess whether or not the frequency distribution of a categorical variable is different from our expectations. There is, however, another version of the test, the chi-square test of independence, which is used to compare the frequency distribution exhibited by two or more groups, and it is this version of the chi-square test that is used in Section 1.5 to assess managerial performance. Notwithstanding this, the mathematics underlying both versions of the test are similar, with both utilising a contingency table and the frequency distribution to compute the chi-square statistic.

1.6 Making Predictions

Most football fans enjoy making predictions, with supporters spending many happy hours discussing and arguing over the likely outcome of forthcoming matches. TV and newspaper pundits thrive on this because they know that every fan has an opinion, with most happy to predict who will win the league, be relegated, or win the cup final, etc. Of course, the predictions made by the fans and pundits may not be very good, but that is not the point, because they do the job of stimulating interest, which ultimately drives up TV audiences, sales of match tickets and newspapers, etc. By contrast, for gamblers and those involved in the sports betting industry, making predictions is a much more serious matter because real money is at stake. Therefore, for bookmakers and gamblers alike, predictions need to be as accurate as possible if money is to be made and not lost.

Over the years, a number of mathematical models have been developed to help predict match outcomes and end-of-season (EoS) points totals. These models, which are discussed in detail in Chapters 5 and 6, are widely used in the sports betting industry and are used by bookmakers along with other information to set the odds for match and league outcomes. R is an excellent vehicle for building soccer prediction tools because it allows users to develop bespoke models that can then be tested and adapted using historical data downloaded from the Internet. As such, R provides the perfect 'virtual laboratory' for building, testing, and refining such models in order to ensure that the predictions produced are as accurate as possible.

One class of prediction tool that is widely used in soccer analytics is the linear regression model (see Key Concept Box 1.2 for more details). This is a statistical technique that describes the relationship between one or more predictor variables (sometimes called independent variables) and a dependent response variable. This type of regression model generally uses a least-squares approach, which minimises the sum of the squared errors of the residuals to produce a best-fit line through the data. Therefore, the technique is often referred to as ordinary least-squares (OLS) regression. OLS regression is a powerful technique because, once the variable coefficients and best-fit line have been established, they can be used to predict future performance, which, of course, is something that many people interested in soccer would like to do. While linear regression is discussed in detail in Chapter 10, we include a short example here as a taster because many readers of this book will be interested in making predictions and may want to know how R can be used to do this.

In order to illustrate the power of R as a prediction tool, the following code is included here, which presents a simple linear regression model. This model utilises shots attempted and points awarded for the EPL during season 2020–2021 to predict the likely EoS points totals for clubs in season 2021–2022. As we will see, despite being a very simple OLS regression model, for the most part it produces surprisingly accurate predictions, which is why such models are extremely popular in soccer analytics.

Before we can build our prediction model, we first need to input the 2020–2021 EoS points totals for the various teams in the EPL, together with the total number of shots attempted. These can be entered as vectors into R as follows:

```
Points <- c(61, 55, 41, 39, 67, 44, 59, 28, 59, 66, 69, 86, 74, 45, 23, 43, 62, 26, 65, 45)
Shots <- c(455,518,476,383,553,346,395,440,524,472,600,590,517,387,319,417,442,
        336,462,462)
```

Having entered the data for season 2020–2021, we can combine the vectors into a data frame object and use the 'head' function in R to display the first six rows as follows:

```
dat2020 <- cbind.data.frame(Points, Shots)
head(dat2020)  # NB. The 'head' function displays the first 6 rows of the data frame.
```

```
##
  Points Shots
1     61   455
2     55   518
3     41   476
4     39   383
5     67   553
6     44   346
```

Now we can build the regression model for season 2020–2021 and display the results. We do this by using the 'lm' function in R to produce a model called 'mod2020'. Here, when we write 'Points ~ Shots' in the code, we are telling R that 'Shots' is the predictor variable and 'Points' is the response variable. To display the results of the model, we use the 'summary' command in R.

```
# Build OLS regression models for season 2020-21
# Using Shots
mod2020 <- lm(Points ~ Shots, data = data.frame(dat2020))
summary(mod2020)
```

This produces:

```
## Call:
lm(formula = Points ~ Shots, data = data.frame(dat2020))

Residuals:
    Min      1Q  Median      3Q     Max
-22.430  -7.576  -2.039  10.446  15.977

Coefficients:
            Estimate Std. Error t value Pr(>|t|)
(Intercept) -21.9996    14.3042  -1.538    0.141
Shots         0.1646     0.0310   5.309 4.78e-05 ***
---
Signif. codes:  0 '***' 0.001 '**' 0.01 '*' 0.05 '.' 0.1 ' ' 1

Residual standard error: 10.83 on 18 degrees of freedom
Multiple R-squared:  0.6103,     Adjusted R-squared:  0.5886
F-statistic: 28.19 on 1 and 18 DF,  p-value: 4.776e-05
```

This summary tells us that for season 2020–2021, the regression model with 'shots' as the predictor was able to explain 61% of the variance in the points total, which is pretty good. This can be illustrated using a scatter plot (Figure 1.5) with a least-squares best-fit line, which can be produced using the following code in R:

```
# Scatter plot with bets-fit lines
plot(dat2020$Shots, dat2020$Points, pch=20, col="black", xlim=c(0,800),
    ylim=c(0,100), ylab="Points", xlab="Shots")
abline(lm(dat2020$Points ~ dat2020$Shots), lty=1)
```

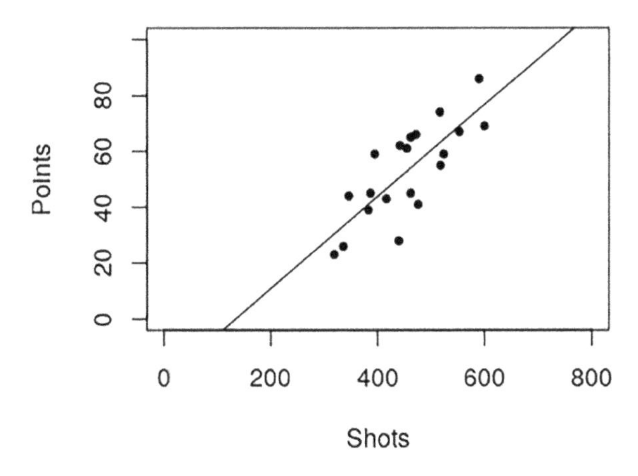

FIGURE 1.5
Scatter plot with best-fit line of EoS shots attempted and league points awarded for EPL season 2020–2021.

Having built the model using 2020–2021 data, we can now apply it to the 2021–2022 season in order to make predictions for specific clubs. Predictions can be made from the 'mod2020' model using the 'predict' function, as follows:

First, we will predict the points total for Chelsea, who finished third and attempted 585 shots on goal in 2021–2022.

```
# Chelsea (who attempted 585 shots and achieved 74 points in season 2021-22)
ChelPts.2021 <- predict(mod2020, list(Shots=585))
print(ChelPts.2021)
```

```
##       1
74.2991
```

In season 2021–2022, Chelsea actually earned 74 points, which is remarkably close to the total of 74.3 points predicted by the model.

Next, we will look at the champions, Manchester City, who attempted 704 shots in 2021–2022.

```
# Manchester City (who attempted 704 shots and achieved 93 points in season 2021-22)
MCPts.2021 <- predict(mod2020, list(Shots=704))
print(MCPts.2021)
```

```
##        1
93.88807
```

As we can see, the model predicts a total of 93.9 points for Manchester City, which is again remarkably close to the 93 points that they actually achieved in season 2021–22.

Finally, we turn our attention to Norwich City, who finished bottom of the league in 2021–22 and who only attempted 374 shots in the whole season.

```
# Norwich City (who attempted 374 shots and achieved 22 points in season 2021-22)
NCPts.2021 <- predict(mod2020, list(Shots=374))
print(NCPts.2021)
```

```
##  1
39.56571
```

Here, the model prediction is not so good because Norwich City only earned 22 points in the whole of that season. As we can see, the model overpredicted the actual point total by 17.6 points, which suggests that the current model might be rather oversimplistic and that a more complex model with more predictor variables might be necessary. Having said this, this simple prediction model highlights the close relationship that exists in the EPL between shots attempted and points awarded. Clearly, the more shots on goal that a team attempts, the more likely it is that some will get through and be converted to goals – hence the close relationship between the two.

While we all think that we know what a shot in football looks like (i.e., when a player strikes the ball hard with their foot in an attempt to score a goal, as in Figure 1.6), it is worth noting that the soccer data companies generally use a looser definition. For example, Opta Sports [17] defines a shot on target as any goal attempt that:

- Goes into the net regardless of intent.
- Is a clear attempt to score that would have gone into the net but for being saved by the goalkeeper or being stopped by a player who is the last man with the goalkeeper having no chance of preventing the goal (i.e., a last line block).

Shots off target are defined as any clear attempt to score that:

- Goes over or wide of the goal without making contact with another player.
- Would have gone over or wide of the goal but for being stopped by a goalkeeper's save or by an outfield player.
- Directly hits the frame of the goal, and a goal is not scored.
- Is blocked by another player who is not the last man, even if the shot is heading towards the goal.

Of course, the total number of shots in any game is the sum of the shots on target and those off target.

FIGURE 1.6
A typical shot in a soccer match is not necessarily the same thing as a shot in soccer analytics.

Source: Shutterstock Editorial. With permission.

KEY CONCEPT BOX 1.2: ORDINARY LEAST-SQUARES REGRESSION

Linear regression models are widely used in statistics and data science to explain and predict phenomena. Such models have the general form:

$$\text{Linear regression model: } y = b_0 + b_1x_1 + b_2x_2 + \ldots + e \qquad (1.2)$$

where y is the response variable; x_1, x_2, … are the predictor variables; b_0, b_1, b_2, … are coefficients applied to the intercept and the predictor variables; and e is the residual error.

So for example, y might be the amount of energy consumed per day by an office building, and x_1 and x_2 could be the average daily outside air temperature and hours of daylight, respectively. Given this, it's not difficult to see that the value of y will be influenced by both x_1 and x_2, because buildings consume more energy on heating and lighting when it is cold and dark outside. However, the energy consumption of the building cannot be fully explained simply by how cold or dark it is outside because other factors such as occupancy pattern and behaviour also play a role. Therefore, Equation 1.2 incorporates a residual error term, e, which is a sort of 'fiddle factor' used to make up for the shortcomings of the predictor variables, x_1 and x_2.

OLS regression is so-called because it uses a least-squares approach to calculate the values of the respective variable coefficients, b_0, b_1, and b_2, in Equation 1.2. A full discussion of the linear algebra underpinning this is beyond the scope of the current text, but suffice to say that the OLS approach minimises the sum of the squared errors of the residuals to produce a best-fit line through the data, which can be described using Equation 1.3. In Equation 1.3, the residual error term, e, is omitted, and the y term is replaced with a $\hat{y}$ term, which is the predicted value of the response variable. Importantly, all the $\hat{y}$ values predicted using Equation 1.3 lie along the best-fit line through the observed data.

$$\text{Multiple linear regression}: \hat{y} = b_0 + b_1x_1 + b_2x_2 + \dots \tag{1.3}$$

When there are several predictor variables, as in Equation 1.3, we call it multiple linear regression analysis. However, when there is only one predictor variable (as shown in Equation 1.4), we generally use the term simple linear regression.

$$\text{Simple linear regression}: \hat{y} = b_0 + b_1x_1 \tag{1.4}$$

When Equation 1.4 is applied and plotted on a scatter plot, the predicted (fitted) values lie along a least-squares best-fit line that runs straight through the middle of the observed data, as illustrated in Figure 1.7, where the intercept at $x=0$ is b_0 and b_1 is the gradient of the best-fit line.

If the model is a good predictor of the observed data, then the residual errors will be small, whereas if it is not that good at prediction, the errors will be large. The goodness of fit of a linear regression model can be assessed using the coefficient of determination, R^2, which is a measure of the amount of variance in the observed response variable that is explained by the model. The R^2 value is in fact identical to the square of the correlation coefficient, r. In the example shown in Figure 1.7, $R^2=0.602$ and the r-value is 0.776, which is the square root of R^2. This

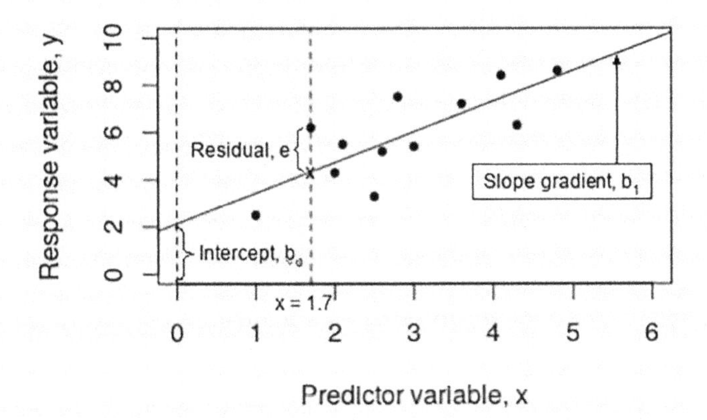

FIGURE 1.7
Scatter plot of x and y for a simple linear regression model with one predictor variable.

implies that the model is able to explain 60.2% of the variation in the value of the response variable.

Another useful measure of goodness of fit that is often used is the mean absolute error, which is the average of the absolute values of the residual errors produced by the model. It is necessary to use the absolute value because some of the residual errors are positive and some are negative, and they would otherwise just cancel themselves out mathematically.

1.7 Match Odds

Those interested in gambling will probably know that one of the key tenets of a successful betting strategy is finding value in the match odds offered by bookmakers. While the subject of betting is covered in detail in Chapter 7, to illustrate how R can be used to assist in deciding where to place bets, let's consider the match that occurred on October 3, 2021, between Tottenham and Aston Villa in the EPL. Tottenham were the home team and won the match by two goals to one.

The pre-match decimal odds offered by William Hill were 2.15 for a home win, 3.30 for a draw, and 3.50 for an away win. By comparison, Pinnacle offered odds of 2.13 (home win), 3.61 (draw), and 3.64 (away win) for the same match. Now, Pinnacle has a reputation within the sports betting industry for setting very keen odds (i.e., odds that are closer to the true calculated odds). Therefore, because the odds offered by William Hill for a home win were greater than those offered by Pinnacle, it suggests that there might have been some value in this bet. The following code shows how R could have been used to evaluate a bet on Tottenham to win at home.

First, let's input the match odds offered by the two bookmakers and save these in a data frame, which is a special class of data object used by R (see Chapter 2 for further details).

```
# William Hill match odds
wh_hwodds <- 2.15 # Odds offered by bookmaker for a home win
wh_dodds <- 3.30 # Odds offered by bookmaker for a draw
wh_awodds <- 3.50 # Odds offered by bookmaker for a away win

# Pinnacle match odds
p_hwodds <- 2.13 # Odds offered by bookmaker for a home win
p_dodds <- 3.61 # Odds offered by bookmaker for a draw
p_awodds <- 3.64 # Odds offered by bookmaker for a away win

# Compile the data frame
WH_odds <- c(wh_hwodds,wh_dodds,wh_awodds) # William Hill
Pin_odds <- c(p_hwodds,p_dodds,p_awodds) # Pinnacle
bet.dat <- cbind.data.frame(WH_odds,Pin_odds)
```

Next, we compute the implied probabilities for the three possible match outcomes and add these to the data frame. The implied probabilities can be computed from these using the following equation:

$$\text{Computing implied probability from decimal odds} : p = \frac{1}{\text{Odds}_d} \tag{1.5}$$

where Odds_d is the decimal odds, and p is the implied probability.

```
# Compute implied probabilities
bet.dat$WH_prob <- round(1/bet.dat$WH_odds,3)
bet.dat$Pin_prob <- round(1/bet.dat$Pin_odds,3)
rownames(bet.dat) <- c("Home win","Draw","Away win")
print(bet.dat)
```

This displays the 'bet.dat' data frame as follows.

```
##
          WH_odds Pin_odds WH_prob Pin_prob
Home win     2.15     2.13   0.465    0.469
Draw         3.30     3.61   0.303    0.277
Away win     3.50     3.64   0.286    0.275
```

From this, we can see that Pinnacle thought that there was a slightly higher chance of Tottenham winning compared with William Hill. However, it is important to note that the quoted implied probabilities are not the true match probabilities because each bookmaker has to add on their profit, also called over-round. The amount of over-round added by each bookmaker can be easily calculated by computing the sum of the implied probabilities using the following code:

```
# William Hill's over-round
WH_or <- sum(bet.dat$WH_prob)-1
print(WH_or)
```

```
## [1] 0.054
```

```
# Pinnacle's over-round
Pin_or <- sum(bet.dat$Pin_prob)-1
print(Pin_or)
```

```
## [1] 0.021
```

From this, we see that William Hill built in 5.4% profit into their quoted match odds, whereas Pinnacle was happy to make just 2.1% profit.

Now, if we had placed a £10 bet on Tottenham to win with William Hill, our profit would have been:

```
# Bet profits
wager <- 10 # £10 wager with WH on Tottenham to win
profit <- wager * (wh_hwodds-1)
print(profit)
```

```
## [1] 11.5
```

So we can see that we would have made £11.50 profit from the £10 wager with William Hill, whereas if the same bet had been taken with Pinnacle, this would result in only £11.30 profit. While for this particular match, the 'value' in the home bet with William Hill was pretty marginal, it nonetheless illustrates the general approach that can be taken.

1.8 Simulating the FA Cup Draw

The FA Cup in England, which was first played in 1871–72, is the oldest domestic knockout football competition in the world. One exciting feature of this competition is that the teams are not seeded, with fixtures determined randomly by drawing numbered balls from a bag – an event that attracts much attention from fans and is often televised live. This means that in the fourth round of the competition, Manchester City could, in theory, just as easily be drawn against lowly Mansfield Town from the fourth tier of English football (at the time of writing in 2022) as Manchester United from the EPL. Given this uncertainty, it is not difficult to see why the televised FA Cup draw attracts a large amount of interest.

One of the most useful features of R is that it can perform stochastic simulations by randomly selecting numbers from a larger population using a random number generator. Stochastic modelling forecasts the probability of various outcomes under different conditions using randomly selected variables, and as such, it is popular in the financial sector for making investment decisions. However, it can also be useful in soccer for simulating outcomes. For example, with reference to the FA Cup draw, the following R code simulates the random draw for the last 16 teams in the competition (i.e., the fifth round of competition). To do this, we first need to create a vector called 'teams,' which contains the unique ID numbers (1–16) for the last 16 teams in the competition.

```
teams <- c(1:16)
print(teams)
```

```
## [1]  1  2  3  4  5  6  7  8  9 10 11 12 13 14 15 16
```

Having done this, the following code can be used to randomly sample the home and away teams and thus create the fixture list: Here we use the 'sample' function to randomly select eight numbers (i.e., the home teams) from the 'teams' vector. The away teams are selected using the code 'sample(teams[-samp.HT])'.

```
# Create matrix to store results
cup.draw <- as.data.frame(matrix(0,8,2))
colnames(cup.draw) <- c("HomeTeam", "AwayTeam")
```

```
# Randomly sample eight home teams
#set.seed(123) # This makes the draw results repeatable
samp.HT <- sample(teams, size=8, replace=FALSE)
samp.AT <- sample(teams[-samp.HT])

cup.draw$HomeTeam <- samp.HT
cup.draw$AwayTeam <- samp.AT
print(cup.draw)
```

This produces the following fixtures:

```
##
  HomeTeam AwayTeam
1       15        8
2       16       11
3        3       12
4       14        1
5       10        4
6        2        7
7        6       13
8        5        9
```

(NB. The fixture list produced may vary depending on your operating system and your version of R, for reasons that are too complex to discuss here.)

From this, we can see that by using a few lines of code in R, we can replicate the FA Cup draw. Mind you, using R is not as exciting as taking numbered balls from a bag live on TV.

If you try this code out for yourself, you may find that you get a different answer than the one printed above. This is because R uses a random number generator, which gives a different set of random numbers each time the program is run. While this is fine, it can cause problems when we want to make things repeatable. In which case, we can add a 'set.seed' command (e.g., set. seed(123)) in order to make things repeatable. If a seed is set, then the same answer will be returned every time the program is run.

1.9 R Has Much More to Offer

If you have read this far, then hopefully you are interested in learning about R and how it can be used to analyse football data. The examples above have been designed to give the reader a taste of what R can do. Although R is considered by many to be mainly a statistical analysis tool, it is in fact much more than this, being a full-blown programming language in its own right. As such, it is packed with many advanced features that can be used to manipulate, edit, and analyse data, and even, if required, to produce reports. Indeed, its capabilities far exceed the scope of this book, with many add-on packages that can be utilised to facilitate advanced machine learning and signal processing techniques. Notwithstanding this, the chapters in this book have been structured in a way that allows the reader to explore various aspects of soccer analytics (e.g., predicting match results, formulating betting strategies, analysing match performance data, etc.) using R. The aim of each chapter is not to produce the definitive predictive model or analytical solution, but rather to equip the reader with knowledge and understanding of the subject that allows them to progress and develop their own soccer analytical skills.

Because R is a free open-source software package, a lot of people use it, especially those in university statistics and data science departments. This means that there are lots of useful tutorials on the Internet and on YouTube that you can use to develop your analytical skills. You are therefore encouraged to copy the R scripts presented in this book and to experiment with them, as this is the best way to learn. More is said about this in Chapter 2, but for now it is enough to know that the best way to learn to program is to copy code and experiment with it – perhaps trying it out on your own data to see what happens. As you play with the code and understand how to use it, your analytical skills will develop, and you may surprise yourself with how much you have learned.

Anyway, now it is over to you. Enjoy learning from the rest of this book.

References

1. Anderson C, Sally D: *The Numbers Game: Why Everything You Know About Soccer Is Wrong:* Penguin Publishing Group; 2013.
2. FB: The global sports analytics market is projected to grow from $2.22 billion in 2022 to $12.60 billion by 2029, at a CAGR of 28.1% in forecast period, 2022–2029. In: *Fortune Business Insights.* https://wwwfortunebusinessinsightscom/sports-analytics-market-102217; 2022.
3. Ihaka R: R: Past and future history. *Computing Science and Statistics* 1998, 392396.
4. Till K, Jones BL, Cobley S, Morley D, O'Hara J, Chapman C, Cooke C, Beggs CB: Identifying talent in youth sport: A novel methodology using higher-dimensional analysis. *PLoS One* 2016, 11(5):e0155047.
5. Beggs CB, Shepherd SJ, Emmonds S: Jones B: A novel application of PageRank and user preference algorithms for assessing the relative performance of track athletes in competition. *PLoS One* 2017, 12(6):e0178458.
6. Weaving D, Jones B, Ireton M, Whitehead S, Till K, Beggs CB: Overcoming the problem of multicollinearity in sports performance data: A novel application of partial least squares correlation analysis. *PLoS One* 2019, 14(2):e0211776.

7. Sportsmail Reporter: Bill Shankly: The top 10 quotes of a Liverpool legend 50 years to the day since he took over. MailOnline. https://www.dailymail.co.uk/sport/football/article-1232318/Bill-Shankly-The-quotes-Liverpool-legend-50-years-day-took-over.html; 1st December 2009.
8. Ziegler M: Premier League's broadcast income to reach £10bn over next three seasons. In: *The Times.* https://www.thetimes.co.uk/article/premier-league-s-broadcast-income-to-reach-10bn-over-next-three-seasons-cm077wpvk; 2022.
9. Lock S: Global amount wagered on European soccer 2019–2020, by league. In: *Statistica;* https://www.statista.com/statistics/1263462/value-betting-on-european-soccer/; 2021.
10. Savvas A: A short history of data analysis in football. In: *The Register.* https://www.theregister.com/2022/06/29/a_short_history_of_data/; 2022.
11. Hytner D: Arsenal's 'secret' signing: Club buys £2m revolutionary data company. In: *The Guardian.* https://www.theguardian.com/football/2014/oct/17/arsenal-place-trust-arsene-wenger-army-statdna-data-analysts; 2014.
12. Dalton-Barron N, Whitehead S, Roe G, Cummins C, Beggs C, Jones B: Time to embrace the complexity when analysing GPS data? A systematic review of contextual factors on match running in rugby league. *Journal of Sports Sciences* 2020, 38(10):1161–1180.
13. Audas R, Dobson S, Goddard J: The impact of managerial change on team performance in professional sports. *Journal of Economics and Business* 2002, 54(6):633–650.
14. Bruinshoofd A, Ter Weel B: Manager to go? Performance dips reconsidered with evidence from Dutch football. *European Journal of Operational Research* 2003, 148(2):233–246.
15. Heuer A, Muller C, Rubner O, Hagemann N, Strauss B: Usefulness of dismissing and changing the coach in professional soccer. *PLoS One* 2011, 6(3):e17664.
16. Beggs C, Bond AJ: A CUSUM tool for retrospectively evaluating team performance: The case of the English Premier League. *Sport, Business and Management: An International Journal* 2020, 10(3):263–289.
17. StatsPerform: Opta event definitions. In: https://www.statsperform.com/opta-event-definitions/; 2022.

2

Getting Started with R

In Chapter 1, we briefly looked at the capabilities of R as a tool for analysing soccer data. In this chapter, we build on that introduction and look further at the practicalities of coding in R. Through a series of easy-to-follow tutorial examples, we will learn how to program in R and produce robust working code. Hopefully, by the end of the chapter, you should be up and running and be able to write simple R code. Hopefully, you will also be confident enough to be able to experiment with and adapt R code that others have written and posted on the Internet, which is a great way to learn.

Of course, some people reading this book will already have experience coding with R and be familiar with the concepts associated with statistical computing and machine learning. If this is the case for you, then you are advised to skip this chapter and move on to Chapter 3. Others, however, may have no background in coding or data science and therefore be in need of a quick tutorial on how to use R. If this applies to you, then this chapter is one that you should read. Its aim is simply to get you up and running with R in as short a time as possible and to make you familiar with a few key concepts that you will find helpful when analysing soccer data. In addition, we will look at how to install R and RStudio and how to load data into R.

2.1 R and RStudio

R is a free, open-source computer programming language, which was developed in the 1990s specifically for statistical and data analysis purposes [1]. As such, it has become the principal tool used by departments of statistics in universities around the world to analyse data. In recent years, its popularity has increased greatly in professional sport and sports science as people have recognised its potential as a tool to analyse the performance of athletes and teams [2–4]. Because it is open-source software, R incorporates a plethora of library packages that can be freely downloaded and used to perform a wide variety of statistical and machine learning-related tasks, making it an extremely powerful tool with which to analyse data. As such, R is an ideal tool for analysing complex datasets associated with soccer.

Although an excellent programming language, the basic version of R is a cumbersome piece of software, which is somewhat difficult to use on its own.

DOI: 10.1201/9781003328568-2

Therefore, it is strongly advised that those wanting to learn R should use an integrated development environment (IDE), which is a wrapper around R that makes it quick and easy to use. Although several IDEs exist, the most popular is RStudio (part of the Posit ecosystem since 2022), which is a multi-paradigm numerical computing environment designed specifically to run R. RStudio has a simple user interface that allows the user to both enter code and run programs in R quickly and easily. Therefore, in this book, we will assume RStudio to be the preferred IDE in which to run R.

2.2 Installing R and RStudio

Before you learn to use R, it is necessary to first install both R and RStudio (both of which are free open-source software packages) on your laptop or computer. As stated above, R is a computer programming language, which was developed in the 1990s for statistical and data analysis purposes [1], while RStudio is a numerical computing environment designed specifically to run R [5].

You should install R first. Details of all the mirror URLs where R can be downloaded are found at https://cran.r-project.org/mirrors.html. It is advisable to choose a mirror location that is close to you, and this will direct you to the appropriate download site. You can then decide which version of R you wish to install (i.e., there are versions for Windows, Macintosh, and Linux operating systems).

Having installed R, you should then install RStudio from Posit, which can be found at:

https://posit.co/download/rstudio-desktop/

There are several versions of RStudio, but most users opt for the free version, which is the one recommended for use with this book. Once loaded onto your computer, the RStudio interface (together with R) should automatically start up when you click on the RStudio icon on the desktop of your computer.

2.3 The RStudio Interface

The RStudio user interface is shown in Figure 2.1. From this, it can be seen that the interface has several windows that perform different tasks. Window 1 is the 'Editor,' and this is where you write your program code; this is the window that you will probably use most. Window 2 is the 'Command window,' and this is where the results of your code will be displayed when you run any programs or scripts that you have created. In addition, if you ask

R to produce graph plots, these will be displayed in Window 4. Details of any library packages that are installed will also be displayed in Window 4. Window 3 allows you to see in detail the data that is stored in the various vectors, matrices, and data frames that have been created by your programs (or scripts, as they are often called).

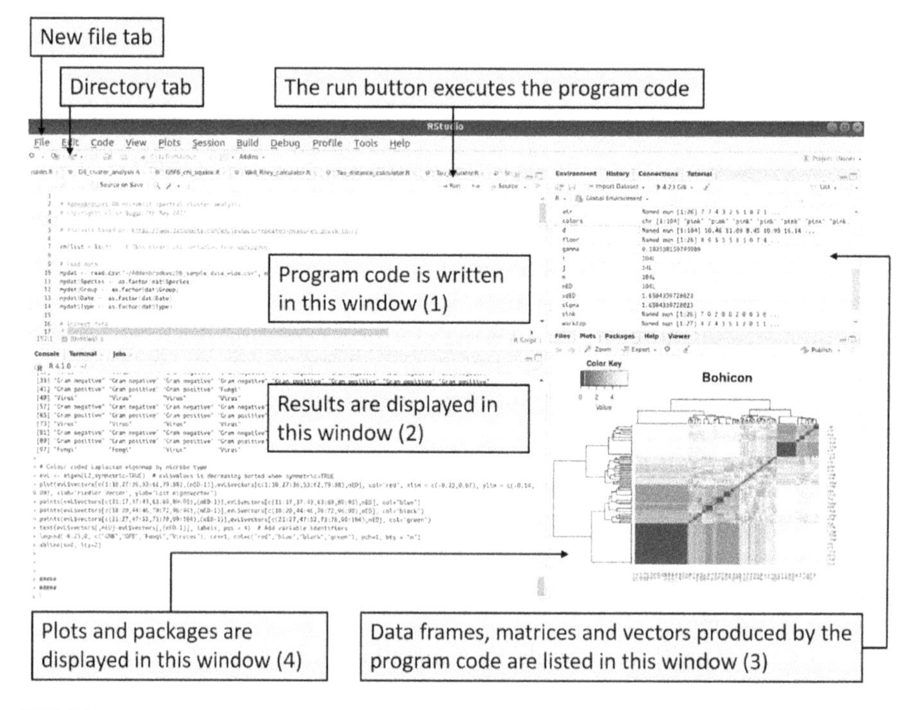

FIGURE 2.1
The RStudio interface with descriptions of the various windows and icons.

Before you start to write code in R using RStudio, there are a few things that are important to know. These are as follows:

1. In R, the '#' symbol is the 'remark' symbol. It enables you to insert comments and remarks into your program without R thinking that your remarks are part of the computing code. As such, any comment prefixed by '#' will have no effect on your computer program. It is very good practice to add comments to your code because this will help you remember what you have done and why you have done it. It also helps others understand your R code when they come to read it.
2. In R, the symbol '<-' is equivalent to '='. You can use '=', but '<-' is preferred because it instantly distinguishes R code from that of other programming languages.

3. In RStudio, if you want to run all or any part of your code, you must first highlight the code before you press the 'run' tab (located above Window 1 in Figure 2.1), as shown in Figure 2.2. If you don't do this, nothing will happen when you press the 'run' button. Alternatively, if you want to run a single line of code, then you can place the cursor on the selected line of code and press the 'run' button.

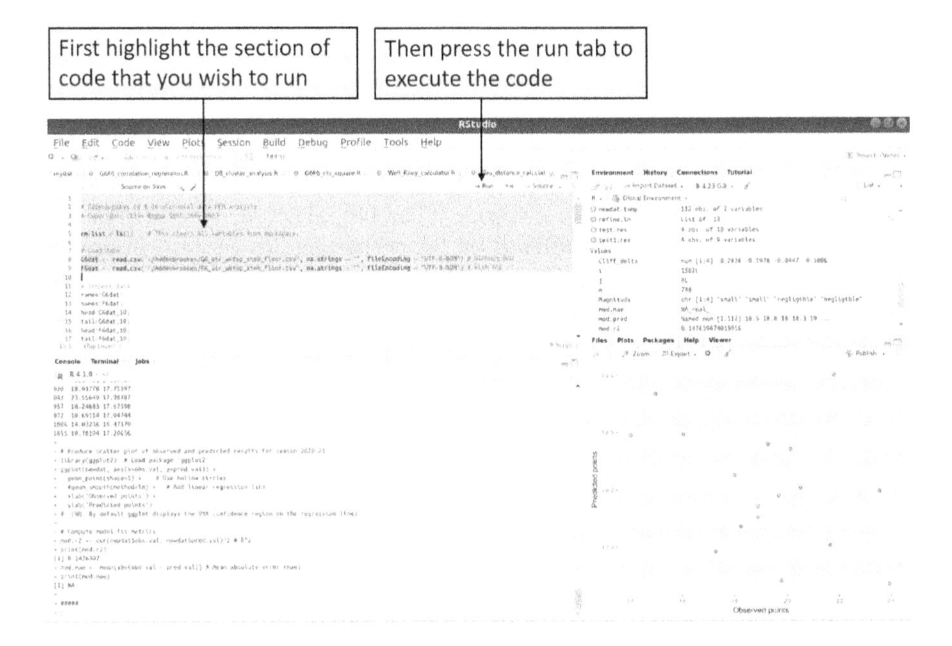

FIGURE 2.2
When using RStudio, the section of code that you wish to execute must first be highlighted (as shown) before it can be run.

2.4 Learning to Code the Easy Way

Many people feel daunted at the thought of learning to code (i.e., write computer programs). Where do you start, and how do you remember all the rules and syntax of the computer language? But in reality, learning to code in a new computer language is actually quite easy if you approach things in the right way. Consider, for example, how a young child learns a language. They do so by copying and experimenting in an iterative process that involves making lots of mistakes. At first, they copy simple sounds, which they frequently hear, but as they learn and become confident, they adapt and synthesise words to form sentences, hopefully learning some grammar along the way. Although this might seem like a rather random process, it is in fact a very successful strategy that enables young children to master complex languages relatively quickly.

Similarly, the best way to learn a computer language like R is to copy code that other more experienced programmers have written and experiment with it. Don't start your learning process by trying to write completely new computer code from scratch; this is the slow way to learn! Instead, find some existing R code that is both interesting and accessible and simply copy it directly into the Editor window (Window 1 in Figure 2.1) of RStudio. Then run the program and see what happens. If it runs successfully, you should see some output (e.g., results) displayed in the RStudio command window (Window 2 in Figure 2.1), and possibly a graphical plot might appear in the plot window (Window 4 in Figure 2.1). Once you have the code running, examine it carefully to see why it works. It can be particularly helpful to create remarks and comments prefixed by '#' in the code to explain things so that you understand (and remember) why the code works. Remember, rather than simply knowing that a piece of R code works, it is more important that you understand why it works.

This book assumes that the best way to learn to program in R is to copy code that others have written and adapt it to suit your own needs. To this end, the example codes presented in this book have been structured in such a way as to teach you how to program in R as well as enable you to analyse soccer data. You are therefore encouraged to copy the various R code examples in this book and to load these into RStudio to see how they run. To make things easy for you, all the R scripts and data files used in this book have been deposited in the GitHub repository (which can be found at: https://github.com/cbbeggs/SoccerAnalytics), and readers are encouraged to copy and adapt these to suit their own needs.

However, there is no point in blindly copying code like an unthinking automaton! In order for the copying process to be beneficial, it is important that you annotate the copied code with remarks explaining how and why the code works. You will then learn from the process, and it will help you remember things when you review the code at a later date. Also, you are encouraged to adapt and improve the example codes in this book, as this will help you develop your programming skills. However, remember to annotate your adaptations with remarks to remind you of why you wrote your code in a particular way. This is important because when you look back at your code in, say, a year or two's time, you may not remember why you took the approach that you did, and so the annotated remarks will help to refresh your memory.

The overall aim of this chapter is not to make you an expert at programming in R but rather to give you insights into the capabilities of R and enable you to become someone who can confidently code in R. Once you become confident and are able to stand on your own two feet with R, like a child developing their language skills, you will be in a position to tackle the more advanced problems presented in the other chapters of this book.

2.5 Starting to Code in R

Given that children learn languages by trial and error, before we look at any syntax, let's dive straight into R and have a go at some coding, as this is the best way to learn. To do this, you should copy the R code presented in the following examples into RStudio and then run it. Please note that throughout this book, lines of R code are denoted with a shaded background, which identifies that it is a line containing executable code. To execute this code, simply copy it into the editor window (Window 1 in Figure 2.1) in the RStudio interface and then run it. Once the code has run, the output it produces will appear in the command window (Window 2 in Figure 2.1). This will be denoted in the text of this book with a '##' symbol at the start of the line or table.

In this first example (Example 2.1), we will consider a soccer team that has played ten matches. Because the example is for teaching purposes only, we will simply make up the data by creating two lists (so-called vectors) of arbitrary numbers, one containing the goals scored and the other the goals conceded by a fictitious soccer team. In Example 2.1, we will also introduce the concept of data frames, which are useful data objects widely used in R, and produce some simple descriptive statistics (see Key Concept Boxes 2.1 and 2.2 for an overview of data frames and descriptive statistics). The aim of this first example is to give you a feel for how data is handled in R.

EXAMPLE 2.1: VECTORS AND DATA FRAMES

Before starting to code it is always good practice to first clear all the existing variables and data from the RStudio workspace, as these might interfere with the new script. This can be easily done using the following line of R code.

```
rm(list = ls())  # This clears all existing variables and data from the workspace.
```

Now that we have cleared the workspace and have a 'blank slate', we can proceed with our first task, which is to create a vector containing the match ID numbers, which in this case is simply 1–10. This represents the ID numbers for ten consecutive matches played by the study team. In R, a vector is simply a list of numbers separated by commas, which is enclosed in brackets and has 'c' at the front. So, we can create a vector containing the match ID numbers 1–10 using the following code:

```
MatchID <- c(1:10) # This creates a vector (1,2,3,4,5,6,7,8,9,10) called 'MatchID'.
```

NB. Here 1:10 denotes a sequence from 1 to 10 in R.
To display this vector we use the command 'print' as follows:

```
print(MatchID) # This displays the vector called 'MatchID'
```

```
## [1]  1  2  3  4  5  6  7  8  9 10    # This is the RStudio output.
```

NB. You can ignore the '[1]' term in the line above, as this simply denotes the line number in the RStudio output.

Next, we will create two new vectors, one containing the goals scored and the other the goals conceded. Note that we use the same format as before, but this time we enter the actual number of goals scored and conceded in each match separated by commas. Again, the outputs produced by RStudio are also shown.

```
GoalsFor <- c(0,2,4,1,3,0,2,2,3,1)  # This is the vector containing the goals scored.
print(GoalsFor) # This displays the goals scored vector.
```

```
## [1] 0 2 4 1 3 0 2 2 3 1   # This is the RStudio output.
```

```
GoalsAgainst <- c(1,1,3,3,0,0,1,1,1,0) # This is a vector containing the goals conceded.
print(GoalsAgainst) # This displays the goals conceded vector.
```

```
## [1] 1 1 3 3 0 0 1 1 1 0   # This is the RStudio output.
```

Having entered the data in the respective vectors, we can then calculate the goal difference for each match and store this in a vector called 'GoalDiff' as follows:

```
GoalDiff <- GoalsFor-GoalsAgainst  # This is a vector containing the goal differences.
print(GoalDiff)  # This displays the goal difference vector.
```

```
## [1] -1  1  1 -2  3  0  1  1  2  1   # This is the RStudio output.
```

Once the respective vectors have been created we can then use them to produce some descriptive statistics as follows:

Goals scored:

```
mean(GoalsFor) # Computes the mean.
```

```
## [1] 1.8
```

```
median(GoalsFor) # Computes the median.
```

```
## [1] 2
```

```
sd(GoalsFor) # Computes the standard deviation.
```

```
## [1] 1.316561
```

```
var(GoalsFor) # Computes the variance.
```

```
## [1] 1.733333
```

Notice that the number of decimal places in the output is not regulated by the above code. However, regulation of the decimal places can easily be done using the 'round' function, which we now apply to produce the rest of the descriptive statistics. (NB. Here we instruct R to restrict the display of the results to 3 decimal places.)

Goals conceded:

```
round(mean(GoalsAgainst),3) # Computes the mean to 3 DP.
```

```
## [1] 1.1
```

```
round(median(GoalsAgainst),3) # Computes the median to 3 DP.
```

```
## [1] 1
```

```
round(sd(GoalsAgainst),3) # Computes the standard deviation to 3 DP.
```

```
## [1] 1.101
```

```
round(var(GoalsAgainst),3) # Computes the variance to 3 DP.
```

```
## [1] 1.211
```

Goal difference:

```
round(mean(GoalDiff),3) # Computes the mean to 3 DP.
```

```
## [1] 0.7
```

```
round(median(GoalDiff),3) # Computes the median to 3 DP.
```

```
## [1] 1
```

```
round(sd(GoalDiff),3) # Computes the standard deviation to 3 DP.
```

```
## [1] 1.418
```

```
round(var(GoalDiff),3) # Computes the variance to 3 DP.
```

```
## [1] 2.011
```

Finally, we can combine the various individual vectors into a single data object called a data frame in R. Data frames are the fundamental data storage object used in R. For most general purposes, they can be considered tables of data, although there are subtle differences (see Key Concept Box 2.1 for more details of vectors, matrices, and data frames). So, now let's combine the four vectors that we have created into a single data frame called 'goals_dat', as follows. We could use the 'cbind' function to combine the vectors, but here we use the related function 'cbind.data.frame' instead because it ensures that a data frame is also created.

```
goals_dat <- cbind.data.frame(MatchID, GoalsFor, GoalsAgainst, GoalDiff)
# NB. The 'cbind.data.frame' command creates a data frame
print(goals_dat) # Displays data frame
```

Which produces:

```
##
  MatchID GoalsFor GoalsAgainst GoalDiff
1       1        0            1       -1
2       2        2            1        1
3       3        4            3        1
4       4        1            3       -2
5       5        3            0        3
```

```
6    6    0    0    0
7    7    2    1    1
8    8    2    1    1
9    9    3    1    2
10   10   1    0    1
```

NB. The first column of the output simply contains the line numbers in RStudio.

Inspection of 'goals_dat' reveals that the four columns (also called variables) in the data frame are actually the four vectors that we created, and that the vector names we specified are now the variable names in the data frame. In R, vectors are assumed to be column vectors, which can be combined (concatenated) horizontally using the 'cbind' function. However, if we simply use the 'cbind' command, then we get a matrix, whereas if we use 'cbind.data.frame', this combines the vectors to produce a data frame.

KEY CONCEPT BOX 2.1: VECTORS, MATRICES AND DATA FRAMES IN R

In mathematics, we often refer to vectors, matrices, and scalar quantities. A matrix is simply a rectangular array of numbers bounded by brackets, like the example below, which shows a [5 × 3] matrix (i.e., 5 rows and 3 columns).

$$\textit{Example of a } 5 \times 3 \textit{ matrix:} \quad \begin{bmatrix} 2.5 & -1.0 & 0 \\ 1.0 & 0 & 1.4 \\ -2.2 & 2.3 & 0.2 \\ 4.2 & -2.3 & -3.5 \\ -1 & 0 & -2.4 \end{bmatrix}$$

If we consider this matrix, we can see that it could be described as 3 separate columns containing 5 numbers (i.e., column vectors) that have been joined together (i.e., horizontally concatenated) to form a matrix. In linear algebra, a vector is simply a single column of numbers, or, in other words, a $[n \times 1]$ matrix, where n is the number of elements in the vector. Each individual element of a vector is called a scalar value, which can be considered equivalent to a $[1 \times 1]$ matrix.

In R, individual vectors can be created using the syntax c(element 1, element 2, element 3, …) as follows:

```
vector.1 <- c(2, 4, 5, 0, 4)
vector.2 <- c(-2, 5, 6, 1, 3)
```

Once created, these two vectors can then be combined (horizontally concatenated) using the 'cbind' command to produce the following [5 × 2] matrix (matrix.1). This can be done using the following R code:

```
matrix.1 <- cbind(vector.1, vector.2)
```

Alternatively, we could vertically concatenate vector.1 and vector.2 using the 'rbind' command in R to produce the following [2×5] matrix (matrix.2).

$$matrix.1: \begin{bmatrix} 2 & -2 \\ 4 & 5 \\ 5 & 6 \\ 0 & 1 \\ 4 & 3 \end{bmatrix} \qquad matrix.2: \begin{bmatrix} 2 & 4 & 5 & 0 & 4 \\ -2 & 5 & 6 & 1 & 3 \end{bmatrix}$$

When dealing with matrices, it is important to remember that it is only possible to add and subtract matrices of the same size and shape (i.e., the same dimensions). So for example, we can add a [5 × 2] matrix to another [5 × 2] matrix but not to a [3 × 2] matrix. When you add (or subtract) two matrices with the same dimensions, all we do is add (or subtract) each individual element in one matrix to (or from) its counterpart in the second matrix. In R, the dimensions of a matrix can easily be identified using the 'nrow' and 'ncol' functions, which give the number of rows and columns, respectively.

Sometimes we want to transpose a matrix and turn it on its side (e.g., convert it from matrix.1 to matrix.2). This can be easily done in R using the following code:

```
matrix.2 <- t(matrix.1)
```

The code for the reverse action is:

```
matrix.1 <- t(matrix.2)
```

Although R uses matrices to perform linear algebra, it is often much more useful to think of matrices as tables of data. To this end, R utilises a special class of data object called a data frame, which is a two-dimensional array-like structure. Unlike matrices, with data frames, the columns and rows can be assigned names, which can be extremely useful when navigating complex datasets. Data frames can also include columns containing words (i.e., strings) rather than numbers. This makes data frames a very useful type of data object because complex data sets often include a mixture of different data types (e.g., integers, floating-point numbers, dates, strings). It is possible to convert a matrix to a data frame using the *'as.data.frame'* command in R. It is also possible to determine the structure of a data frame and list its variables using the *'str'* and *'names'* commands, respectively.

While in Example 2.1 it is easy to display the data frame 'goals_dat' in its entirety, with many larger data frames, some of which may contain thousands of rows (observations) and hundreds of columns (variables), this is impossible. So, it is useful to know a few techniques that can be used to interrogate data frames in order to find out what is in them. Example 2.2 shows a few simple techniques in R that can be used to quickly interrogate data. For continuity, we will continue to use the 'goal_dat' data frame.

EXAMPLE 2.2: INTERROGATING DATA

One of the most useful commands in R is the 'names' function, which simply lists all the variable (column) names in a data frame. Using this command instantly enables you to sce what is in your data set, as follows:

```
names(goals_dat)
```

```
## [1] "MatchID"      "GoalsFor"      "GoalsAgainst" "GoalDiff"
```

From this, we can see how useful the 'names' command is because it enables variable names to be displayed at any point in an R script.

Another useful command with which to quickly inspect a data frame is the 'head' function, which displays the first six rows in the data frame.

```
head(goals_dat)
```

```
##
  MatchID GoalsFor GoalsAgainst GoalDiff
1       1        0            1       -1
2       2        2            1        1
3       3        4            3        1
4       4        1            3       -2
5       5        3            0        3
6       6        0            0        0
```

By default, the 'head' function displays just six rows. However, this can be altered as follows, allowing the user to specify the number of rows to be displayed. In this case, we display the first eight rows.

```
head(goals_dat, 8) # This tells R to display the first eight rows.
```

```
##
   MatchID GoalsFor GoalsAgainst GoalDiff
1        1        0            1       -1
2        2        2            1        1
3        3        4            3        1
4        4        1            3       -2
5        5        3            0        3
6        6        0            0        0
7        7        2            1        1
8        8        2            1        1
```

Likewise, the 'tail' command can be used to tell R to display the last six rows in the data frame.

```
tail(goals_dat)
```

```
##
   MatchID GoalsFor GoalsAgainst GoalDiff
5        5        3            0        3
6        6        0            0        0
7        7        2            1        1
8        8        2            1        1
9        9        3            1        2
10      10        1            0        1
```

Often, it is useful to know the size of a data frame. This can be done using two commands, 'nrow' and 'ncol', as follows:

```
nrow(goals_dat) # This gives the number of rows in the data frame.
```

```
##          [1] 10
```

```
ncol(goals_dat) # This gives the number of columns in the data frame.
```

```
##          [1] 4
```

We can also use the 'dim' function to perform the same task, as follows:

```
dim(goals_dat) # This gives the dimensions of the data frame.
```

```
##          [1] 10  4
```

So from this, we can see that the data frame 'goals_dat' comprises 10 rows (or observations) and 4 columns (or variables). In other words, it has dimensions [10 × 4].

We can also find out a lot about the structure of the data frame by using the 'str' command, as follows:

```
str(goals_dat) # Displays the structure of the data.
```

```
##
'data.frame':  10 obs. of  4 variables:
 $ MatchID     : int  1 2 3 4 5 6 7 8 9 10
 $ GoalsFor    : num  0 2 4 1 3 0 2 2 3 1
 $ GoalsAgainst: num  1 1 3 3 0 0 1 1 1 0
 $ GoalDiff    : num  -1 1 1 -2 3 0 1 1 2 1
```

This tells us that the 'goals_dat' data frame comprises 10 observations and 4 variables. Furthermore, it tells us that the first variable, 'MatchID,' comprises wholly of integers, whereas the other variables contain floating-point numbers (i.e., numbers that have a decimal place).

Often when coding, we want to select or display just part of a data frame, say, for example, the data in the second column (i.e., the 'GoalsFor' variable). In R, we can do this in two ways:

Method 1: Specifying the variable name using the '$' symbol.

```
goals_dat$GoalsFor # Here, $ denotes that 'GoalsFor' is a variable belonging to 'goals_dat'.
```

```
##   [1] 0 2 4 1 3 0 2 2 3 1
```

Method 2: Specifying the variable position using square brackets.

```
goals_dat[ ,2] # Here, the square brackets are used to select the second column.
```

```
##   [1] 0 2 4 1 3 0 2 2 3 1
```

Although both methods produce identical results, the first method uses the variable name prefixed by $, while the second uses square brackets to specify the location of the variable in the data frame (i.e., the second column in this case). Note that by specifying [,2] we are telling R that we want only the second column, whereas we are not controlling the number of rows that are selected. In other words, this tells R that we want all the rows to be included.

By extension of Method 2, we could use [,c(2,3)] to display both the second and third columns in 'goals_dat'.

```
goals_dat[,c(2,3)] # This selects the second and third variables in 'goals_dat'.
```

```
##
   GoalsFor GoalsAgainst
1         0            1
2         2            1
3         4            3
4         1            3
5         3            0
6         0            0
7         2            1
8         2            1
9         3            1
10        1            0
```

Similarly, if we want to select specific rows, we can do so as follows:

```
goals_dat[c(3,4,5),] # This selects the third, fourth and fifth rows in 'goals_dat'.
```

This selects the third, fourth and fifth rows in 'goals_dat' and produces:

```
##
  MatchID GoalsFor GoalsAgainst GoalDiff
3       3        4            3        1
4       4        1            3       -2
5       5        3            0        3
```

2.6 For Loops and If Statements

Now that we understand the concept of data frames and have a feel for how data is handled in R, it is perhaps worth considering how we can use the information contained in the data to make decisions or determine outcomes. For example, we might want to add a new column to the data frame 'goals_dat' that contains the outcome (i.e., win, lose, or draw) of the individual matches. While this could be done by hand, it would be a slow process (especially if the data frame was large), and therefore it is much quicker to use the specialised functions in R designed to perform repetitive tasks, as illustrated in Example 2.3. These functions allow the user to perform repeated tasks very rapidly, so they are widely used in data analytics.

In Example 2.3, we apply two methods to the 'goals_dat' data frame to determine match outcomes: one using the 'ifelse' function and another using a 'for loop' in conjunction with the 'if' statement. 'For loops' are widely used in coding to execute repetitive tasks in large data sets, and as such, are a technique that is well worth learning. Both methods can be useful, but as you will see, the 'ifelse' function is best suited to situations requiring dichotomous outputs (e.g., yes or no; win or lose; etc.), while 'for loops' are more flexible. (NB. There are other functions in R (e.g., 'lapply', 'sapply') that can also be used to perform repetitive tasks more rapidly than 'for loops', but these are not covered here.)

EXAMPLE 2.3: USING 'FOR LOOPS' AND 'IF' STATEMENTS TO PERFORM REPEATED TASKS

Let's look again at the 'goals_dat' data frame from Example 2.1. We can remind ourselves of the variables in this data frame using the command:

```
names(goals_dat)
```

which produces the output:

```
## [1] "MatchID"      "GoalsFor"     "GoalsAgainst" "GoalDiff"
```

(NB. Before working with a data frame, it is good practice to first have a look at it using the commands outlined in Example 2.2, as this will help you when writing new code.)

From this, we see that nowhere in the data frame does it tell us the outcome of the individual matches (i.e., win, lose, or draw). So in this example, we will use the existing goals scored and conceded data to create a new categorical vector containing the match outcomes.

From the 'GoalsFor' and 'GoalsAgainst' variables, it is not difficult to work out the match outcomes by hand. For any given match, if the GoalsFor is greater than the GoalsAgainst, then our team wins. Likewise, if the GoalsFor is less than the GoalsAgainst, then the team lost. Finally, if the GoalsFor equals the GoalsAgainst, then the match was a draw. However, while this is very easy to compute in our heads or on a piece of paper, it is actually somewhat tricky to do when coding it in R, particularly if the functions used are best suited to dichotomous outcomes. So, let's consider two of the ways in which we might approach this problem.

Method 1. Use the 'ifelse' function to specify the match outcome. This inspects the specified vectors using some test criteria to determine whether or not the test is met. As such, it is an excellent method to use if a dichotomous output is required.

```
outcome1 <- ifelse(goals_dat$GoalsFor > goals_dat$GoalsAgainst, "Win", "Did not win")
print(outcome1)
```

```
## [1] "Did not win" "Win"  "Win"  "Did not win" "Win" "Did not win" "Win"
 [8] "Win"   "Win"   "Win"
```

What the 'ifelse' function is doing here is comparing the observations in the GoalsFor and GoalsAgainst columns and applying the specified test. So, if GoalsFor is greater than GoalsAgainst, then "Win" is inserted in the 'outcome1' vector; otherwise "Did not win" is inserted. From this, we can see that while the 'ifelse' method is helpful, it does not get us to where we want to be because we actually want an output vector containing three categories (i.e., win, lose, and draw) rather than just two. Nevertheless, the 'ifelse' function would have been an excellent tool to use had we wanted a dichotomous outcome.

Method 2. Use the 'if' function in conjunction with a 'for loop' to specify the match outcome. This runs along the specified vectors, one step at a time, and applies the specified tests. As such, it is a much more flexible approach compared with Method 1 above.

Unlike Method 1, in Method 2, we have to first create an empty vector (called in this case 'outcome2') in which to store the results produced by the 'for loop'. This can be done as follows:

```
outcome2 <- c() # This creates an empty vector in which to store the results.
```

Next, we want to know how many iterations (repeated actions) to specify in the 'for loop', which is actually the number of matches (i.e., 10 in our case). This can be done by finding out the number of rows in the 'goal_dat' data frame using the 'nrow' function, as follows:

```
n <- nrow(goals_dat)
```

Here we assign the variable name '*n*' to the number of iterations in the 'for loop'. Having done this, we can then write the 'for loop' code as follows:

```
for(i in 1:n){
  if(goals_dat$GoalsFor[i] > goals_dat$GoalsAgainst[i]){outcome2[i] <- "Win"}
  if(goals_dat$GoalsFor[i] < goals_dat$GoalsAgainst[i]){outcome2[i] <- "Lose"}
  if(goals_dat$GoalsFor[i] == goals_dat$GoalsAgainst[i]){outcome2[i] <- "Draw"}
}
```

NB. The symbol '==' means "is identical to" or "is exactly the same as" in R. It should be used instead of '=' when writing 'if' statements.

```
print(outcome2)
```

This produces the following vector, which contains the desired three categories.

```
## [1] "Lose" "Win"  "Win"  "Lose" "Win" "Draw" "Win"  "Win" Win"  "Win"
```

Reviewing the 'for loop' R code, we can see that it comprises two main parts:

- A 'for' statement that has the general form 'for(i in 1:n){…}'; and
- A series of 'if' statements contained within the curly brackets of the 'for' statement.

First, let's look at the 'for(i in 1:n){…}' bit of the code in more detail. What this code actually does is create a temporary variable called 'i' that counts, one step at a time, along a series from 1 to n (i.e., from 1 to 10 in this case), telling R to perform some sort of repeated action (task), which is specified in the curly brackets. It starts at i=1, which is the first element in the vectors 'GoalsFor' and 'GoalsAgainst', and runs along until it reaches i=10, which is the last element in the respective vectors. When i=10, the 'for loop' stops.

Now, let's look at the 'if' code within the curly brackets. The first thing to notice is that the 'if' statement has the same basic form as the 'for' statement (i.e., a command applied to a statement or test contained in brackets, which tells R to do something specified in curly brackets). Let's consider, for example:

```
if(goals_dat$GoalsFor[i] > goals_dat$GoalsAgainst[i]){outcome2[i] <- "Win"}
```

This code is telling R that if the ith element in the 'GoalsFor' vector is greater than the ith element in the 'GoalsAgainst' vector, then insert "Win" in the 'outcome2' vector at the ith position. Here '[i]' is used to denote the position of the counting variable, i, on the respective vectors.

In this example, to keep things simple, we have just used three 'if' statements to create the desired three outcomes. However, in other applications, it is often helpful to incorporate an 'else' statement into the 'for loop', which denotes that after a specific test has been applied, everything else should be whatever is specified in the curly brackets.

Notice also that in the third 'if' statement we used '==' rather than '=' in the test. This is because R uses relational operators to compare vector elements (see Table 2.2), and '==' is the relational operator for equal to.

Having created the 'outcome2' vector, we can now combine this with the 'goals_dat' data frame as follows to create a new data frame called 'match_dat'.

```
match_dat <- cbind.data.frame(goals_dat, outcome2)
names(match_dat)
```

```
## [1] "MatchID"  "GoalsFor"  "GoalsAgainst"  "GoalDiff"   "outcome2"
```

But "outcome2" is perhaps not the best name to use for the match outcome variable. So, we can easily change it to, say "Result", in R using the following code:

```
colnames(match_dat)[colnames(match_dat) == 'outcome2'] <- 'Result'
print(match_dat)
```

Which produces the output:

```
##
   MatchID GoalsFor GoalsAgainst GoalDiff Result
1        1        0            1       -1   Lose
2        2        2            1        1    Win
3        3        4            3        1    Win
4        4        1            3       -2   Lose
5        5        3            0        3    Win
6        6        0            0        0   Draw
7        7        2            1        1    Win
8        8        2            1        1    Win
9        9        3            1        2    Win
10      10        1            0        1    Win
```

Often, when editing large data sets, we want to pick out observations that meet some specific criteria. So, to illustrate how this can be done in R, we will use the following code that selects all the games in which our study team won:

```
# Find all matches where the team won
winResults <- match_dat[match_dat$Result == "Win",]   # Selects the games won
print(winResults)
```

This instantly identifies all the matches that meet the selection criteria (i.e., won match) and displays them in the following table:

```
##
   MatchID GoalsFor GoalsAgainst GoalDiff Result
2        2        2            1        1    Win
3        3        4            3        1    Win
5        5        3            0        3    Win
7        7        2            1        1    Win
8        8        2            1        1    Win
9        9        3            1        2    Win
10      10        1            0        1    Win
```

2.7 Useful Syntax in R

Now that you have gained a feel for R and can do a little coding, it is perhaps a good time to look at some of the syntax used in R. To this end, the following section provides a useful overview of some of the syntax commonly used in R.

First, let's review the arithmetic operators used in R. These are listed in Table 2.1, from which it can be seen that most of them are pretty self-explanatory and familiar to the average user. The only exception is the matrix multiplication operator '%*%', which is an advanced technique that is used in linear algebra.

TABLE 2.1

Useful Arithmetic Operators Used in R

Arithmetic Operator	Description
x+y	$x+y$
x−y	$x-y$
x/y	$x \div y$
x * y	$x \times y$
X %*% Y	Matrix multiplication of X and Y
x^2	x^2
sqrt(x)	$\sqrt{x}$

From Example 2.3, we can see that relational operators (e.g. >, <, ==) are often used in 'for loops' to compare values. Table 2.2 presents a list of some useful relational and logical operators that are frequently used in R.

TABLE 2.2

Some Useful Relational and Logical Operators Used in R

Logical Operator	Description
<	Less than
<=	Less than or equal to
>	Greater than
>=	Greater than or equal to
==	Exactly equal to
!=	Not equal to
!a	Not a
a&b	a and b
a|b	a or b

2.8 Library Packages

Because it is open-source software, R incorporates a plethora of library packages that can be freely downloaded and employed to perform a wide variety of advanced statistical tasks, making its capabilities immense. These library packages can easily be downloaded and installed into RStudio using the 'Install' tab on the user interface (see Figure 2.3). Installed packages can be viewed using the 'Packages' tab on the RStudio interface (see Figure 2.3).

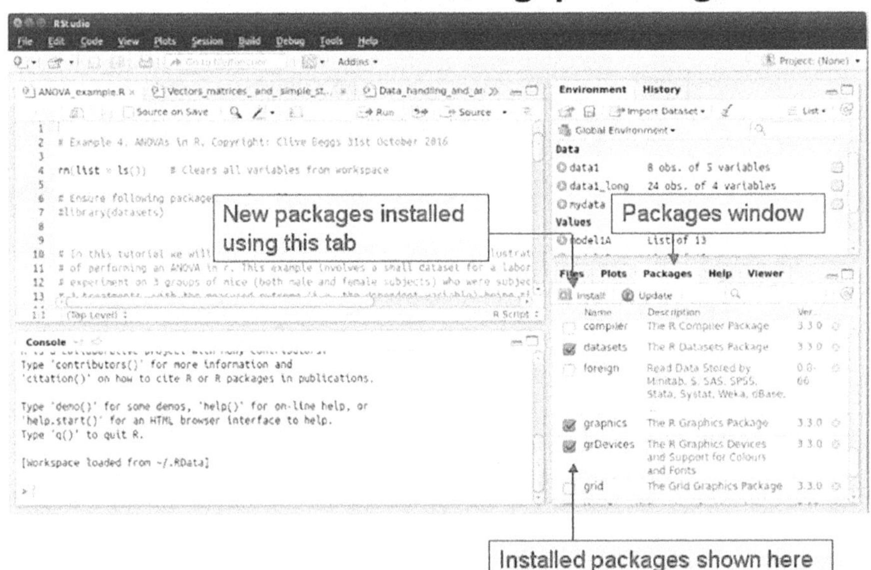

FIGURE 2.3
The RStudio interface with descriptions of the windows and tabs used to download library packages.

Because there are thousands of free library packages available in R, no attempt will be made here to describe them all; rather, we will content ourselves here with using one package, 'psych', to produce some sophisticated descriptive statistics. In Example 2.4, we will apply the 'describeBy' function in the 'psych' package to produce some descriptive statistics for the 'match_dat' data frame, which we created in Example 2.3.

EXAMPLE 2.4: USING THE 'PSYCH' LIBRARY PACKAGE TO PRODUCE DESCRIPTIVE STATISTICS

As before, let's first familiarise ourselves with the 'match_dat' data frame.

```
names(match_dat)
```

which produces:

```
##      [1] "MatchID"    "GoalsFor"   "GoalsAgainst" "GoalDiff"   "Result"
```

In R, we can produce a summary of descriptive statistics simply by applying the 'summary' function to a given data frame, as follows:

```
summary(match_dat)
```

which produces:

```
##
    MatchID         GoalsFor      GoalsAgainst      GoalDiff
 Min.   : 1.00   Min.   :0.00   Min.   :0.00   Min.   :-2.00
 1st Qu.: 3.25   1st Qu.:1.00   1st Qu.:0.25   1st Qu.: 0.25
 Median : 5.50   Median :2.00   Median :1.00   Median : 1.00
 Mean   : 5.50   Mean   :1.80   Mean   :1.10   Mean   : 0.70
 3rd Qu.: 7.75   3rd Qu.:2.75   3rd Qu.:1.00   3rd Qu.: 1.00
 Max.   :10.00   Max.   :4.00   Max.   :3.00   Max.   : 3.00
    Result
 Length:10
 Class :character
 Mode  :character
```

While the 'summary' command is extremely useful, it is noticeable that the function does not produce any standard deviation or variance results, which is a glaring omission. So, we need to turn to the 'describeBy' in the 'psych' package to produce more sophisticated results.

To do this, we first have to install the 'psych' library package, which can be done using either the 'Install' tab on the RStudio interface (as described in Section 2.8) or by typing the command:

```
install.packages("psych") # This installs the 'psych' package.
```

This command only needs to be executed once to install the package. Thereafter, the 'psych' library can be called using the command:

```
library(psych)  # This loads the psych library into R.
```

Having loaded the 'psych' package into RStudio, we can now start to use it. The command that we need is the 'describeBy' function, which, as you will see, is extremely powerful. It can be applied using:

```
describeBy(match_dat)
```

which produces:

```
##
             vars  n mean   sd median trimmed  mad min max range  skew
MatchID         1 10  5.5 3.03    5.5    5.50 3.71   1  10     9  0.00
GoalsFor        2 10  1.8 1.32    2.0    1.75 1.48   0   4     4  0.06
GoalsAgainst    3 10  1.1 1.10    1.0    1.00 0.74   0   3     3  0.73
GoalDiff        4 10  0.7 1.42    1.0    0.75 0.74  -2   3     5 -0.37
Result*         5 10  2.6 0.70    3.0    2.75 0.00   1   3     2 -1.19
             kurtosis   se
MatchID         -1.56 0.96
GoalsFor        -1.36 0.42
GoalsAgainst    -0.92 0.35
GoalDiff        -0.71 0.45
Result*         -0.07 0.22
```

From this, it can be seen that the 'describeBy' function produces a data frame of comprehensive results containing the: variable name; variable ID number; number of valid observations; the mean; standard deviation; trimmed mean (with trim defaulting to .1); the median; the median absolute deviation (mad); the minimum and maximum values; the skewness; kurtosis; and the standard error. However, closer inspection reveals two shortcomings in the output:

- The last variable, 'Result' is marked by an asterisk, which indicates that it is a categorical variable and that the results displayed for this variable are therefore not valid.
- The first variable, 'MatchID', although numeric, actually contains ID numbers and therefore should not be included in the descriptive statistics.

Therefore, we need to modify the code as follows to obtain just the descriptive results for variables two to four, which we have saved as a data frame called 'des_res'.

```
des_res <- describeBy(match_dat[,c(2:4)])
print(des_res)
```

which produces:

```
##
             vars  n mean   sd median trimmed  mad min max range  skew
GoalsFor        1 10  1.8 1.32      2    1.75 1.48   0   4     4  0.06
GoalsAgainst    2 10  1.1 1.10      1    1.00 0.74   0   3     3  0.73
GoalDiff        3 10  0.7 1.42      1    0.75 0.74  -2   3     5 -0.37
             kurtosis   se
GoalsFor        -1.36 0.42
GoalsAgainst    -0.92 0.35
GoalDiff        -0.71 0.45
```

KEY CONCEPT BOX 2.2: DESCRIPTIVE STATISTICS

Sporting data can be complex and difficult to understand, and so when communicating such data, we often want to condense it down into a form that can be easily understood. The standard way to do this is to produce so-called 'descriptive statistics'. These are statistics (metrics) that capture the essence of the data in such a way that it can be described to others in a manner that they can easily understand. However, when producing descriptive statistics, it is important to know the type of data being analysed; otherwise, errors can easily occur. For example, we can describe the average height of players on a soccer team, but we cannot describe their average eye colour. This is because there is no such thing as an average eye colour. So, with eye colour, all we can do is produce a frequency count (e.g., three players have blue eyes; five have brown eyes; two have hazel eyes; and one has green eyes) to describe the data.

Often, when assessing the performance of individual athletes or teams, we record various types of data. For example, we might record the number of goals scored by a team in each match, which will result in a vector containing a list of integers (whole numbers), because nobody can score 0.5 of a goal. We might also measure the amount of possession (expressed as a percentage) that a team has in each match, in which case the vector produced will be what is called a continuous variable, containing a list of floating-point (decimal-point) numbers (e.g., 45.2, 57.6, 35.0). Finally, we might want to record events that fall into specific categories, such as match results that can only be either a win, lose, or draw. This will produce what is known as a categorical variable that contains a sequence of results drawn from these three categories (e.g., win, win, draw, win, lose, lose). Each of these variable types contains its own peculiarities, which means that they need to be treated differently from each other when being analysed. So for example, while we can compute the average number of points that a team earns per match in a season (e.g., 1.85 points), we cannot talk about the average outcome of the team's matches; all we can do is say how many matches they won, lost, and drew.

So how do we produce descriptive statistics? Well, the usual way is to identify those metrics (variables) that are important and then decide whether or not they are categorical. If they contain specific categories (e.g., win, lose, draw), then we need to report the number of events or observations that fall into each category (e.g., 20 wins, 11 losses, and 7 draws). However, if the data is non-categorical and contains either floating-point numbers (i.e. a continuous variable) or integers (e.g., points per match (i.e., 0, 1, or 3) – so-called interval data), then it is possible to compute average values for the respective variables. Interestingly, the average (mean) value will always be a floating-point value, even if the data has specific intervals, as is the case with points awarded per match. So, it is perfectly OK to report that in a given season, a team earned an average of 1.76 points per match at home but only 1.23 away from home.

In statistics, the average value is the mean, which in the case of a single variable (i.e., a vector) is the sum of all the numbers (i.e., elements) in that vector divided by the number of elements. The mean value, $\bar{x}$, can be represented mathematically as follows: where x_i are the elements in the vector, and n is the count (i.e., the number of elements in the vector).

$$\text{Mean}: \quad \bar{x} = \frac{\sum_{i=1}^{n} x_i}{n} \tag{2.1}$$

Another measure of central tendency commonly used in statistics is the median value, which is the central number of a data set. This number divides the data exactly in half, so that 50% of the data in a vector is above this value and 50% is below it. Often, the mean and median values are close to each other, but this is not always the case, especially if the data is heavily skewed.

Often, we also want to communicate how widely the data is distributed around the mean. This can be done by computing the variance (σ^2), which can be calculated using:

$$\text{Variance: } \sigma^2 = \frac{\sum_{i=1}^{n} \left(x_i - \bar{x}\right)^2}{n} \tag{2.2}$$

However, it is often more convenient to express how widely the data is distributed in terms of the standard deviation (σ), which is the square root of the variance. This is because approximately two standard deviations (actually 1.96) from the mean account for 95% of the variance in the data in many data sets (i.e., data sets that have a normal distribution).

2.9 Loading Data into R

When working with R, one of the first challenges that beginners encounter is simply how to get data into R so that it can be analysed. Fortunately, this process is relatively straightforward if the data is in the correct format. Probably the easiest way to input data into R is to put it on a spreadsheet and then save it as a comma-separated values (CSV) file – something that can easily be done in *Microsoft Excel* or *OpenOffice Calc*. Once saved in CSV format, the data can be inputted into R using the 'read.csv' function. This command inputs a text file, which has a comma as a separator and a dot as a decimal point by default. Unless stated otherwise, the first line of the file is assumed to contain the column headers (i.e., the variable names).

The general way to apply the 'read.csv' function is as follows, where "input.csv" is the CSV file being loaded. In this case, for example, the code creates a data frame called 'dat'.

```
dat <- read.csv("input.csv")
```

In practice, however, it is likely that the "input.csv" file will be stored in a directory somewhere on the computer. So for example, if the CSV file is stored in a directory called say 'Datasets', then the file path would have to be included in the 'read.csv' command, as follows:

For a PC computer:
```
dat <- read.csv("C:/Datasets/input.csv")
```
For an Apple computer:
```
dat <- read.csv("~/Desktop/Datasets/input.csv")
```

Alternatively, the 'Import' button on the top right of the RStudio console can be used to import directly CSV files as well as Excel and SPSS files. The 'read_excel' function in the "readxl" package can also be utilised as follows:

```
install.packages("readxl") # This is required to install package.
```

This command only needs to be executed once to install the package. Thereafter, the 'readxl' library can be called using the command:

```
library("readxl") # This calls the readxl library.

# For xls files
dat <- read_excel("C:/Datasets/input.xls")

# For xlsx files
dat <- read_excel("C:/Datasets/input.xlsx")
```

Example 2.5 illustrates how the 'read.csv' function can be used to input a CSV file containing goals scored (GF) and goals conceded (GA) data for Arsenal and Chelsea over the seasons 2011–2020 in the English Premier League (EPL). The file containing the data is called "Arsenal_Chelsea_comparison.csv" and we will assume that it is stored in a directory called 'Datasets'. The 'Arsenal_Chelsea_comparison.csv' file can be downloaded from GitHub at the following address: https://github.com/cbbeggs/SoccerAnalytics. Once downloaded, you should save it in a directory called 'Datasets', which you should create on your computer's hard drive.

EXAMPLE 2.5: IMPORTING DATA AND EXPORTING RESULTS TO AND FROM R

This example illustrates how to import and export data to and from R using CSV files. Before starting to code, let's first clear all the existing variables and data from the RStudio workspace.

```
rm(list = ls())  # This clears all variables and data from the workspace.
```

Now the data can be loaded into R and displayed using the following code. (NB. The data file (Arsenal_Chelsea_comparison.csv) can be found at: https://github.com/cbbeggs/SoccerAnalytics and is assumed here to be located in the 'Datasets' directory.)

```
dat <- read.csv("C:/Datasets/Arsenal_Chelsea_comparison.csv")
print(dat)
```

(NB. Here we are assuming that the reader is using a Windows-based PC. When using an Apple Mac computer the code should be adjusted accordingly in keeping with the comments made in Section 2.9.)

This displays the 'dat' data frame as follows, from which it can be seen that the total goals for (GF) and goals against (GA), together with the points awarded in the EPL, are collated for Arsenal and Chelsea for the seasons 2011–12 to 2020–21.

```
##
   Season Arsenal_GF Arsenal_GA Arsenal_points Chelsea_GF Chelsea_GA
1  2011-12        74         49             70         65         46
2  2012-13        72         37             73         75         39
3  2013-14        68         41             79         71         27
4  2014-15        71         36             75         73         32
5  2015-16        65         36             71         59         53
6  2016-17        77         44             75         85         33
7  2017-18        74         51             63         62         38
8  2018-19        73         51             70         63         39
9  2019-20        56         48             56         69         54
10 2020-21        55         39             61         58         36
   Chelsea_points
1              64
2              75
3              82
4              87
5              50
6              93
7              70
8              72
9              66
10             67
```

Having loaded the data into R and created the 'dat' data frame, we can now do something with it, which in this case we shall do by producing some descriptive statistics using the 'describeBy' function from the 'psych' package, as follows:

```
library(psych)  # This loads the psych library into R.

des_results <- describeBy(dat[,c(2:7)])
print(des_results)
```

which displays:

```
##
               vars  n mean    sd median trimmed  mad min max range  skew
Arsenal_GF        1 10 68.5  7.62   71.5   69.12 4.45  55  77    22 -0.76
Arsenal_GA        2 10 43.2  6.18   42.5   43.12 8.90  36  51    15  0.08
Arsenal_points    3 10 69.3  7.17   70.5   69.75 6.67  56  79    23 -0.50
Chelsea_GF        4 10 68.0  8.33   67.0   67.12 8.15  58  85    27  0.57
Chelsea_GA        5 10 39.7  8.84   38.5   39.50 8.90  27  54    27  0.39
Chelsea_points    6 10 72.6 12.40   71.0   72.88 8.90  50  93    43 -0.02
               kurtosis   se
Arsenal_GF        -1.06 2.41
Arsenal_GA        -1.88 1.95
Arsenal_points    -1.15 2.27
Chelsea_GF        -0.81 2.63
Chelsea_GA        -1.26 2.80
Chelsea_points    -0.92 3.92
```

Having produced a data frame containing the descriptive results called "des_results", we can save this to a directory, which in this case we have called

"AnalysisResults". (NB. You can store it in any directory on your computer by substituting the appropriate directory name.)

```
write.csv(des_results, "C:/AnalysisResults/descriptive_results.csv")
```

Note that with the write.csv command, the first term in brackets is the name of the file being saved, and the second term in quotes is the path to where the file is being saved on your computer. (NB. The data frame name does not have to be the same as the CSV file name on your computer.)

2.10 Plotting Data

While descriptive statistics are useful, when communicating information to others, it is often helpful to produce plots, as these are generally easier to understand. These can easily be produced in R using a few simple commands. To illustrate how this can be done, consider Example 2.6, which shows how a simple time-series line plot can be produced using the 'plot' function. In R, the 'plot' function has the general form plot(*x*, *y*, ...) and can be used to produce both scatter plots and line plots. With line plots, it is generally used to specify the first line in the plot, with subsequent lines added using the 'lines' function, as follows:

```
plot(x, y, ...)  # This specifies the first line in the plot.
lines(x, y, ...) # This is used for subsequent additional lines.
```

When applying either function, it is necessary to specify the characteristics of the lines and data points used in the plot. This is done by incorporating arguments into the 'plot' and 'lines' statements, as summarised in Table 2.3.

TABLE 2.3

Arguments Used to Construct Plots in R

Description	Argument	Popular Options
Line configuration	type="..."	"p" for points; "l" for lines; "o" for over-plotted
Line type	lty=...	1 for a solid line; 2 for a dashed line; 3 for a dotted line; 4 for a dot-dashed line
Line thickness	lwd=...	1, 2, 3, etc. (NB. lwd = 1 is default)
Line colour	col="..."	"black" (default colour); "blue"; "red"; "green"; "magenta"; "cyan"; "orange"; "yellow"; "gray"
Data point character	pch=...	20 for a solid point; 1 for a hollow circle; 2 for a hollow triangle; 3 for a "+"; 4 for a "×"; 5 for a hollow diamond; 15 for a solid square; 17 for a solid triangle; 18 for a solid diamond
X-axis label	xlab="..."	Specify the units used in between the inverted commas
Y-axis label	ylab="..."	Specify the units used in between the inverted commas
Limits of X-axis	xlim=c(...,...)	Specify the lower and upper limits of the X-axis. For example xlim=c(0,10)
Limits of Y-axis	ylim=c(...,...)	Specify the lower and upper limits of the Y-axis. For example ylim=c(–10,10)

EXAMPLE 2.6: PRODUCING A LINE PLOT

In this example, we use the data in the 'dat' data frame to produce a time-series line plot of the goals scored and conceded by Arsenal and Chelsea over ten seasons in the EPL. To do this, we use the following code, in which the normal *x*-axis labels are replaced with a vector of characters containing the years in which the respective seasons commenced.

```
seasons <- c("2011","2012","2013","2014","2015","2016","2017","2018","2019","2020")

plot(seasons, dat$Arsenal_GF, type="o", lty=1, pch=20, col="black", ylim=c(0,140),
     ylab="Goals", xlab="Season")
lines(seasons, dat$Chelsea_GF, type="o", lty=2, pch=20)
lines(seasons, dat$Arsenal_GA, type="o", lty=1, pch=4)
lines(seasons, dat$Chelsea_GA, type="o", lty=2, pch=4)
legend(2011,145, c("Arsenal goals for","Arsenal goals against","Chelsea goals for",
"Chelsea goals against"), cex=0.8, col=c("black","black","black","black"),
lty=c(1,1,2,2), pch=c(20,4,20,4), bty = "n")
title("Arsenal and Chelsea comparison")
```

In this code, the 'plot' function is used to specify the first line, with 'lines' used thereafter and type="o" specified to indicate that the data point characters are overlaid onto the lines. Note that because we have not specified the line colour in the respective 'lines' statements, R automatically defaults to using black. So, when producing black-and-white plots, there is no real need to specify line colour. The code also includes two other commands, 'title' and 'legend'. The 'title' function is pretty self-explanatory and inserts a title above the plot, while the legend function inserts a legend at a specified location, which in this case is (2011, 145), signifying the position of the top left-hand corner of the legend box.

The above code produces the black-and-white line plot shown in Figure 2.4.

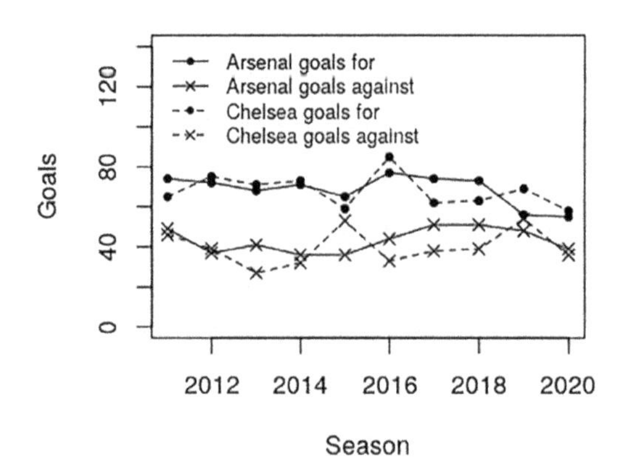

FIGURE 2.4
Line plot of goals scored and conceded by Arsenal and Chelsea over ten seasons.

From Figure 2.4, it can be seen that specifying (2011, 145) in the legend command positions the legend nicely on the plot. Here we have specified bty="n", which indicates that there is no boundary box around the legend (the default setting is to draw a boundary box). The argument 'cex=0.8' indicates that the lettering in the legend is only 80% of its default size. Notice also that in the 'legend' command we specify the descriptor labels, line types, and point characters as vectors having the general form c(...).

Box plots (sometimes called box and whisker plots) can also be produced very easily using as little as a single line of R code, as illustrated in Example 2.7. Box plots are an excellent method for graphically demonstrating the centrality, spread, and skewness of groups of numerical data. They comprise a box drawn from the first quartile to the third quartile, which is divided by the median value. In addition, whiskers go from each quartile to the minimum or maximum values.

EXAMPLE 2.7: PRODUCING A BOX PLOT

In this example, we will use the 'dat' data frame to produce a simple box plot of the goals scored and conceded by Arsenal and Chelsea, which we can do using just a single line of code.

```
boxplot(dat[,c(2,3,5,6)], ylab="Goals")
```

This produces the box plot in Figure 2.5, which gives the median, quartile and range values for the goals scored and conceded by both teams over the ten seasons.

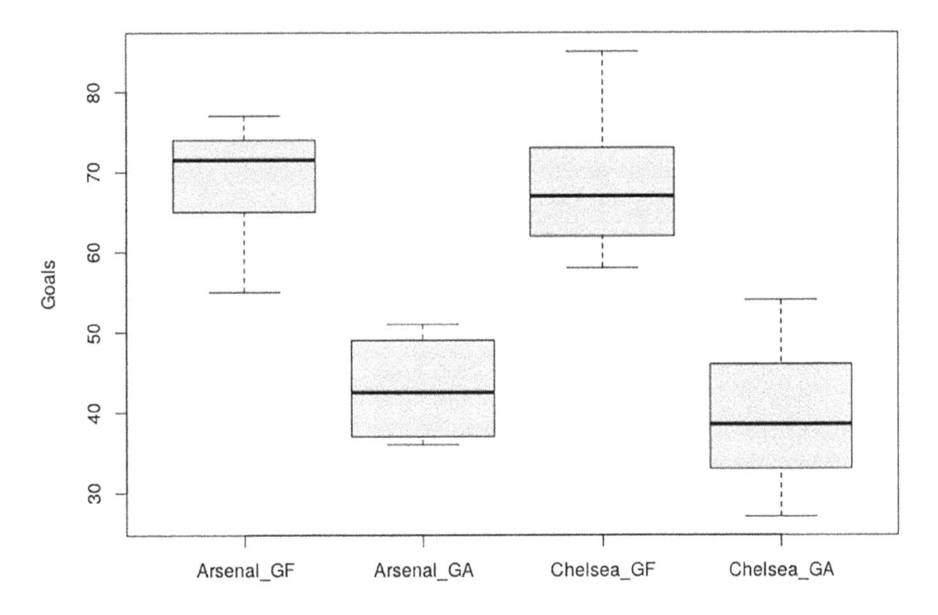

FIGURE 2.5
Box plot of goals scored and conceded by Arsenal and Chelsea over ten seasons.

Notice that the box plot mirrors the output produced using the 'summary' command, as follows:

```
summary(dat[,c(2,3,5,6)])
```

```
##
   Arsenal_GF      Arsenal_GA      Chelsea_GF      Chelsea_GA
 Min.   :55.00   Min.   :36.00   Min.   :58.00   Min.   :27.00
 1st Qu.:65.75   1st Qu.:37.50   1st Qu.:62.25   1st Qu.:33.75
 Median :71.50   Median :42.50   Median :67.00   Median :38.50
```

```
Mean   :68.50   Mean   :43.20   Mean   :68.00   Mean   :39.70
3rd Qu.:73.75   3rd Qu.:48.75   3rd Qu.:72.50   3rd Qu.:44.25
Max.   :77.00   Max.   :51.00   Max.   :85.00   Max.   :54.00
```

We can also produce scatter plots to show the relationships between the variables in the data. For example, we can compare goals conceded against points awarded for both teams using the code in Example 2.8.

EXAMPLE 2.8: PRODUCING A SCATTER PLOT

In this example, we use the 'dat' data frame to produce a black-and-white scatter plot of the goals conceded by Arsenal and Chelsea and the respective points awarded. This can be done using the following code:

```
plot(dat$Chelsea_GA, dat$Chelsea_points, pch=20, col="black", xlim=c(0,60),
     ylim=c(0,100), ylab="Points", xlab="Goals conceded")
points(dat$Arsenal_GA, dat$Chelsea_points, pch=4)
abline(lm(dat$Chelsea_points ~ dat$Chelsea_GA), lty=1)
abline(lm(dat$Arsenal_points ~ dat$Arsenal_GA), lty=2)
legend(5,40, c("Arsenal goals conceded","Arsenal bestfit line",
               "Chelsea goals conceded","Chelsea bestfit line"),
               cex=0.8, col=c("black","black","black","black"),
               lty=c(0,1,0,2), pch=c(4,NA,20,NA), bty = "n")
```

Notice that although we are using the 'plot' function, because we are creating a scatter plot, there is no need to specify the line configuration (i.e., no need to use type ="..."). Also, because we are producing a scatter plot, in the second line of the code, we use the 'points' function rather than the 'lines' function. Other than this, the arguments used with each command are very similar to those used in Example 2.6.

The above code produces the black-and-white scatter plot shown in Figure 2.6, which incorporates two least-squares best-fit lines through the data. The code for the best-fit lines utilises the 'abline' function together with the linear model function 'lm'.

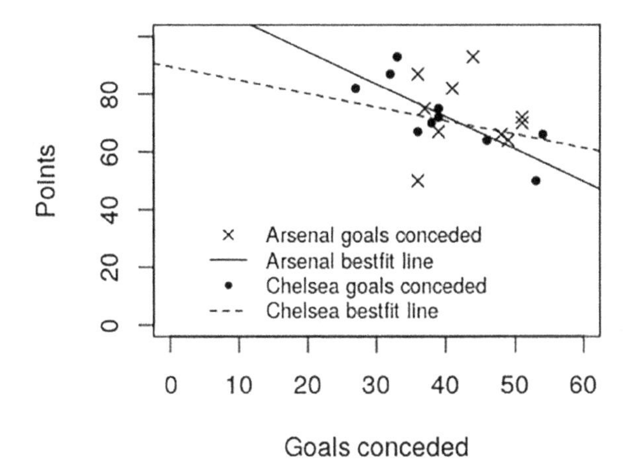

FIGURE 2.6
Scatter plot of goals conceded and points awarded for Arsenal and Chelsea over ten seasons.

2.11 Statistical Tests

When analysing data, it is important to remember that we often don't have the full picture. For example, if we wanted to compare the running speeds of soccer players in the EPL with those in, say, the German Bundesliga, we might perform a sprint experiment involving 20 players from each league. Having done this, if we observed that the tested German players were on average faster than their English counterparts, we might be tempted to conclude that players from the Bundesliga are faster than those from the EPL. But how do we know that this is actually the case? For example, we could have inadvertently selected 20 very fast players from the Bundesliga and 20 atypically slow players from the EPL, with the result that the observed difference between the sample groups is biased and not representative of players in the two leagues. So, this is where we have to rely on statistical tests, because we cannot measure the speed of every player in each league. With this in mind, when considering data, it often important to ask two simple questions:

- *"Is there a real difference (effect) between the two observed groups, or is the apparent difference just down to natural variation in the data (i.e., chance)?"*
- *"Is there a real relationship (correlation) between two observed measures (variables), or is the apparent relationship just down to chance?"*

The first of these questions could very well be applied to the box plot in Figure 2.5. Yes, Arsenal scored and conceded more goals than Chelsea over the ten seasons in the data set, but is this difference really significant? In other words, is there a real effect, or could the observed difference just be down to natural variation (i.e., chance)? (See Key Concept Box 2.3.) Likewise, when we consider the scatter plots in Figure 2.6, we can see that a broadly inverse relationship exists between goals conceded and the number of points awarded, but is this correlation a real effect, and if so, is it statistically significant for both teams?

It is very easy to answer both of these questions using R. The first can be answered using a paired *t*-test and the second using Pearson correlation analysis, as shown in Example 2.9. (N.B. More information on statistical hypothesis testing can be found in Key Concept Box 2.3.)

EXAMPLE 2.9: STATISTICAL TESTING

Again, we refer to the 'dat' data frame. Say we wanted to know if there was any statistical difference between (i) the goals scored by Arsenal and Chelsea and (ii) the goals conceded by Arsenal and Chelsea. To determine this, we could use a paired *t*-test, which can be applied using the following code for Welch's *t*-test. (NB. Welch's *t*-test is a widely used modification of Student's *t*-test, which is robust when the variances of the two samples are not equal to each other.)

```
t.test(dat$Arsenal_GF, dat$Chelsea_GF, paired=TRUE)
t.test(dat$Arsenal_GA, dat$Chelsea_GA, paired=TRUE)
```

In this example, the paired argument is set to TRUE because the data for Arsenal and Chelsea is matched. However, had the data not been matched, then we would set 'paired=FALSE', which is the default condition.

The outputs of these two tests are as follows:

```
## For goals scored:
     Paired t-test

data:  dat$Arsenal_GF and dat$Chelsea_GF
t = 0.19012, df = 9, p-value = 0.8534
alternative hypothesis: true difference in means is not equal to 0
95 percent confidence interval:
 -5.449373  6.449373
sample estimates:
mean of the differences
                    0.5

## For goals conceded:
     Paired t-test

data:  dat$Arsenal_GA and dat$Chelsea_GA
t = 1.1254, df = 9, p-value = 0.2895
```

```
alternative hypothesis: true difference in means is not equal to 0
95 percent confidence interval:
 -3.535353 10.535353
sample estimates:
mean of the differences
                    3.5
```

From this analysis, it can be seen that the p-values are 0.853 and 0.290 for goals scored and conceded, respectively, indicating that for both of these variables, there is no statistical evidence to indicate a difference between the performances of Arsenal and Chelsea over the ten seasons.

In order to determine whether or not the correlations shown in Figure 2.6 are significant, a Pearson correlation analysis needs to be performed to establish the correlation between goals conceded and points awarded. This can be done using the 'cor.test' function as follows:

```
# Arsenal
cor.test(dat$Arsenal_GA,dat$Arsenal_points)
```

```
##    Pearson's product-moment correlation

data:  dat$Arsenal_GA and dat$Arsenal_points
t = -1.2549, df = 8, p-value = 0.2449
alternative hypothesis: true correlation is not equal to 0
95 percent confidence interval:
 -0.8246159  0.3009105
sample estimates:
       cor
-0.4055522
```

```
# Chelsea
cor.test(dat$Chelsea_GA,dat$Chelsea_points)
```

```
##    Pearson's product-moment correlation

data:  dat$Chelsea_GA and dat$Chelsea_points
t = -3.8028, df = 8, p-value = 0.005216
alternative hypothesis: true correlation is not equal to 0
95 percent confidence interval:
 -0.9513764 -0.3491690
sample estimates:
       cor
-0.8023925
```

This reveals that for Arsenal, the correlation between goals conceded and points awarded is r=–0.406 (p=0.245, which is not significant), whereas for Chelsea, the relationship is much stronger, with $r=-0.802$ and $p=0.005$, which is highly significant. As such, the statistical results confirm the visually discernible trend in the scatter plot shown in Figure 2.6. There is a clear inverse relationship between goals conceded and points awarded in the case of Chelsea, which did not reach significance for Arsenal.

KEY CONCEPT BOX 2.3: STATISTICAL HYPOTHESIS TESTING

Often, when analysing data, we want to know things like:

- Is an observed event exceptional (possibly due to some underlying effect), or is it something that could have occurred due to random variation (i.e., just by chance)?
- Has an intervention (e.g., a new training regime or a change of manager) actually worked, or could the observed effect (i.e., the improvement in team performance) have occurred just by chance?
- Is there a real difference between the two groups from which samples are taken?

Answering these questions is not as easy because, in most situations, we don't have perfect knowledge. For example, when the owner of a soccer club sacks the team manager and appoints a new one, they cannot be sure that the observed improvement in the team's performance is actually due to the managerial change. Perhaps the improvement would also have occurred under the leadership of the old manager. After all, the form of soccer teams often dips, only to improve some time later when the results revert back to the mean. So how do we know when an intervention (e.g., a change of manager or the purchase of a new player) has had an impact on team performance that could not otherwise have occurred by chance? Well, to answer this type of question, statisticians use a variety of hypothesis-testing techniques.

The key idea underpinning hypothesis testing is being able to tell whether or not an observed event is likely to have occurred by chance. So, if we can say that an observed event is unlikely to have occurred by chance, then we can be pretty confident that it must be due to some underlying real effect. In statistics, the 'null distribution' of all the possible outcomes that could occur just by chance (i.e., when there is no underlying real effect) is assumed to be a normal distribution with a mean value of zero and a standardised x-axis (Z) as shown in Figure 2.7.

If the observed event lies in the tails of the null distribution (i.e., $Z < -1.96$ or $Z > 1.96$, which correspond to the 2.5% and 97.5% percentiles), then it is considered highly unlikely that the event has occurred by chance, and therefore it is assumed that some underlying effect must be at work. In which case we deem the event to be 'significant' at $p < 0.05$. Which is shorthand for saying that it is a highly unlikely event with a probability of occurring that is less than 5% if the 'null hypothesis' (i.e., the null distribution) is correct.

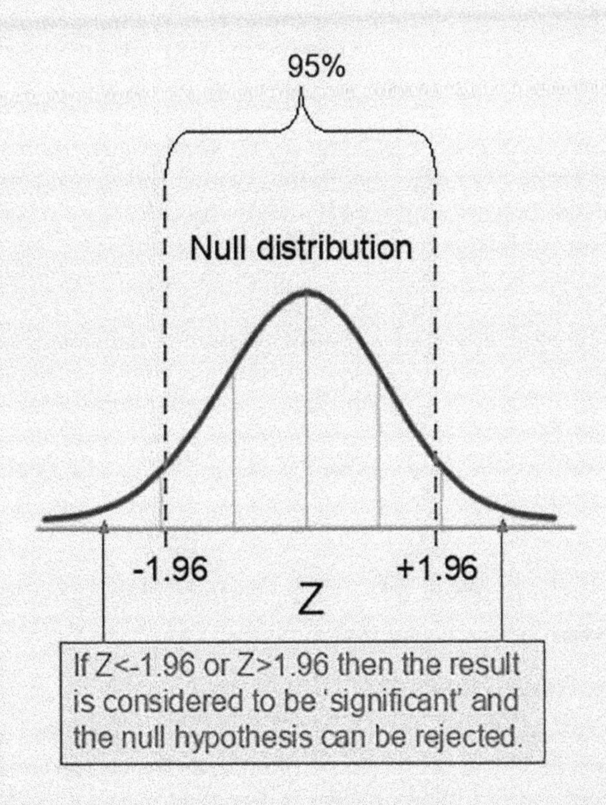

FIGURE 2.7
Null distribution with limits of significance shown as vertical dashed lines.

Say we wanted to know if there was any statistical difference between the heights of players in the EPL and the German Bundesliga. To answer this question we could sample 50 players from each league and measure their respective heights. In which case, the 'effect' would be the difference between the means of the two sample groups, and the Z-score would be computed using the following equation, in which the numerator is the effect and the denominator is the standard error of the difference between the means – a combined measure of the variance in the two sample groups.

$$Z = \frac{\bar{x}_1 - \bar{x}_2}{\sqrt{\frac{\sigma_1^2}{n_1} + \frac{\sigma_2^2}{n_2}}} \quad (2.3)$$

where $\bar{x}_1$ is the mean of the first sample; $\bar{x}_2$ is the mean of the second sample; σ_1 is the standard deviation of the first sample; σ_2 is the standard deviation of the second sample; and n_1 and n_2 are the numbers in each sample.

If the computed Z-score is less than –1.96 or greater than 1.96, then we can be pretty confident that there is a real difference between the heights of the players in the EPL and the Bundesliga. In which case we reject the 'null hypothesis' that there is no difference between the heights of the players in the two leagues in favour of the 'alternative hypothesis' that a real difference exists.

This is the basic idea underpinning all parametric statistical testing, and tests such as Welch's independent *t*-test work on this principle, although with *t*-tests, a *t*-distribution is assumed rather than a normal distribution. Non-parametric tests such as the chi-square test work in a different way, but they too rely on the concept of rejecting or accepting the null hypothesis.

2.12 Linear Regression

In Chapter 1, we saw how linear regression analysis can be used to make predictions in soccer. Linear regression is often used to analyse the relationship between a response variable and several predictor variables. While a full discussion of linear models is well beyond the scope of this chapter (see Key Concept Box 1.2 and Chapter 10 for a fuller discussion of linear regression models), it is perhaps worthwhile taking a few moments here to illustrate how R can be used to quickly and easily produce linear regression models. For convenience, in Example 2.10, we will use the Arsenal-Chelsea data set introduced in Example 2.5 to produce two linear regression models – one for Arsenal and one for Chelsea. We will then use these models to evaluate whether or not the goals scored and conceded by the respective teams can be used to predict the total number of points awarded in each season. In addition, we will use Example 2.10 as an opportunity to introduce the 'ggplot2' package, which is very popular with users of R because it allows high-quality graphical plots to be easily produced.

Linear regression models with multiple predictor variables are widely used in statistics, both to explain observed phenomena and to make predictions. Such models have the general form:

$$\text{Multiple linear regression model}: y = b_0 + b_1 x_1 + b_2 x_2 + \ldots + e \quad (2.4)$$

where y is the response variable; x_1, x_2, ... are the predictor variables; b_0, b_1, b_2, ... are coefficients applied to the intercept and the predictor variables; and e is the residual error.

Basically, what Equation 2.4 says is that the observed data (i.e., the response variable, y) can be 100% replicated mathematically using a linear weighted combination of the predictor variables, x_1, x_2, … , plus some residual error, e. Many people get confused by the concept of residual errors, but in essence they are simply a 'fiddle-factor' vector that makes up the difference between the linear weighted combination of the predictor variables, $\hat{y}$, and the observed values, y, hence the term 'error'. Given this, we can use the regression model (minus the residual error values) to predict the values in the response variable using Equation 2.5.

$$\text{Predicted values for response variable}: \hat{y} = b_0 + b_1 x_1 + b_2 x_2 + \ldots \quad (2.5)$$

The clever thing about such regression models (see Key Concept Box 1.2) is that when plotted on a scatter plot, the predicted values lie along a least-squares best-fit line that runs straight through the observed data, which is a very neat idea! So if the model is a good predictor of the observed data, then the residual errors will be small, whereas if it is not so good at prediction, the errors will be large. The goodness of fit of a linear regression model can be assessed using the coefficient of determination (R^2), which is a measure of the amount of variance in the observed data that is explained by the model.

While this might all sound very complex, in practice, linear regression is quite easy to implement in R. Furthermore, once implemented in R, it becomes much easier to understand how linear regression models work. So, let's explore the simple linear regression model presented in Example 2.10, which highlights the main concepts associated with the technique.

EXAMPLE 2.10: SIMPLE MULTIPLE LINEAR REGRESSION MODEL

In this example, we will use the Arsenal-Chelsea data set introduced in Example 2.5 to produce two linear regression models – one for Arsenal and one for Chelsea. We will then use these models to evaluate whether or not the goals scored and conceded by the respective teams can be used to predict the total number of points awarded in each season. In addition, we will use the 'ggplot2' package to produce high-quality regression plots.

First, we need to create two linear regression models, one for Arsenal and one for Chelsea. In these models, the predictor variables are the goals conceded and the goals scored, and the points total is the response variable. This can be done very easily using the 'lm' command as follows:

```
# Arsenal
Arsenal.lm <- lm(Arsenal_points ~ Arsenal_GA + Arsenal_GF, data = dat)
summary(Arsenal.lm) # This produces a summary.
```

This produces the following summary of the model results for Arsenal:

```
## Call:
lm(formula = Arsenal_points ~ Arsenal_GA + Arsenal_GF, data = dat)

Residuals:
    Min      1Q  Median      3Q     Max
-5.0048 -1.2063 -0.7429  0.4706  8.5622

Coefficients:
            Estimate Std. Error t value Pr(>|t|)
(Intercept)  48.5241    14.2488   3.405  0.01136 *
Arsenal_GA   -0.6842     0.2296  -2.980  0.02052 *
Arsenal_GF    0.7348     0.1862   3.946  0.00556 **
---
Signif. codes:  0 '***' 0.001 '**' 0.01 '*' 0.05 '.' 0.1 ' ' 1

Residual standard error: 4.136 on 7 degrees of freedom
Multiple R-squared:  0.7409,     Adjusted R-squared:  0.6669
F-statistic: 10.01 on 2 and 7 DF,  p-value: 0.008852
```

Now let's run the same model for Chelsea.

```
# Chelsea
Chelsea.lm <- lm(Chelsea_points ~ Chelsea_GA + Chelsea_GF, data = dat)
summary(Chelsea.lm) # This produces a summary.
```

```
## Call:
lm(formula = Chelsea_points ~ Chelsea_GA + Chelsea_GF, data = dat)

Residuals:
    Min      1Q  Median      3Q     Max
-4.3079 -2.6929  0.2186  2.7669  3.9266

Coefficients:
            Estimate Std. Error t value Pr(>|t|)
(Intercept)  44.1345    14.1690   3.115 0.016965 *
Chelsea_GA   -0.7892     0.1479  -5.336 0.001080 **
Chelsea_GF    0.8793     0.1571   5.598 0.000818 ***
---
Signif. codes:  0 '***' 0.001 '**' 0.01 '*' 0.05 '.' 0.1 ' ' 1

Residual standard error: 3.586 on 7 degrees of freedom
Multiple R-squared:  0.935, Adjusted R-squared:  0.9164
F-statistic: 50.32 on 2 and 7 DF,  p-value: 7.015e-05
```

From this, we can see that R gives us lots of information, much of which is beyond the scope of this current example. However, the important things to note are:

- That for both Arsenal and Chelsea, all the predictor variables are significant ($p<0.05$). This means that both goals scored and goals conceded can be used to explain the end-of-season points totals accumulated by the two clubs. This, of course, is completely unsurprising, as one would expect goals scored and conceded to be strongly associated with points accumulated.
- While the Chelsea model explains 93.5% of the variance in the observed point data, the Arsenal model explains only 74.1% of the variance. This is a rather unexpected finding, which suggests that compared with Chelsea, Arsenal were involved in more matches where they either won or lost heavily, with the result that the relationship between goals scored/conceded and points awarded is weaker for them compared with Chelsea.
- The 'Estimate' column in the results tables contains the linear coefficients (i.e., b_0, b_1, and b_2) used in the respective models for Arsenal and Chelsea to predict the points awarded.

Having built the two linear regression models, we can now use them to predict (compute) the expected number of points awarded for each season, as follows:

```
# Arsenal
Arsenal.pred <- round(predict(Arsenal.lm, data = dat, type="response"),1)

# Chelsea
Chelsea.pred <- round(predict(Chelsea.lm, data = dat, type="response"),1)

# Create new data frame
new.dat <- cbind.data.frame(dat,Arsenal.pred,Chelsea.pred)
print(new.dat)
```

```
##
    Season Arsenal_GF Arsenal_GA Arsenal_points Chelsea_GF Chelsea_GA
1  2011-12         74         49             70         65         46
2  2012-13         72         37             73         75         39
3  2013-14         68         41             79         71         27
4  2014-15         71         36             75         73         32
5  2015-16         65         36             71         59         53
6  2016-17         77         44             75         85         33
7  2017-18         74         51             63         62         38
8  2018-19         73         51             70         63         39
9  2019-20         56         48             56         69         54
10 2020-21         55         39             61         58         36
   Chelsea_points Arsenal.pred Chelsea.pred
1              64         69.4         65.0
2              75         76.1         79.3
3              82         70.4         85.3
4              87         76.1         83.1
5              50         71.7         54.2
```

```
6           93          75.0        92.8
7           70          68.0        68.7
8           72          67.3        68.8
9           66          56.8        62.2
10          67          62.3        66.7
```

From this, we can instantly see that both models produce predictions that are reasonably close to the observed point values. However, it can be a bit difficult to get a handle on these, and a much better way to assess the accuracy of the models is to produce regression plots. These can be easily produced using the 'ggplot2' package. However, before using 'ggplot2' it is helpful to first create some new vectors as follows:

```
# Arsenal
obs.Arsenal <- new.dat$Arsenal_points
pred.Arsenal <- new.dat$Arsenal.pred

# Chelsea
obs.Chelsea <- new.dat$Chelsea_points
pred.Chelsea <- new.dat$Chelsea.pred
```

Now we can produce regression plots for Arsenal and Chelsea using the following code:

```
install.packages("ggplot2") # NB. Use once to install ggplot2.

library(ggplot2)  # Call the 'ggplot2' package.

# Produce scatter plot of observed and predicted results for Arsenal
ggplot(new.dat, aes(x=pred.Arsenal, y=obs.Arsenal)) +
  geom_point(color="black", size=3) +   # Specify data point size and colour
  geom_smooth(method=lm, se = FALSE) +  # Add linear regression line
  xlab("Predicted points") +
  ylab("Observed points")

# Produce scatter plot of observed and predicted results for Chelsea
ggplot(new.dat, aes(x=pred.Chelsea, y=obs.Chelsea)) +
  geom_point(color="black", size=3) +   # Specify data point size and colour
  geom_smooth(method=lm, se = FALSE) +  # Add linear regression line
  xlab("Predicted points") +
  ylab("Observed points")
```

This produces Figures 2.8 and 2.9, which are scatter plots of the observed and predicted values for Arsenal and Chelsea, respectively. The diagonal line through the centre of the plot is the ordinary least-squares best-fit line. In both cases, the best-fit line coincides with the line of perfect prediction when the observed and predicted values are identical, indicating that the models are likely to be reasonably good predictors of the end-of-season points total. It should also be noted that the closer the predicted values are to the best-fit line, the better the regression model.

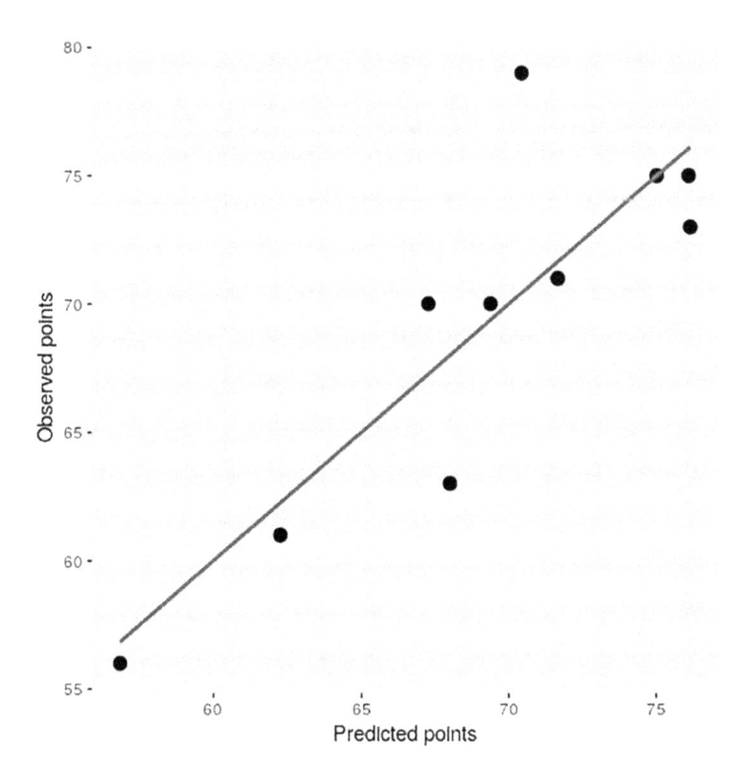

FIGURE 2.8
Scatter plot of observed and predicted points awarded to Arsenal, with best-fit regression line (R^2=0.741).

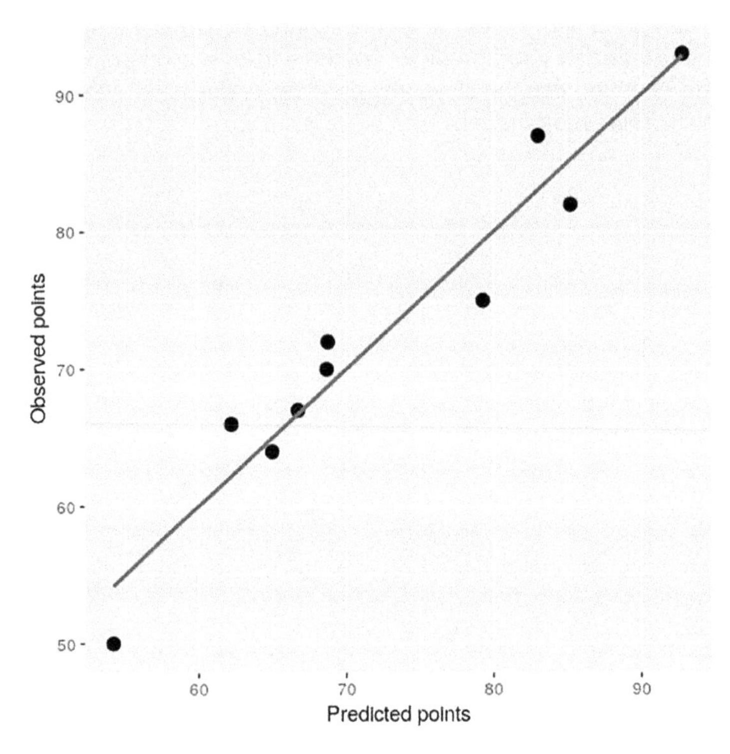

FIGURE 2.9
Scatter plot of observed and predicted points awarded to Chelsea, with best-fit regression line (R^2=0.935).

2.13 Tidyverse

The tutorials in this chapter are mostly written in what is called 'base-R' – that is, the core set of functions and data structures that are built into R and that are available by default when R is installed. These are excellent and easy to use, and they allow the base-R user to perform a wide range of data manipulation, statistical analysis, and visualisation tasks. However, because R is an open-access platform in which users often develop add-on functions and packages, it means that over the years, base-R has become somewhat fragmented, with the result that some data analysis tasks involving multiple steps can be problematic because of inconsistencies between packages created by different developers. In response to this, the Tidyverse ecosystem was developed [6] in an attempt to provide a more consistent approach to data analysis in R.

While a full discussion of Tidyverse is beyond the scope of this book, it is perhaps worth outlining here what Tidyverse is, so that the reader is aware of its potential. Unlike base-R, which is a bit of a free-for-all, Tidyverse is a curated ecosystem containing a collection of packages that are designed to work together seamlessly and provide a more consistent approach to data analysis. As such, Tidyverse emphasises the principle of 'tidy data', which is a standard way of organising and structuring data in order to make it easier to work with. Tidyverse includes popular R packages such as 'dplyr', 'tidyr', 'ggplot2', and 'tibble', and as such provides a 'one-stop' cohesive toolkit for data manipulation, visualisation, and analysis.

The Tidyverse approach is intended to be more user-friendly and intuitive than base-R. As such, it is a useful vehicle when performing standard tasks such as data cleaning, visualisation, and common statistics. However, Tidyverse tends to have some limitations when performing non-standard work such as package development and advanced statistical procedures – in these cases, base-R is preferable because it works more from first principles. One important stylistic feature that is encouraged in Tidyverse rather than in base-R is 'piping', where functions are piped together using '%>%' to emphasise a sequence of actions rather than specifying the object on which the actions are being performed.

2.14 Copy, Experiment, and Adapt

If you have been able to get the example R scripts in this chapter to run and have gained some understanding of how they work, then you are well equipped to tackle the rest of this book. Congratulations! You are now up and running in R. All you need to do now is build on what you have already learned, like a child learning a language. To do this, you should experiment by copying code produced by others (there is lots of this on the Internet) and seeing how it runs in RStudio. When you get the code to execute correctly, try adapting it or applying it to new data. That way, you will learn quickly. But as you experiment, it is important to annotate your code with detailed comments explaining how and why the code works. This will help you learn quickly and also be able to remember what you did when you revisit your code in a couple of years.

References

1. Ihaka R: R: Past and future history. *Computing Science and Statistics* 1998, 392396.
2. Till K, Jones BL, Cobley S, Morley D, O'Hara J, Chapman C, Cooke C, Beggs CB: Identifying talent in youth sport: A novel methodology using higher-dimensional analysis. *PLoS One* 2016, 11(5):e0155047.
3. Beggs CB, Shepherd SJ, Emmonds S, Jones B: A novel application of PageRank and user preference algorithms for assessing the relative performance of track athletes in competition. *PLoS One* 2017, 12(6):e0178458.
4. Weaving D, Jones B, Ireton M, Whitehead S, Till K, Beggs CB: Overcoming the problem of multicollinearity in sports performance data: A novel application of partial least squares correlation analysis. *PLoS One* 2019, 14(2):e0211776.
5. Racine JS: RStudio: A platform-independent IDE for R and Sweave. *Journal of Applied Econometrics* 2012, 27: 167–172. https://onlinelibrary.wiley.com/doi/abs/10.1002/jae.1278
6. Wickham H, Averick M, Bryan J, Chang W, McGowan LDA, Francois R, Grolemund G, Hayes A, Henry L, Hester J: Welcome to the Tidyverse. *Journal of Open Source Software* 2019, 4(43):1686.

3

Using R to Harvest and Process Soccer Data

When working with R, one of the first challenges many people encounter is simply how to get data into R so that it can be analysed. There is a huge amount of soccer-related data on the Internet that is freely available, but how do we get it off the web into R without spending many tedious hours copying it by hand onto spreadsheets? Fortunately, R has many useful tools that can be employed to harvest and process data from the Internet. In this chapter, we will investigate how to get data into R so that it can be analysed and processed. The methods outlined in this chapter are by no means exhaustive; other methods exist that are not covered, but hopefully, when you have worked through the examples presented here, you will be able to not only import soccer data from the Internet but also perform simple processing tasks to analyse the data that you have harvested.

3.1 Importing and Editing Data Files in R

The subject of importing a comma-separated values (CSV) file into R is briefly covered in Chapter 2, but here we will cover things in a little more detail. Probably the easiest way to input data into R is to put it on a spreadsheet and then save it as a CSV file – something that can easily be done in *Microsoft Excel* or *OpenOffice Calc*. Once saved in CSV format, the data can be loaded into R using the 'read.csv' function. This command inputs a text file, which has a comma as a separator and a dot as a decimal point by default. Unless stated otherwise, the first line of the file is assumed to contain the column headers (i.e., the variable names).

The general way to apply the 'read.csv' function is as follows, where "input.csv" is the CSV file being loaded. In this case, the code creates a data frame called 'dat'.

```
dat <- read.csv("input.csv")
```

In practice, however, it is likely that the "input.csv" file will be stored in a directory somewhere on the computer. So for example, if the CSV file is

DOI: 10.1201/9781003328568-3

stored in a directory called 'Datasets', then the file path would have to be included in the 'read.csv' command, as follows:

For a PC computer:
```
dat <- read.csv("C:/Datasets/input.csv")
```
For an Apple computer:
```
dat <- read.csv("~/Desktop/Datasets/input.csv")
```

Alternatively, the 'Import' button on the top right of the RStudio console (see Figure 3.1, shown with a dropdown menu) can be used to import directly CSV files as well as Excel and SPSS files. When using this method to import files, you should select the 'Text (base)' option to import CSV files.

The 'read_excel' function in the "readxl" package can also be utilised to import Excel spreadsheet files as follows:

```
install.packages("readxl") # Required to install the package.

library("readxl") # This calls the readxl library.

# For xls files
dat <- read_excel("C:/Datasets/input.xls")

# For xlsx files
dat <- read_excel("C:/Datasets/input.xlsx")
```

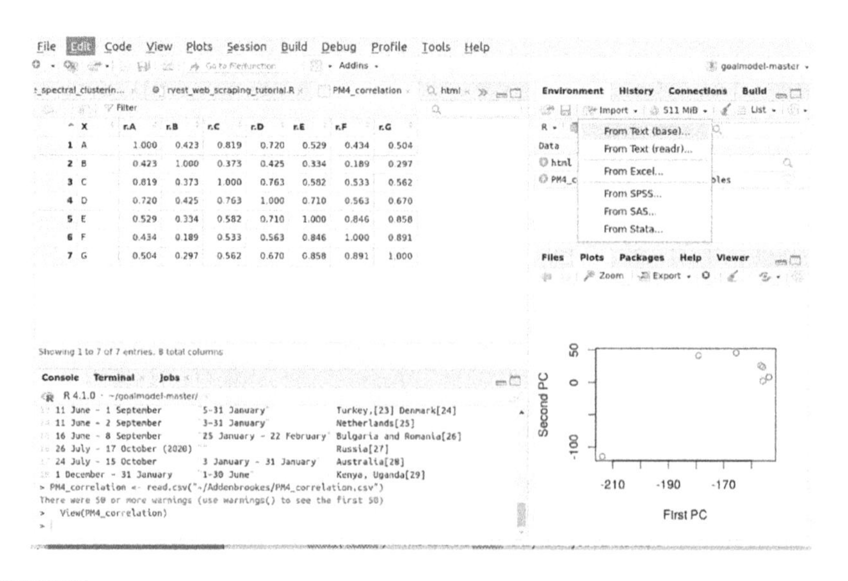

FIGURE 3.1
Use the 'Import' button and then select from the dropdown menu to import data files into RStudio.

When importing data files into R, it is good practice to follow the general strategy outlined in Figure 3.2. This involves:

1. Importing the data file into R.
2. Inspecting the data to get a feel for it (see Chapter 2 for details of how to do this).
3. Making a working copy of the imported data so that the original cannot be altered or corrupted in any way. (NB. Sometimes it is possible to accidentally corrupt data when working with original files, and so it is good practice to work on copies of the original.)
4. Cleaning and editing the working copy of the data. This might involve, for example, removing incomplete or corrupted data, selecting part of the data set, and adding new variables computed from the existing data.
5. Saving the cleaned and edited version of the data and exporting it as a CSV file.

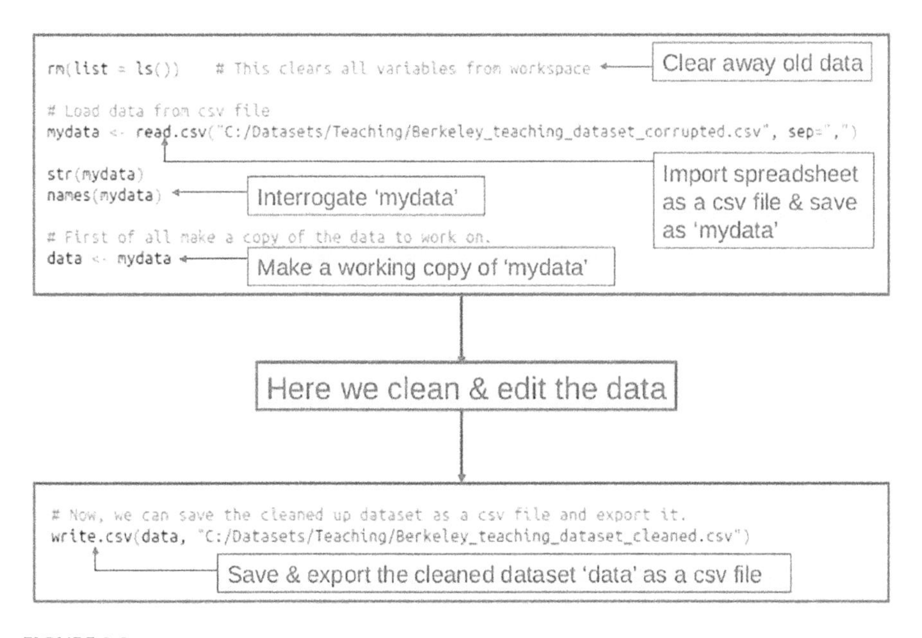

FIGURE 3.2
General strategy for importing, editing, and exporting cleaned data in R.

Example 3.1 shows how this strategy can be executed in practice using a small data set containing the results for Arsenal's home matches in the English Premier League (EPL) during the 2020–2021 season. The file containing the data is called "Arsenal_home_2020.csv," and we will assume that it is stored in a directory called 'Datasets'. In this example, we use the 'read.csv' function to load the data into R and the 'write.csv' command to export the edited data frame. Example 3.1 also shows how to add new derived variables to an existing data frame. The 'Arsenal_home_2020.csv' file can be downloaded from GitHub at the following address: https://github.com/cbbeggs/SoccerAnalytics. Once downloaded, you should save it in a directory called 'Datasets', which you should create on your computer's hard drive.

EXAMPLE 3.1: IMPORTING A CSV FILE INTO R AND EDITING THE DATA FRAME

This example illustrates how to import and modify data in R using files in CSV format. Before starting to code, it is always good practice to first clear all the existing variables and data from the RStudio workspace, as these might interfere with the new script. This can be easily done using the following line of R code:

```
rm(list = ls())   # This clears all variables from the workspace
```

Now that we have a blank slate, the data can be imported into R in CSV format and displayed using the following code. (NB. Here we assume the use of a Windows PC. If you are using an Apple Mac computer, then the code should be amended as described in Section 3.1 above.). The CSV data file used in this example (Arsenal_home_2020.csv), which contains data from Arsenal's home matches for season 2020–2021, can be downloaded from https://github.com/cbbeggs/SoccerAnalytics. The downloaded CSV file should be saved in a directory called 'Datasets'. Once stored, the data can be loaded into R using the 'read.csv' function as follows:

```
ArsenalHome <- read.csv("C:/Datasets/Arsenal_home_2020.csv")
print(ArsenalHome)
```

```
##
        Date HomeTeam          AwayTeam HG AG Result HS AS HST AST
1 19/09/2020  Arsenal          West Ham  2  1      W  7 14   3   3
2 04/10/2020  Arsenal Sheffield United  2  1      W  6  6   5   2
3 25/10/2020  Arsenal         Leicester  0  1      L 12  6   4   2
4 08/11/2020  Arsenal       Aston Villa  0  3      L 13 15   2   6
5 29/11/2020  Arsenal            Wolves  1  2      L 13 11   2   5
6 13/12/2020  Arsenal           Burnley  0  1      L 18 10   6   2
7 16/12/2020  Arsenal       Southampton  1  1      D  9 13   4   3
```

```
8  26/12/2020  Arsenal          Chelsea  3  1      W 15 19   7   3
9  14/01/2021  Arsenal   Crystal Palace  0  0      D 11 12   4   2
10 18/01/2021  Arsenal        Newcastle  3  0      W 20  4   6   1
11 30/01/2021  Arsenal       Man United  0  0      D 17 14   3   3
12 14/02/2021  Arsenal            Leeds  4  2      W 13  9   5   5
13 21/02/2021  Arsenal         Man City  0  1      L  7 15   1   3
14 14/03/2021  Arsenal        Tottenham  2  1      W 13  6   3   3
15 03/04/2021  Arsenal        Liverpool  0  3      L  3 16   2   7
16 18/04/2021  Arsenal           Fulham  1  1      D 18  3   5   1
17 23/04/2021  Arsenal          Everton  0  1      L 14  8   3   1
18 09/05/2021  Arsenal        West Brom  3  1      W 15 12   7   1
19 23/05/2021  Arsenal         Brighton  2  0      W 16  5   5   1
```

This displays the 'ArsenalHome' data frame, which presents data for all Arsenal's EPL home matches for the season 2020–2021. The data includes the date on which the match was played, the number of goals scored by Arsenal (HG), the number of goals conceded by Arsenal (AG), the match result, the number of shots made by Arsenal (HS), the number of shots conceded by Arsenal (AS), the number of shots on target made by Arsenal (HST), and the number of shots on target conceded by Arsenal (AST).

Having loaded the data into R and created a data frame called 'ArsenalHome', we will now inspect the data.

```
names(ArsenalHome) # This displays the variable names
```

```
##
[1] "Date" "HomeTeam" "AwayTeam" "HG" "AG" "Result" "HS" "AS" "HST" "AST"
```

```
str(ArsenalHome) # This displays the structure of the data frame.
```

```
##
'data.frame':    19 obs. of  10 variables:
 $ Date    : chr  "19/09/2020" "04/10/2020" "25/10/2020" "08/11/2020" ...
 $ HomeTeam: chr  "Arsenal" "Arsenal" "Arsenal" "Arsenal" ...
 $ AwayTeam: chr  "West Ham" "Sheffield United" "Leicester" ...
 $ HG      : int  2 2 0 0 1 0 1 3 0 3 ...
 $ AG      : int  1 1 1 3 2 1 1 1 0 0 ...
 $ Result  : chr  "W" "W" "L" "L" ...
 $ HS      : int  7 6 12 13 13 18 9 15 11 20 ...
 $ AS      : int  14 6 6 15 11 10 13 19 12 4 ...
 $ HST     : int  3 5 4 2 2 6 4 7 4 6 ...
 $ AST     : int  3 2 2 6 5 2 3 3 2 1 ...
```

This gives the variable names and tells us which variables contain integers and which contain characters.

In order to avoid corruption of the original data, it is good practice to make a working copy of the data, which in this case we will call 'Ar_dat'.

```
Ar_dat <- ArsenalHome   # This makes a working copy of the data frame.
```

Now we can add some new derived variables to the data frame 'Ar_dat'. In this example, we will add four new variables: goal difference (GD); total goals (TG); total shots ratio for the home team (HTSR); and total shots ratio for the away team (ATSR). The total shots ratio (TSR) is a widely used metric of how well teams perform during matches and is calculated using:

$$TSR = \frac{SF}{SF + SA} \tag{3.1}$$

where SF is the number of shots attempted by a team during a match, and SA is the number of shots conceded by the team.

We can add the new derived variables in two steps: (i) Step 1, which involves creating some new columns in the data frame that are temporarily populated with NAs (NA means the data is not available); and (ii) Step 2, which involves populating the new columns with calculated values derived from the other variables.

```
# Step 1 – Create the new empty variables populated with NAs.
Ar_dat["GD"] <- NA # This creates a new column to store goal difference.
Ar_dat["TG"] <- NA # This creates the new column to store total goals scored.
Ar_dat["HTSR"] <- NA # This creates the new column to store home team shots ratio.
Ar_dat["ATSR"] <- NA # This creates the new column to store away team shots ratio.

# Step 2 - Populate the new columns with the calculated values.
Ar_dat$GD <- Ar_dat$HG - Ar_dat$AG  # Goal difference
Ar_dat$TG <- Ar_dat$HG + Ar_dat$AG  # Total goals
Ar_dat$HTSR <- round(Ar_dat$HS/(Ar_dat$HS + Ar_dat$AS),3) # HTSR rounded to 3dp
Ar_dat$ATSR <- round(Ar_dat$AS/(Ar_dat$AS + Ar_dat$HS),3)  # ATSR rounded to 3dp
```

Having done this, we can inspect the new data frame to check that it is correct.

```
names(Ar_dat)
```

```
##
[1] "Date" "HomeTeam" "AwayTeam" "HG" "AG" "Result" "HS" "AS" "HST" "AST"
[11] "GD"   "TG"   "HTSR"   "ATSR"
```

```
head(Ar_dat,8) # This displays the first eight rows of the modified data frame.
```

```
##
        Date HomeTeam          AwayTeam HG AG Result HS AS HST AST GD TG
1 19/09/2020  Arsenal          West Ham  2  1      W  7 14   3   3  1  3
2 04/10/2020  Arsenal Sheffield United  2  1      W  6  6   5   2  1  3
3 25/10/2020  Arsenal         Leicester  0  1      L 12  6   4   2 -1  1
4 08/11/2020  Arsenal       Aston Villa  0  3      L 13 15   2   6 -3  3
5 29/11/2020  Arsenal            Wolves  1  2      L 13 11   2   5 -1  3
6 13/12/2020  Arsenal           Burnley  0  1      L 18 10   6   2 -1  1
7 16/12/2020  Arsenal       Southampton  1  1      D  9 13   4   3  0  2
```

```
8 26/12/2020  Arsenal            Chelsea  3  1      W 15 19   7   3  2  4
   HTSR  ATSR
1 0.333 0.667
2 0.500 0.500
3 0.667 0.333
4 0.464 0.536
5 0.542 0.458
6 0.643 0.357
7 0.409 0.591
8 0.441 0.559
```

From this, we can see that the four new columns have been added to the existing data frame.

Now, all that is left to do is export the results as a CSV file, as follows:

```
write.csv(Ar_dat, "C:/AnalysisResults/Arsenal_home_shots_ratio.csv")
```

NB. Here we are saving the file to a directory called 'AnalysisResults'. Alternatively, the file can be saved to any directory, provided that the path is specified.

3.2 Dividing Data into Subgroups

When editing and processing data, it is often the case that we want to divide the data set into subgroups. For example, when looking at a team's performance, we might want to separate out all the matches won, lost, and drawn into three separate subgroups. This can be easily done in R using the code in Example 3.2, which is applied to the 'Ar_dat' data frame that we created in Example 3.1.

EXAMPLE 3.2: DIVIDING DATA INTO SUBGROUPS

In this example, we use the 'Ar_dat' data frame that we created in Example 3.1 to illustrate how the data can be divided into subgroups, each of which is a data frame. Here we group the matches according to match outcome (i.e., win, lose, or draw).

First of all, it is good practice to review the data.

```
head(Ar_dat)
```

```
##
        Date HomeTeam          AwayTeam HG AG Result HS AS HST AST GD TG
1 19/09/2020  Arsenal          West Ham  2  1      W  7 14   3   3  1  3
2 04/10/2020  Arsenal Sheffield United  2  1      W  6  6   5   2  1  3
3 25/10/2020  Arsenal         Leicester  0  1      L 12  6   4   2 -1  1
4 08/11/2020  Arsenal       Aston Villa  0  3      L 13 15   2   6 -3  3
5 29/11/2020  Arsenal            Wolves  1  2      L 13 11   2   5 -1  3
6 13/12/2020  Arsenal           Burnley  0  1      L 18 10   6   2 -1  1
   HTSR  ATSR
1 0.333 0.667
2 0.500 0.500
3 0.667 0.333
4 0.464 0.536
5 0.542 0.458
6 0.643 0.357
```

From this, we can see that the 'Result' variable reports the outcome of the matches played. So, we can use this to sub-divide the data as follows:

```
# This splits the data into separate win, lose and draw data frames.
win <- Ar_dat[Ar_dat$Result == "W",]   # Wins
lose <- Ar_dat[Ar_dat$Result == "L",]   # Lose
draw <- Ar_dat[Ar_dat$Result == "D",]   # Draw
```

To display the results, we simply use the following code:

```
print(win)
```

```
##
         Date HomeTeam          AwayTeam HG AG Result HS AS HST AST GD TG
1  19/09/2020  Arsenal          West Ham  2  1      W  7 14   3   3  1  3
2  04/10/2020  Arsenal Sheffield United  2  1      W  6  6   5   2  1  3
8  26/12/2020  Arsenal           Chelsea  3  1      W 15 19   7   3  2  4
10 18/01/2021  Arsenal         Newcastle  3  0      W 20  4   6   1  3  3
12 14/02/2021  Arsenal             Leeds  4  2      W 13  9   5   5  2  6
14 14/03/2021  Arsenal         Tottenham  2  1      W 13  6   3   3  1  3
18 09/05/2021  Arsenal         West Brom  3  1      W 15 12   7   1  2  4
19 23/05/2021  Arsenal          Brighton  2  0      W 16  5   5   1  2  2
    HTSR  ATSR
1  0.333 0.667
2  0.500 0.500
8  0.441 0.559
10 0.833 0.167
12 0.591 0.409
14 0.684 0.316
18 0.556 0.444
19 0.762 0.238
```

```
print(lose)
```

```
##
         Date HomeTeam      AwayTeam HG AG Result HS AS HST AST GD TG  HTSR
3  25/10/2020  Arsenal     Leicester  0  1      L 12  6   4   2 -1  1 0.667
4  08/11/2020  Arsenal   Aston Villa  0  3      L 13 15   2   6 -3  3 0.464
5  29/11/2020  Arsenal        Wolves  1  2      L 13 11   2   5 -1  3 0.542
6  13/12/2020  Arsenal       Burnley  0  1      L 18 10   6   2 -1  1 0.643
13 21/02/2021  Arsenal      Man City  0  1      L  7 15   1   3 -1  1 0.318
15 03/04/2021  Arsenal     Liverpool  0  3      L  3 16   2   7 -3  3 0.158
17 23/04/2021  Arsenal       Everton  0  1      L 14  8   3   1 -1  1 0.636
    ATSR
3  0.333
4  0.536
5  0.458
6  0.357
13 0.682
15 0.842
17 0.364
```

```
print(draw)
```

```
##
         Date HomeTeam        AwayTeam HG AG Result HS AS HST AST GD TG
7  16/12/2020  Arsenal     Southampton  1  1      D  9 13   4   3  0  2
9  14/01/2021  Arsenal Crystal Palace  0  0      D 11 12   4   2  0  0
11 30/01/2021  Arsenal     Man United  0  0      D 17 14   3   3  0  0
16 18/04/2021  Arsenal          Fulham  1  1      D 18  3   5   1  0  2
    HTSR  ATSR
7  0.409 0.591
9  0.478 0.522
11 0.548 0.452
16 0.857 0.143
```

3.3 Missing Data

Data sets can often be incomplete, with data entries missing for some variables. This can cause problems when analysing the data because certain functions struggle when data is missing. For example, if missing data points (denoted by 'NA' in R) are present, then the 'mean' function will not work unless it is modified. To illustrate this, let's consider a simple example (Example 3.3) using made-up data for a few soccer players.

EXAMPLE 3.3: HOW TO DEAL WITH MISSING DATA

Let's first create a data frame containing the average number of shots, goals, passes, and tackles per match for eight players. Importantly, this data set is not complete because some of the data points are missing. In R, these are denoted by 'NA', which means that the data is not available.

```
# Create some data with missing data entries
players <- c("Paul","Stephen","James","Kevin","Tom","Edward","John","David") # Players
shots <- c(2.4,3.6,0.3,1.1,4.2,2.3,NA,0.6) # Average number of shots per game
goals <- c(0.2,0.6,0.0,0.1,0.7,0.3,0.1,0.0) # Average number of goals per game
passes <- c(23.1,NA,39.2,25.5,18.6,37.4,28.3,28.3) # Average number of passes per game
tackles <- c(6.3,4.5,10.6,9.8,4.1,5.3,11.2,7.8) # Average number of tackles per game

# Create data frame
perf_dat <- cbind.data.frame(players,shots,goals,passes,tackles) # Creates data frame
print(perf_dat)
```

This produces the following data frame, which reveals that two data points are missing.

```
##
  players shots goals passes tackles
1    Paul   2.4   0.2   23.1     6.3
2 Stephen   3.6   0.6     NA     4.5
3   James   0.3   0.0   39.2    10.6
4   Kevin   1.1   0.1   25.5     9.8
5     Tom   4.2   0.7   18.6     4.1
6  Edward   2.3   0.3   37.4     5.3
7    John    NA   0.1   28.3    11.2
8   David   0.6   0.0   28.3     7.8
```

So, how can we deal with this? One approach that is sometimes taken is to simply remove all the subjects or observations that contain any NAs. This can be done in R by using the command 'na.omit' as follows:

```
# Lines containing NAs can be completely removed if so desired.
na.omit(perf_dat)
```

This produces the following reduced data set, which is OK but contains 25% less data.

```
##
  players shots goals passes tackles
1    Paul   2.4   0.2   23.1     6.3
3   James   0.3   0.0   39.2    10.6
4   Kevin   1.1   0.1   25.5     9.8
5     Tom   4.2   0.7   18.6     4.1
6  Edward   2.3   0.3   37.4     5.3
8   David   0.6   0.0   28.3     7.8
```

So, in order to prevent excess loss of important information (data), it is often better to modify the 'mean', 'median', etc. functions by adding a 'na.rm' (NA remove) argument. If this is set to TRUE, then any NA data points are ignored when computing the relevant descriptive statistic. Without adding 'na.rm=TRUE', the 'mean' function will not compute if NAs are present, as illustrated below:

```
# Try 'mean' function
mean(perf_dat$shots) # This does not work.
```

```
## [1] NA
```

```
mean(perf_dat$goals) # This works.
```

```
## [1] 0.25
```

```
mean(perf_dat$passes) # This does not work.
```

```
## [1] NA
```

```
mean(perf_dat$tackles) # This works.
```

```
## [1] 7.45
```

```
# Now adding na.rm = TRUE
mean(perf_dat$shots, na.rm = TRUE) # This now works.
```

```
## [1] 2.071429
```

```
mean(perf_dat$goals, na.rm = TRUE) # This now works.
```

```
## [1] 0.25
```

```
mean(perf_dat$passes, na.rm = TRUE) # This now works.
```

```
## [1] 28.62857
```

```
mean(perf_dat$tackles, na.rm = TRUE) # This now works.
```

```
## [1] 7.45
```

Alternatively, you can use the 'describeBy' function in the "psych" package, which automatically ignores any NAs that might be present. This can be applied as follows:

```
install.packages("psych") # This installs the 'psych' package.
# NB. This command only needs to be executed once to install the package.
# Thereafter, the 'psych' library can be called using the command.

library(psych)
describeBy(perf_dat[,c(2:5)])
```

which produces:

```
##
        vars n  mean   sd median trimmed  mad  min  max range skew kurtosis
shots      1 7  2.07 1.49   2.30    2.07 1.93  0.3  4.2   3.9 0.16    -1.79
goals      2 8  0.25 0.27   0.15    0.25 0.22  0.0  0.7   0.7 0.63    -1.41
passes     3 7 28.63 7.41  28.30   28.63 7.71 18.6 39.2  20.6 0.23    -1.60
tackles    4 8  7.45 2.82   7.05    7.45 3.93  4.1 11.2   7.1 0.12    -1.90
          se
shots   0.56
goals   0.09
passes  2.80
tackles 1.00
```

3.4 Importing Data from the Internet

While there is an abundance of soccer data on the Internet, getting that data into the correct format so that it can be analysed using R can be a major challenge. Copying data by hand from tables on the web into spreadsheets is a long and tedious process, which is best avoided if possible. Fortunately, some websites offer access to data that is already in CSV format. For example, the Football-Data.co.uk website (https://www.football-data.co.uk/data.php) contains historical soccer match data in CSV format from leagues around the world. This data can be easily accessed with a few lines of code in R, as illustrated in Example 3.4.

EXAMPLE 3.4: ACCESSING CSV DATA FILES FROM THE INTERNET

This example illustrates how to access CSV files from the Internet. In this case we will access historical match data for the English Premier League (EPL) for season 2020–2021 from https://www.football-data.co.uk/mmz4281/2021/E0.csv.

First of all, we clear all the existing variables and data from the workspace so that they do not interfere with the new code.

```
rm(list=ls()) # This clear existing variables and data from the workspace.
```

We can then access the CSV file from the website and load it into R as follows:

```
EPL2020_dat <- read.csv('https://www.football-data.co.uk/mmz4281/2021/E0.csv')
```

This produces a data frame called "EPL2020_dat," which contains 380 observations (matches) and, in this case, 106 variables containing match performance data and betting data. However, often we don't actually want all these variables, so it is advisable to use the following code instead, which restricts the number of variables selected. Here the function 'head' is used to select all 380 EPL matches in season 2020–2021. (NB. For other leagues, it might be necessary to use an alternative value rather than 380.) In addition, here we select only the first sixteen columns in the data set.

```
# Load data
EPL2020_dat <- head(read.csv('https://www.football-data.co.uk/mmz4281/2021/E0.csv'),
            380)[,1:16]
```

To review the concise version of the 'EPL2020_dat' data frame, we can use the following commands:

```
names(EPL2020_dat) # This lists the names of the variables in the data frame.
```

```
##
[1] "Div"   "Date"  "Time"  "HomeTeam" "AwayTeam" "FTHG"  "FTAG"   "FTR"
[9] "HTHG"  "HTAG"  "HTR"   "Referee"  "HS"    "AS"      "HST"    "AST"
```

```
head(EPL2020_dat,10) # This displays the first ten lines of the data frame.
```

This produces:

```
##
   Div       Date  Time          HomeTeam    AwayTeam FTHG FTAG FTR HTHG
1   E0 12/09/2020 12:30            Fulham     Arsenal    0    3   A    0
2   E0 12/09/2020 15:00   Crystal Palace Southampton    1    0   H    1
3   E0 12/09/2020 17:30         Liverpool       Leeds    4    3   H    3
4   E0 12/09/2020 20:00          West Ham   Newcastle    0    2   A    0
5   E0 13/09/2020 14:00         West Brom   Leicester    0    3   A    0
6   E0 13/09/2020 16:30         Tottenham     Everton    0    1   A    0
7   E0 14/09/2020 20:15          Brighton     Chelsea    1    3   A    0
8   E0 14/09/2020 18:00 Sheffield United      Wolves    0    2   A    0
9   E0 19/09/2020 12:30           Everton   West Brom    5    2   H    2
10  E0 19/09/2020 15:00             Leeds      Fulham    4    3   H    2
   HTAG HTR    Referee HS AS HST AST
1     1   A C Kavanagh  5 13   2   6
2     0   H     J Moss  5  9   3   5
3     2   H   M Oliver 22  6   6   3
4     0   D  S Attwell 15 15   3   2
5     0   D   A Taylor  7 13   1   7
```

```
6     0   D M Atkinson  9 15   5   4
7     1   A   C Pawson 13 10   3   5
8     2   A     M Dean  9 11   2   4
9     1   H     M Dean 17  6   7   4
10    1   H   A Taylor 10 14   7   6
```

Where Div is the division (i.e. E0 is the EPL); Date is the date of the match; Time is the time the match started; HomeTeam and AwayTeam are the home and away teams, respectively; FTHG is the full-time home goals scored; FTAG is the full-time away goals scored; FTR is the full-time result (i.e., H=home win; A=away win; or D=draw); HTHG, HTAG, and HTR are the half-time equivalents of the full-time variables; Referee is the name of the match referee; HS is home team shots; AS is away team shots; HST is home team shots on target; and AST is away team shots on target.

Now we can export the 'EPL2020_dat' data frame to a file called "EPL_results_2021.csv", which we store in the directory called "AnalysisResults." (NB. As stated above, you can store the file in any directory on your computer by specifying the appropriate directory name.)

```
write.csv(EPL2020_dat, "C:/AnalysisResults/EPL_results_2021.csv")
```

If data for multiple seasons is required, then the code in Example 3.5 can be used. This download EPL match results for the five seasons from 2016–2017 to 2020–2021. (NB. This code can easily be adapted to suit other leagues, divisions, and seasons.)

EXAMPLE 3.5: ACCESSING SOCCER DATA FROM THE INTERNET FOR MULTIPLE SEASONS

```
# Clear existing variables and data from the workspace.
rm(list=ls())

# Download results from website for the 5 seasons 2016-2020.
seasons <- c(rep("1617",1), rep("1718",1), rep("1819",1), rep("1920",1), rep("2021",1))
division <- c(rep(c("E0"),5)) # NB. "E0" is the EPL and 5 refers to the number of seasons

urls = paste(seasons, division, sep="/")
urls = paste("https://www.football-data.co.uk/mmz4281", urls, sep="/")

# Load all the data using a loop and selecting just a few variables.
download_data = NULL
for(i in 1:length(urls)){
temp = read.csv(urls[i])
temp = temp[,c("Div","Date","HomeTeam","AwayTeam","FTHG","FTAG","FTR")]
download_data = rbind(download_data, temp)
}
```

This produces a data frame (download_data) with 1900 observations and 7 variables, which can be visualised using:

```
# Visualise data frame
head(download_data,10) # This displays the first 10 rows.
```

```
##
   Div     Date       HomeTeam   AwayTeam FTHG FTAG FTR
1   E0 13/08/16        Burnley    Swansea    0    1   A
2   E0 13/08/16 Crystal Palace  West Brom    0    1   A
3   E0 13/08/16        Everton  Tottenham    1    1   D
4   E0 13/08/16           Hull  Leicester    2    1   H
5   E0 13/08/16       Man City Sunderland    2    1   H
6   E0 13/08/16  Middlesbrough      Stoke    1    1   D
7   E0 13/08/16    Southampton    Watford    1    1   D
8   E0 14/08/16        Arsenal  Liverpool    3    4   A
9   E0 14/08/16    Bournemouth Man United    1    3   A
10  E0 15/08/16        Chelsea   West Ham    2    1   H
##
```

```
tail(download_data,10) # This displays the last 10 rows.
```

```
     Div       Date          HomeTeam      AwayTeam FTHG FTAG FTR
1891  E0 23/05/2021           Arsenal      Brighton    2    0   H
1892  E0 23/05/2021       Aston Villa       Chelsea    2    1   H
1893  E0 23/05/2021            Fulham     Newcastle    0    2   A
1894  E0 23/05/2021             Leeds     West Brom    3    1   H
1895  E0 23/05/2021         Leicester     Tottenham    2    4   A
1896  E0 23/05/2021         Liverpool Crystal Palace   2    0   H
1897  E0 23/05/2021          Man City       Everton    5    0   H
1898  E0 23/05/2021 Sheffield United       Burnley    1    0   H
1899  E0 23/05/2021          West Ham   Southampton    3    0   H
1900  E0 23/05/2021            Wolves    Man United    1    2   A
```

From this, it can be seen that the code selects all the EPL matches for the five seasons from 2016–2017 to 2020–2021.

3.5 Harvesting Soccer Data from the Internet

While it is relatively easy to load data from the Internet that is conveniently packaged as CSV files, it is much more difficult to obtain data if it is just contained in tables on web pages. In order to capture this data in this format, it is necessary to use an R package to harvest and scrape the data from the Internet. One such bespoke R package is the 'worldfootballR' library package [1] (https://jaseziv.github.io/worldfootballR/), which is designed to extract soccer data from FBref.com (https://fbref.com/en/). The FBref.com website contains a wealth of soccer-related data about both players and teams in all the major leagues around the world. As such, it has much useful data that can be used to analyse both player and team performance.

The 'worldfootballR' library package can be found on GitHub. However, in order to access packages from GitHub, it is first necessary to install the devtools package, which will enable the installation of worldfootballR. This can be done as follows:

```
# First install the devtools package, which enables packages to be installed from GitHub.
install.packages("devtools")

# Then install worldfootballR from GitHub.
devtools::install_github("JaseZiv/worldfootballR", ref = "main")
```

Alternatively, 'worldfootballR' can be downloaded from *the Comprehensive R Archive Network* (CRAN).

After the worldfootballR package has been successfully installed, it is only necessary thereafter to load the library every time you want to use it. This can be done as follows:

```
library(worldfootballR) # This calls up the library package.
```

When seeking to analyse match data using the 'worldfootballR' package, there are two functions, 'fb_match_urls' and 'fb_match_summary,' that are particularly useful. The first of these allows us to obtain the URLs for all the match reports for the specified league in a given season, while the second function enables summary reports to be produced for specified matches.

To obtain, for example, a list of match urls from FBref.com for the EPL during season 2020–2021, we can use the code in Example 3.6, where: ENG denotes the country (i.e., England); 1st denotes the first tier (i.e., the EPL); M denotes males; and the 'season_end_year' is 2021.

EXAMPLE 3.6: ACCESSING MATCH REPORT DATA FROM THE INTERNET

```
# Clear existing variables and data from the workspace.
rm(list=ls())

# Load library
library(worldfootballR) # This calls up the library package.
match_urls <- fb_match_urls(country = "ENG",gender = "M",tier = "1st",
                            season_end_year = c(2021))
```

We can then display the urls for the first 10 matches in season 2020–2021 using the command:

```
head(match_urls,10)
```

This displays a list of the URLs containing the various match reports. (NB. Because the list of URL addresses is complex, for the sake of simplicity, this list is not replicated here.)

Having got a list of URLs, we can use this to produce a data frame containing a list of the main events in any specified match. So for example, say we wanted to produce a summary for the Fulham-Arsenal match that took place on 12 September 2020, using the URL https://fbref.com/en/matches/bf52349b/Fulham-Arsenal-September-12-2020-Premier-League. This could be done using the 'fb_match_summary' function as follows:

```
match_summary <- fb_match_summary(match_url =
 "https://fbref.com/en/matches/bf52349b/Fulham-Arsenal-September-12-2020-Premier-League")
```

The summary produced gives an overview of the Fulham-Arsenal match, with the timing and nature of the main events (e.g., goal, yellow card, etc.) all listed. We can display the match summary data frame using the line 'print(match_summary)', although this displays much superfluous information. Therefore, it is better to edit the results as follows:

```
print(match_summary[,c(19:25)]) # This displays the variables of interest (i.e. columns 19-25).
```

This produces:

```
##
      Team Home_Away Event_Time Is_Pens Event_Half  Event_Type
1  Arsenal      Away          8   FALSE          1        Goal
```

```
2     Fulham       Home         26    FALSE          1 Yellow Card
3    Arsenal       Away         39    FALSE          1 Yellow Card
4    Arsenal       Away         49    FALSE          2        Goal
5    Arsenal       Away         55    FALSE          2 Yellow Card
6    Arsenal       Away         57    FALSE          2        Goal
7     Fulham       Home         61    FALSE          2 Yellow Card
8     Fulham       Home         63    FALSE          2  Substitute
9     Fulham       Home         63    FALSE          2  Substitute
10    Fulham       Home         75    FALSE          2  Substitute
11   Arsenal       Away         76    FALSE          2  Substitute
12   Arsenal       Away         79    FALSE          2  Substitute
13   Arsenal       Away         87    FALSE          2  Substitute
                                     Event_Players
1                              Alexandre Lacazette
2                                   Michael Hector
3                       Pierre-Emerick Aubameyang
4                Gabriel Dos Santos Assist: Willian
5                                   Héctor Bellerín
6         Pierre-Emerick Aubameyang Assist: Willian
7                                       Tom Cairney
8        Aleksandar Mitrović for Aboubakar Kamara
9   Andre-Frank Zambo Anguissa for Neeskens Kebano
10                       Bobby Reid for Josh Onomah
11                         Nicolas Pépé for Willian
12                   Dani Ceballos for Granit Xhaka
13          Eddie Nketiah for Alexandre Lacazette
```

The 'match_summary' data frame shown above contains a list of chronological events that occurred during the match. From this, it can be seen that goals were scored for Arsenal in the 8th minute by Alexandre Lacazette, the 49th minute by Gabriel Dos Santos, and the 57th minute by Pierre-Emerick Aubameyang. Although not shown here, the summary data also tells us that no penalties were awarded during the match.

The FBref.com website also has a considerable amount of data relating to team squads and their performance. This data can be easily accessed using the 'fb_season_team_stats' function in the 'worldfootballR' package, as illustrated in Example 3.7. When using this function, it is necessary to specify the type of data that you want to harvest. This can be done using the 'stat_type' argument, which has the following options:

- "standard" – General performance data for squads, including expected goals (xG)
- "shooting" – Shooting data
- "passing" – Passing data
- "passing_types" – Type of passes

- "defense" – Defensive data
- "possession" – Possession data
- "goal_shot_creation" – Goal shot creation data
- "playing_time" – Playing time data
- "keeper" – Goalkeeper data
- "keeper_adv" – Advanced goalkeeping data
- "misc" – Squad miscellaneous data

Example 3.7 shows how squad data can be accessed from the FBref.com website.

EXAMPLE 3.7: ACCESSING SQUAD PERFORMANCE DATA FROM THE INTERNET

First we need to ensure that the 'worldfootballR' library is loaded.

```
library(worldfootballR)
```

So, now we can start to import squad statistics from the FBref.com website. To illustrate the technique, we will download general performance data for the squads in the EPL for the 2020–2021 season.

```
EPL_2020_standard <- fb_season_team_stats(country = "ENG", gender = "M",
                        season_end_year = "2021", tier = "1st", stat_type = "standard")
```

The variables in the 'EPL_2020_standard' data frame can be displayed using the 'names' function.

```
names(EPL_2020_standard)
```

```
##
 [1] "Competition_Name"              "Gender"
 [3] "Country"                       "Season_End_Year"
 [5] "Squad"                         "Team_or_Opponent"
 [7] "Num_Players"                   "Age"
 [9] "Poss"                          "MP_Playing_Time"
[11] "Starts_Playing_Time"           "Min_Playing_Time"
[13] "Mins_Per_90_Playing_Time"      "Gls"
[15] "Ast"                           "G_plus_A"
[17] "G_minus_PK"                    "PK"
[19] "PKatt"                         "CrdY"
[21] "CrdR"                          "xG_Expected"
[23] "npxG_Expected"                 "xAG_Expected"
[25] "npxG_plus_xAG_Expected"        "PrgC_Progression"
[27] "PrgP_Progression"              "Gls_Per_Minutes"
```

```
[29] "Ast_Per_Minutes"                "G_plus_A_Per_Minutes"
[31] "G_minus_PK_Per_Minutes"         "G_plus_A_minus_PK_Per_Minutes"
[33] "xG_Per_Minutes"                 "xAG_Per_Minutes"
[35] "xG_plus_xAG_Per_Minutes"        "npxG_Per_Minutes"
[37] "npxG_plus_xAG_Per_Minutes"
```

From this, we see that the data frame comprises 34 variables, which cover various aspects of squad performance. However, for conciseness, we will only display variables 4, 5, 6, 8, 9, 14, and 15. (NB. 'Poss' is possession; 'Gls' is the number of goals; and 'Ast' is the number of assists.)

```
print(EPL_2020_standard[,c(4,5,6,8,9,14,15)])
```

```
##
   Season_End_Year               Squad Team_or_Opponent  Age Poss Gls Ast
1             2021             Arsenal             team 25.9 53.5  53  38
2             2021         Aston Villa             team 25.2 48.5  52  38
3             2021            Brighton             team 25.8 51.1  39  24
4             2021             Burnley             team 28.3 42.1  32  20
5             2021             Chelsea             team 26.0 60.9  56  38
6             2021      Crystal Palace             team 29.1 40.6  39  29
7             2021             Everton             team 26.3 46.4  45  32
8             2021              Fulham             team 25.2 49.7  26  18
9             2021        Leeds United             team 26.1 57.3  60  45
10            2021      Leicester City             team 26.5 54.3  64  45
11            2021           Liverpool             team 26.8 62.1  65  43
12            2021     Manchester City             team 26.1 63.5  82  55
13            2021      Manchester Utd             team 25.6 55.7  70  51
14            2021       Newcastle Utd             team 27.1 38.8  44  26
15            2021       Sheffield Utd             team 26.7 41.7  19  13
16            2021         Southampton             team 26.6 52.0  47  33
17            2021           Tottenham             team 27.2 51.6  66  50
18            2021           West Brom             team 26.4 37.9  33  20
19            2021            West Ham             team 27.8 43.2  60  46
20            2021              Wolves             team 26.3 49.3  34  21
21            2021          vs Arsenal         opponent 26.7 46.2  35  25
22            2021      vs Aston Villa         opponent 26.7 51.9  45  30
23            2021         vs Brighton         opponent 26.4 48.7  44  30
24            2021          vs Burnley         opponent 26.6 58.3  54  47
25            2021          vs Chelsea         opponent 26.3 38.6  35  26
26            2021   vs Crystal Palace         opponent 26.4 59.9  66  47
27            2021          vs Everton         opponent 26.4 53.5  47  33
28            2021           vs Fulham         opponent 26.6 50.1  52  34
29            2021     vs Leeds United         opponent 26.5 42.4  52  36
30            2021   vs Leicester City         opponent 26.5 45.4  48  29
31            2021        vs Liverpool         opponent 26.6 37.6  42  29
32            2021  vs Manchester City         opponent 26.4 36.1  31  17
33            2021   vs Manchester Utd         opponent 26.6 44.2  42  28
34            2021    vs Newcastle Utd         opponent 26.6 61.8  59  41
35            2021    vs Sheffield Utd         opponent 26.7 58.5  60  45
36            2021      vs Southampton         opponent 26.5 47.8  67  44
37            2021        vs Tottenham         opponent 26.5 48.3  42  24
38            2021        vs West Brom         opponent 26.6 62.4  73  52
39            2021         vs West Ham         opponent 26.6 57.1  43  34
40            2021           vs Wolves         opponent 26.6 50.7  49  34
```

Now, we can export the 'EPL_2020_standard' data frame as a csv file.

```
write.csv(EPL_2020_standard, "C:/AnalysisResults/EPL_2020_standard.csv")
```

As well as squad statistics, the FBref.com website also has a wealth of data devoted specifically to individual players. This can be accessed using the 'fb_player_season_stats' function as shown in Example 3.8, which relates to shooting data for Cristiano Ronaldo.

EXAMPLE 3.8: ACCESSING PLAYER PERFORMANCE DATA FROM THE INTERNET

To access the shooting data for Cristiano Ronaldo, we use the following code:

```
library(worldfootballR)
Ronaldo_shooting <-
  fb_player_season_stats("https://fbref.com/en/players/dea698d9/Cristiano->Ronaldo",
                          stat_type = 'shooting')
```

The variables in the 'Ronaldo_shooting' data frame can be displayed using:

```
names(Ronaldo_shooting)
```

```
##
 [1] "player_name"           "player_url"              "Season"
 [4] "Age"                   "Squad"                   "Country"
 [7] "Comp"                  "Mins_Per_90"             "Gls_Standard"
[10] "Sh_Standard"           "SoT_Standard"            "SoT_percent_Standard"
[13] "Sh_per_90_Standard"    "SoT_per_90_Standard"     "G_per_Sh_Standard"
[16] "G_per_SoT_Standard"    "Dist_Standard"           "FK_Standard"
[19] "PK_Standard"           "PKatt_Standard"          "xG_Expected"
[22] "npxG_Expected"         "npxG_per_Sh_Expected"    "G_minus_xG_Expected"
[25] "np:G_minus_xG_Expected"
```

This reveals that the data frame comprises 25 variables.

Selected variables from this data frame can be displayed using:

```
head(Ronaldo_shooting[,c(1,3:5,9,11)], 20) # This displays the top 20 rows.
```

```
##
        player_name    Season Age            Squad Gls_Standard SoT_Standard
1  Cristiano Ronaldo 2002-2003  17     Sporting CP            3           15
2  Cristiano Ronaldo 2002-2003  17     Sporting CP            0           NA
3  Cristiano Ronaldo 2003-2004  18 Manchester Utd            0            4
4  Cristiano Ronaldo 2003-2004  18 Manchester Utd            4           21
```

```
5  Cristiano Ronaldo 2004-2005  19 Manchester Utd           0          9
6  Cristiano Ronaldo 2004-2005  19 Manchester Utd           5         59
7  Cristiano Ronaldo 2005-2006  20 Manchester Utd           0          9
8  Cristiano Ronaldo 2005-2006  20 Manchester Utd           9         47
9  Cristiano Ronaldo 2006-2007  21 Manchester Utd           3         22
10 Cristiano Ronaldo 2006-2007  21 Manchester Utd          17         71
11 Cristiano Ronaldo 2007-2008  22 Manchester Utd           8         21
12 Cristiano Ronaldo 2007-2008  22 Manchester Utd          31         89
13 Cristiano Ronaldo 2008-2009  23 Manchester Utd           4         33
14 Cristiano Ronaldo 2008-2009  23 Manchester Utd          18         66
15 Cristiano Ronaldo 2009-2010  24    Real Madrid           7         17
16 Cristiano Ronaldo 2009-2010  24    Real Madrid          26         92
17 Cristiano Ronaldo 2010-2011  25    Real Madrid           6         26
18 Cristiano Ronaldo 2010-2011  25    Real Madrid          40        101
19 Cristiano Ronaldo 2011-2012  26    Real Madrid          10         28
20 Cristiano Ronaldo 2011-2012  26    Real Madrid          46        113
```

Here, 'Gls_Standard' is goals shored and 'SoT_Standard' is shots on target.

3.6 Scraping Soccer Data from the Internet

While the 'worldfootballR' package is an excellent tool for harvesting soccer data from the FBref.com website, it is not designed for general scraping of the web. Web scraping is the process of extracting data from websites using automated software. It involves writing code to access a website, locate specific pieces of information or data on that site, and then extract the target data for use in another context. So, if we want to extract a specific table, for example, from some random website, we have to turn instead to a more general web scraping tool, like 'rvest' [2], which is a popular package developed for R. rvest can be installed using the 'install' packages button on RStudio or the following code:

```
install.packages("rvest") # NB. This only needs to be run once to install the package.
```

The rvest package is widely used to scrape information and data from web pages and, as such, has many applications, most of which are beyond the scope of this book. So, here we will content ourselves with looking at the relatively simple task of lifting tables of data from single web pages.

Tabular data can be fairly easily scraped from the Internet using the 'read_html' function in the rvest package, as illustrated in Example 3.9. This involves an interactive approach in which we have to browse to the desired web page and locate the target table that we want to extract. In Example 3.9, the target is a table of player transfer windows for various countries, which can be found on Wikipedia (https://en.wikipedia.org/wiki/Transfer_window).

Once the target table has been selected, the next thing to do is to right click on it and choose "inspect element". This splits the page horizontally and displays the html code. As you hover over the page elements in the html code in the bottom window, sections of the web page become highlighted in blue in the top window. The html code that you want is the bit that highlights the whole target table in blue when you hover over it (see Figure 3.3). Once you have found this, right click on the line, then click *Copy>Copy selector* (Firefox: *Copy>CSS selector*; Safari: *Copy>Selector Path*). Finally, return to RStudio, create a variable for your CSS selector, and then paste in the selector that you copied, like this:

```
css_selector <- "copied CSS selector" # Insert copied CSS selector between quotation marks.
```

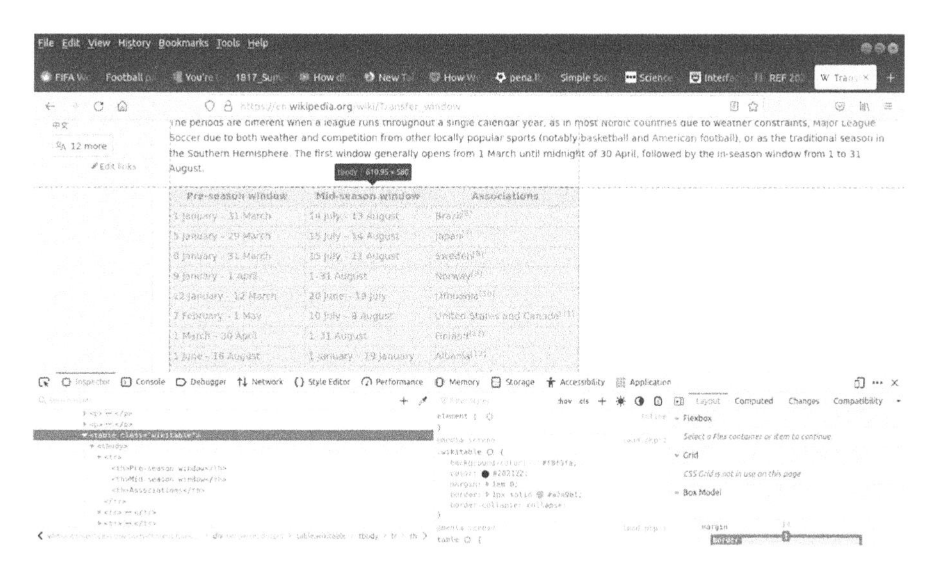

FIGURE 3.3
Screen shot showing the Wikipedia page and the target table identified for scraping.

Having created the 'css_selector' variable, we can use this to scrape the table from the web, as illustrated in Example 3.9.

While web scraping is widely used in data mining and for market research purposes, it is important to note that it can be subject to legal and ethical considerations. Therefore, it is important to exercise caution when web scraping and ensure that your activities are conducted in compliance with applicable laws and best practice.

EXAMPLE 3.9: USING RVEST TO SCRAPE A TABLE FROM A WEB PAGE

To illustrate how rvest can be used to scrape a table from a web page, we will apply it to the Wikipedia page on player transfer windows in soccer (https://en.wikipedia.org/wiki/Transfer_window). The code below enables us to harvest the data from the target table, which lists player transfer windows for various countries.

```
# Read html
library(rvest)
tran_window <- read_html("https://en.wikipedia.org/wiki/Transfer_window")

# Create CSS selector
css_selector <- ".wikitable"

# Scrape table and convert it to a data frame
tw_df <- as.data.frame(tran_window %>%
            html_element(css = css_selector) %>%
        html_table(fill=TRUE))

print(tw_df)
```

NB. Here, %>% is used as an operator that 'pipes' together functions, so that a data object is pushed from one function to another. This helps with readability and makes it easier to follow the flow between multiple functions. In this case, the code is telling R to select the html element ".wikitable" from 'tran_window', and then push (pipe) it through to the 'html_table' function for processing. By using 'fill=TRUE,' we are telling R to fill in the missing cells in tables automatically with NA. Also, by using the 'as.data.frame' function, we are telling R to save the output as a data frame.

The output of this code produces the following table, which lists the player transfer windows for the countries reported.

```
##
              Pre-season window          Mid-season window
1          1 January - 31 March       14 July - 13 August
2          5 January - 29 March       15 July - 14 August
3          8 January - 31 March       15 July - 11 August
4            9 January - 1 April             1-31 August
5         12 January - 12 March         20 June - 19 July
6           10 February - 4 May         7 July - 4 August
7             1 March - 30 April             1-31 August
8             1 June - 18 August            1-19 January
9           10 June - 31 August 18 January - 15 February
10         10 June - 1 September            1-31 January
11             9 June - 31 August           1-31 January
12          1 July - 2 September            3-31 January
13     1 July - 1 September[22]        1-31 January[23]
```

```
14         1 July - 2 September   1 January - 2 February
15         9 June - 1 September   1 January - 1 February
16        11 June - 1 September             5-31 January
17        11 June - 2 September             3-31 January
18        16 June - 8 September 25 January - 22 February
19 26 July - 17 October (2020)
20        24 July - 15 October              3-31 January
21      1 December - 31 January                1-30 June
                            Associations
1                               Brazil[6]
2                                Japan[7]
3                               Sweden[8]
4                               Norway[9]
5                           Lithuania[10]
6  United States and Canada[11][12]
7                             Finland[13]
8                             Albania[14]
9                         Switzerland[15]
10             England,[16] France[17]
11                              India[18]
12                    Italy[19][20][21]
13                                Germany
14                              Spain[24]
15                           Scotland[25]
16             Turkey,[26] Denmark[27]
17                        Netherlands[28]
18             Bulgaria and Romania[29]
19                             Russia[30]
20                          Australia[31]
21                      Kenya, Uganda[32]
```

3.7 Final Comments

The aim of this chapter has been to give a practical overview of some of the techniques that can be used to import soccer data into R so that it can be processed and analysed. However, the techniques shown are by no means exhaustive, and there are plenty of other tools out there that can be used. The 'rvest' package in particular is a very powerful tool, which has a much wider range of applications than those described above. Notwithstanding this, if you have worked through the examples in this chapter, you should be well equipped to acquire and process soccer-related data from a wide variety of sources, and hopefully this will stand you in good stead as you develop your analytical skills.

References

1. Zivkovic J (2023). *worldfootballR: Extract and Clean World Football (Soccer) Data*. R package version 0.6.4.0003, https://github.com/JaseZiv/worldfootballR
2. Wickham H, Wickham MH: Package 'rvest'. https://cranr-project.org/web/packages/rvest/rvest pdf; 2016, https://doi.org/10.1002/wics.147

4

Match Data and League Tables

Much analysis relating to soccer involves gathering statistics from historical match data and predicting end-of-season (EoS) league positions. Data scientists working in professional sport can gain a lot of useful information through the analysis of historical data and spend much time attempting to identify features (metrics) that explain team performance. In this chapter, we will focus on techniques for capturing performance metrics from historical match data, as these can be useful predictors of likely EoS league position. Also, in this chapter, we will show how to compile home and away performance statistics from historical match data, as well as obtain head-to-head (H2H) match statistics for equivalent matches in previous seasons.

4.1 Compiling Team Performance Statistics from Historical Match Data

Historical match data can be an excellent resource with which to assess the performance of soccer teams. However, arranging the data in a form that can be easily analysed is often a tricky task, especially as teams generally play each other both home and away, something that can cause confusion if not handled correctly. Also, we frequently want to augment data sets with performance metrics that can be derived from the match data. So in this section, we will look at how historical match data can be augmented and used to produce performance statistics.

Match performance data sets often include statistics relating to shots attempted and shots on target, and these can be used to produce derived metrics such as the total shots ratio (TSR), which is a widely used indicator of team performance (see Chapter 5 for more details on TSR). TSR is calculated as follows:

$$\text{TSR} = \frac{\text{SF}}{\text{SF} + \text{SA}} \tag{4.1}$$

where SF is the number of shots for the team and SA is the number of shots against it.

DOI: 10.1201/9781003328568-4

In Example 4.1, we will augment match data from the English Premier League (EPL) for the 2020–2021 season by computing the TSR for the home and away teams, as well as the number of points awarded to each team.

EXAMPLE 4.1: AUGMENTING HISTORICAL MATCH DATA

Before starting to code, it is always good practice to first clear existing variables and data from the RStudio workspace, as these might interfere with the new script. This can be easily done using the following line of R code, which, although not very memorable, works very well.

```
rm(list = ls())   # This clears all variables from the workspace
```

Now that we have a blank slate, the data can be imported into R in CSV format and displayed using the code below.

This example uses historical data from: https://www.football-data.co.uk/englandm.php for the EPL season 2020–2021. However, we will not use all the variables in the data set. Here we will just utilise the first 16 variables, which are:

- Div=League division
- Date=Match date (dd/mm/yy)
- HomeTeam=Home team
- AwayTeam=Away team
- FTHG=Full time home team goals
- FTAG=Full time away team goals
- FTR=Full time result (H=Home Win, D=Draw, A=Away Win)
- HTHG=Half time home team goals
- HTAG=Half time away team goals
- HTR=Half time result (H=Home Win, D=Draw, A=Away Win)
- Referee=Match referee
- HS=Home team shots
- AS=Away team shots
- HST=Home team shots on target
- AST=Away team shots on target

The data can be loaded into R using the following code, which selects only the first 16 variables.

```
# Load data (NB. Here we select just the first 16 variables).
EPL2020_data <- head(read.csv('https://www.football-data.co.uk/mmz4281/2021/E0.csv'),
                     380)[,1:16]

# Inspect data
names(EPL2020_data)
```

```
##
[1] "Div"       "Date"      "Time"      "HomeTeam"    "AwayTeam"    "FTHG"
[7] "FTAG"      "FTR"       "HTHG"      "HTAG"        "HTR"         "Referee"
[13] "HS"       "AS"        "HST"       "AST"
```

From this, we can see that we have included some variables that are irrelevant. So, for ease of use, we will remove the variables "Div", "HTHG", "HTAG", "HTR", and "Referee," which are not of interest to us here.

```
dat <- EPL2020_data[,c("Date","HomeTeam","AwayTeam","FTHG","FTAG",
                "FTR","HS","AS","HST","AST")]  # This creates a working data frame

head(dat) # Displays first six rows of dat
```

Which produces:

```
##
        Date       HomeTeam    AwayTeam FTHG FTAG FTR HS AS HST AST
1 12/09/2020         Fulham     Arsenal    0    3   A  5 13   2   6
2 12/09/2020 Crystal Palace Southampton    1    0   H  5  9   3   5
3 12/09/2020      Liverpool       Leeds    4    3   H 22  6   6   3
4 12/09/2020       West Ham   Newcastle    0    2   A 15 15   3   2
5 13/09/2020      West Brom   Leicester    0    3   A  7 13   1   7
6 13/09/2020      Tottenham     Everton    0    1   A  9 15   5   4
```

Now we can augment the data by adding some new derived variables. To do this, we first need to create some new 'empty' variables populated with NAs and zeros in which to store the derived statistics.

```
dat["GD"] <- NA # Creates a new column for goal difference populated with NAs
dat["TG"] <- NA # Creates a new column for total goals scored populated with NAs
dat["HTSR"] <- NA # Creates a home team shots ratio variable populated with NAs
dat["ATSR"] <- NA # Creates a away team shots ratio variable populated with NAs
dat["HPts"] <- 0 # Creates a home team points variable populated with zeros
dat["APts"] <- 0 # Creates an away team points variable populated with zeros
```

Having created the empty variables, we can now compute the derived statistics and populate the variables as follows:

```
dat$GD <- dat$FTHG - dat$FTAG
dat$TG <- dat$FTHG + dat$FTAG
dat$HTSR <- round(dat$HS/(dat$HS+dat$AS),3)  # Here we round to 3 decimal places
dat$ATSR <- round(dat$AS/(dat$AS+dat$HS),3)  # Here we round to 3 decimal places
```

Computing the home and away points awarded per match is a little trickier. But this can be done fairly easily using a 'for loop' and four 'if statements', as follows.

```
for(i in 1:nrow(dat)){
  if(dat$FTR[i] == "H"){dat$HPts[i] <- 3}
  if(dat$FTR[i] == "A"){dat$APts[i] <- 3}
  if(dat$FTR[i] == "D") {dat$HPts[i] <- 1}
  if(dat$FTR[i] == "D") {dat$APts[i] <- 1}
}
```

We can also rename some of the columns in the data frame.

```
colnames(dat)[colnames(dat) == 'FTHG'] <- 'HG'
colnames(dat)[colnames(dat) == 'FTAG'] <- 'AG'
colnames(dat)[colnames(dat) == 'FTR'] <- 'Result'
```

Finally, we can display the new augmented 'dat' data frame.

```
head(dat) # This displays only the first six rows of the data frame.
```

```
##
        Date       HomeTeam    AwayTeam HG AG Result HS AS HST AST GD TG
1 12/09/2020         Fulham     Arsenal  0  3      A  5 13   2   6 -3  3
2 12/09/2020 Crystal Palace Southampton  1  0      H  5  9   3   5  1  1
3 12/09/2020      Liverpool       Leeds  4  3      H 22  6   6   3  1  7
4 12/09/2020       West Ham   Newcastle  0  2      A 15 15   3   2 -2  2
5 13/09/2020      West Brom   Leicester  0  3      A  7 13   1   7 -3  3
6 13/09/2020      Tottenham     Everton  0  1      A  9 15   5   4 -1  1
   HTSR  ATSR HPts APts
1 0.278 0.722    0    3
2 0.357 0.643    3    0
3 0.786 0.214    3    0
4 0.500 0.500    0    3
5 0.350 0.650    0    3
6 0.375 0.625    0    3
```

Having downloaded and augmented the match data as shown in Example 4.1, it is often the case that we want to look at the performance of individual teams in the league. This can be tricky because the teams in the league play both home and away matches, which means that when assessing performance, we need to distinguish between the two. So, when assessing the performance of individual teams, it is helpful to convert the data into 'for' and 'against' rather than 'home' and 'away', as illustrated in Example 4.2, which considers the performance of Liverpool during the 2020–2021 season.

EXAMPLE 4.2: EXTRACTING DATA FOR INDIVIDUAL TEAMS

Here we will assess the performance of Liverpool during the 2020–2021 season. Here, Liverpool is the target (or study) team.

```
study.team <- "Liverpool" # NB. The team name should be in quotes.
```

Next, we extract data for just the target team.

```
home.matches <- dat[dat$HomeTeam == study.team,] # Selects target team's home matches.
away.matches <- dat[dat$AwayTeam == study.team,] # Selects target team's away matches.
```

Now, we need to add a 'status' variable to denote whether a match is home or away.

```
home.matches["Status"] <- "Home" # Creates the new column to denote home matches.
away.matches["Status"] <- "Away" # Creates the new column to denote away matches.
```

Next, we need to change the variable names to be club specific (i.e. for and against rather than home and away).

```
home <- home.matches # Make duplicate copy of the data frame on which to make changes
away <- away.matches # Make duplicate copy of the data frame on which to make changes
```

At this point, it is a good idea to inspect the data frames, just to check that we have got everything right.

```
head(home)
```

```
##
          Date  HomeTeam        AwayTeam HG AG Result HS AS HST AST GD
3   12/09/2020 Liverpool           Leeds  4  3      H 22  6   6   3  1
28  28/09/2020 Liverpool         Arsenal  3  1      H 21  4   8   3  2
53  24/10/2020 Liverpool Sheffield United  2  1      H 17 13   5   2  1
62  31/10/2020 Liverpool        West Ham  2  1      H  9  4   5   3  1
86  22/11/2020 Liverpool       Leicester  3  0      H 24 11  13   4  3
106 06/12/2020 Liverpool          Wolves  4  0      H 11  9   6   3  4
    TG  HTSR  ATSR HPts APts Status
3    7 0.786 0.214    3    0   Home
28   4 0.840 0.160    3    0   Home
53   3 0.567 0.433    3    0   Home
62   3 0.692 0.308    3    0   Home
86   3 0.686 0.314    3    0   Home
106  4 0.550 0.450    3    0   Home
```

```
head(away)
```

```
##
          Date    HomeTeam  AwayTeam HG AG Result HS AS HST AST GD TG
15  20/09/2020     Chelsea Liverpool  0  2      A  5 18   3   6 -2  2
38  04/10/2020 Aston Villa Liverpool  7  2      H 18 14  11   8  5  9
39  17/10/2020     Everton Liverpool  2  2      D 11 22   5   8  0  4
77  08/11/2020    Man City Liverpool  1  1      D  7 10   2   3  0  2
90  28/11/2020    Brighton Liverpool  1  1      D 11  6   3   2  0  2
115 13/12/2020      Fulham Liverpool  1  1      D 10 12   5   6  0  2
     HTSR  ATSR HPts APts Status
15  0.217 0.783    0    3   Away
38  0.562 0.438    3    0   Away
39  0.333 0.667    1    1   Away
77  0.412 0.588    1    1   Away
90  0.647 0.353    1    1   Away
115 0.455 0.545    1    1   Away
```

From this, it can be seen that we are still describing things in terms of home and away goals rather than goals for and against Liverpool, which is much more useful. So, we need to change the names of the variables in the 'home' and 'away' data frames to reflect this. In R, there are lots of different ways in which variable names can be changed. For example, we used the 'colnames' function in Example 4.1 to do this task. However, this approach tends to become cumbersome when there is a need to change multiple variable names at the same time. So, in this example, we will instead use the 'rename' command in the 'dplyr' package, which can be installed as follows:

```
install.packages("dplyr") # This installs the 'dplyr' package.
# NB. This command only needs to be executed once to install the package.
# Thereafter, the 'dplyr' library can be called using the 'library' command.
```

We can use the 'dplyr' package to rename multiple variables, as follows:

```
# First, we change the variable names in the 'home' data frame.
library(dplyr) # Load library package 'dplyr'
home <- rename(home, c("GF"="HG","GA"="AG","SF"="HS","SA"="AS",
                "STF"="HST","STA"="AST","TSRF"="HTSR","TSRA"="ATSR",
                "PF"="HPts","PA"="APts"))
```

We also need to replace the 'H', 'A', and 'D' elements in Results vector with 'W', 'L', and 'D', which can be done using the 'recode' function in 'dplyr', as follows:

```
home$Result <- recode(home$Result, "H"="W", "A"="L", "D"="D")

head(home)
```

```
##
          Date  HomeTeam           AwayTeam GF GA Result SF SA STF STA GD
3   12/09/2020 Liverpool              Leeds  4  3      W 22  6   6   3  1
28  28/09/2020 Liverpool            Arsenal  3  1      W 21  4   8   3  2
53  24/10/2020 Liverpool Sheffield United  2  1      W 17 13   5   2  1
62  31/10/2020 Liverpool           West Ham  2  1      W  9  4   5   3  1
86  22/11/2020 Liverpool          Leicester  3  0      W 24 11  13   4  3
106 06/12/2020 Liverpool             Wolves  4  0      W 11  9   6   3  4
    TG  TSRF  TSRA PF PA Status
3    7 0.786 0.214  3  0   Home
28   4 0.840 0.160  3  0   Home
53   3 0.567 0.433  3  0   Home
62   3 0.692 0.308  3  0   Home
86   3 0.686 0.314  3  0   Home
106  4 0.550 0.450  3  0   Home
```

Now we repeat the process for the 'away' data frame, with the only difference being that we have to change the sign on the match goal differences.

```
away <- rename(away, c("GA"="HG","GF"="AG","SA"="HS","SF"="AS",
                "STA"="HST","STF"="AST","TSRA"="HTSR","TSRF"="ATSR",
                "PA"="HPts","PF"="APts"))

away$GD <- -1*away$GD # Change sign on goal difference to reflect use of 'for' and 'against'.

# Replace elements in Results vector with 'W', 'L', 'D.
away$Result <- recode(away$Result, "H"="L", "A"="W", "D"="D")

head(away)
```

```
##
          Date    HomeTeam  AwayTeam GA GF Result SA SF STA STF GD TG
15  20/09/2020     Chelsea Liverpool  0  2      W  5 18   3   6  2  2
38  04/10/2020 Aston Villa Liverpool  7  2      L 18 14  11   8 -5  9
39  17/10/2020     Everton Liverpool  2  2      D 11 22   5   8  0  4
77  08/11/2020    Man City Liverpool  1  1      D  7 10   2   3  0  2
90  28/11/2020    Brighton Liverpool  1  1      D 11  6   3   2  0  2
115 13/12/2020      Fulham Liverpool  1  1      D 10 12   5   6  0  2
     TSRA  TSRF PA PF Status
15  0.217 0.783  0  3   Away
38  0.562 0.438  3  0   Away
39  0.333 0.667  1  1   Away
77  0.412 0.588  1  1   Away
90  0.647 0.353  1  1   Away
115 0.455 0.545  1  1   Away
```

This gives us exactly what we wanted: a complete breakdown of all Liverpool's matches, which are grouped by their home and away status.

Now that we have the data in the correct format, we can produce some descriptive statistics for Liverpool using the code in Example 4.3, which computes the mean, median, and standard deviation for the various performance metrics.

EXAMPLE 4.3: PRODUCING DESCRIPTIVE STATISTICS

In this example, we shall produce some performance statistics for Liverpool for the 2020–2021 season using the data frames compiled in Example 4.2. However, before we do this, we need to select which variables we want to include in the analysis. This can be done by inspecting the data using the 'names' command and then identifying variables that are redundant.

```
names(home)
```

```
##
[1] "Date"      "HomeTeam" "AwayTeam" "GF"        "GA"        "Result"
[7] "SF"        "SA"        "STF"       "STA"       "GD"        "TG"
[13] "TSRF"      "TSRA"      "PF"        "PA"        "Status"
```

Here we are not interested in the "Date", "HomeTeam", "AwayTeam", and "Result" variables.

To produce the descriptive statistics for home and away matches, we can use the following code, which utilises the 'describeBy' function in the 'psych' library package.

```
library(psych)

# Home matches descriptive statistics
home.temp <- home[,c(4,5, 7:16)] # Variables selected for statistical analysis
H.stats <- describeBy(home.temp) # These are the descriptive statistics
H.sums <- colSums(home.temp) # Column sums
home.des <- cbind.data.frame(H.stats$n, H.stats$mean, H.stats$median, H.stats$sd, H.sums)
colnames(home.des) <- c("Pld", "Mean", "Median", "SD", "Total") # Rename variables
print(round(home.des,3)) # Display the home match descriptive statistics
```

This produces the following descriptive statistics for Liverpool's home matches:

```
##
     Pld   Mean Median    SD   Total
GF    19  1.526   2.00 1.349  29.000
GA    19  1.053   1.00 1.026  20.000
SF    19 16.684  17.00 5.697 317.000
SA    19  8.316   8.00 2.868 158.000
STF   19  5.737   6.00 3.314 109.000
STA   19  3.842   4.00 1.167  73.000
GD    19  0.474   1.00 1.744   9.000
TG    19  2.579   2.00 1.644  49.000
TSRF  19  0.656   0.68 0.129  12.467
TSRA  19  0.344   0.32 0.129   6.533
PF    19  1.737   3.00 1.408  33.000
PA    19  1.105   0.00 1.370  21.000
```

```
# Away matches descriptive statistics
away.temp <- away[,c(4,5, 7:16)] # Variables selected for statistical analysis
A.stats <- describeBy(away.temp) # These are the descriptive statistics
A.sums <- colSums(away.temp) # Column sums
away.des <- cbind.data.frame(A.stats$n, A.stats$mean, A.stats$median, A.stats$sd, A.sums)
colnames(away.des) <- c("Pld", "Mean", "Median", "SD", "Total") # Rename variables
print(round(away.des,3)) # Display the home match descriptive statistics
```

This produces the following descriptive statistics for Liverpool's away matches.

```
##
     Pld   Mean Median    SD  Total
GA    19  1.158  1.000 1.642  22.00
GF    19  2.053  2.000 1.615  39.00
SA    19  9.211 10.000 4.090 175.00
SF    19 15.316 15.000 4.473 291.00
STA   19  3.737  3.000 2.182  71.00
STF   19  5.526  6.000 2.245 105.00
GD    19  0.895  1.000 2.401  17.00
TG    19  3.211  3.000 2.200  61.00
TSRA  19  0.371  0.364 0.128   7.05
TSRF  19  0.629  0.636 0.128  11.95
PA    19  0.789  0.000 1.084  15.00
PF    19  1.895  3.000 1.243  36.00
```

To produce descriptive statistics for all the matches (home and away combined) in the season, we can simply write:

```
all <- rbind(home, away)
# All matches descriptive statistics
all.temp <- all[,c(4,5, 7:16)] # Variables selected for statistical analysis
all.stats <- describeBy(all.temp) # These are the descriptive statistics
all.sums <- colSums(all.temp) # Column sums
all.des <- cbind.data.frame(all.stats$n, all.stats$mean, all.stats$median, all.stats$sd, all.sums)
colnames(all.des) <- c("Pld", "Mean", "Median", "SD", "Total") # Rename variables
print(round(all.des,3)) # Display the home match descriptive statistics
```

Which produces:

```
##
     Pld   Mean Median    SD   Total
GF    38  1.789  2.000 1.492  68.000
GA    38  1.105  1.000 1.351  42.000
SF    38 16.000 16.000 5.099 608.000
SA    38  8.763  8.500 3.514 333.000
STF   38  5.632  6.000 2.794 214.000
STA   38  3.789  3.000 1.727 144.000
GD    38  0.684  1.000 2.081  26.000
```

```
TG     38  2.895  2.500 1.942 110.000
TSRF   38  0.643  0.667 0.127  24.417
TSRA   38  0.357  0.333 0.127  13.583
PF     38  1.816  3.000 1.312  69.000
PA     38  0.947  0.000 1.229  36.000
```

While the methodology outlined above produces tables that can be extremely informative, they can nonetheless be difficult to follow, particularly if lots of information is presented. So, sometimes it is helpful to produce plots from the data, as these are generally easier to understand.

Box plots can be particularly helpful when communicating results to lay audiences who might not be very numerate. So to illustrate how box plots can be constructed in R, we will use the following code to produce a box plot of the shots on target (SoT), both for and against Liverpool during home and away matches.

```
# Produce box plot of home and away shots on target.
HSTF <- home$STF # SoT for at home
HSTA <- home$STA # SoT against at home
ASTF <- away$STF # SoT for away
ASTA <- away$STA # SoT against away
SoT <- cbind.data.frame(HSTF,HSTA,ASTF,ASTA)
boxplot(SoT, ylab="Shots on target")
```

This produces the box plot in Figure 4.1, which shows that for both home and away matches, Liverpool made many more SoT than they conceded.

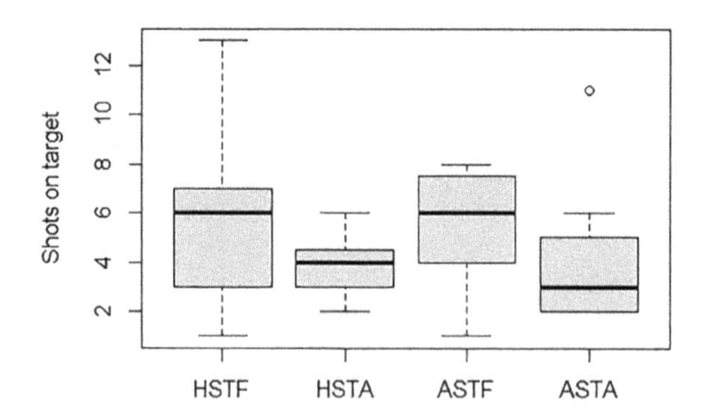

FIGURE 4.1
Box plot of the shots on target, both for and against Liverpool, during their home and away matches in the 2020–2021 season. (HSTF = home shots on target for; HSTA = home shots on target against; ASTF = away shots on target for; ASTA = away shots on target against).

From Examples 4.1–4.3, we can see that it is relatively easy in R to produce detailed performance statistics for any club in any season directly from the historical match data. As such, the code presented in these examples can be utilised and adapted to gain deeper insights into the performance of teams. For example, from the results in Example 4.3, it is interesting to note that although Liverpool were awarded more points per away match (1.895 points) compared with home matches (1.737 points), in terms of TSRF and TSRA per match, they actually performed better at home than away.

4.2 Producing Head-to-Head Statistics from Historical Match Data

Often, when seeking to predict the outcome of a given match, it is helpful to look at the outcomes of the same H2H fixture in previous seasons, as this might provide useful information regarding the likely outcome of the match [1]. However, doing this by hand is a slow and tedious process, so it is much better to let R do the hard work. Example 4.4, which concerns matches between Arsenal and Chelsea in the five seasons from 2016–2017 to 2020–2021, illustrates how this can be done in practice.

EXAMPLE 4.4: PRODUCING HISTORICAL HEAD-TO-HEAD MATCH STATISTICS

This example illustrates how to obtain head-to-head (H2H) historical match data for the Arsenal (home team) versus Chelsea (away team) fixture for the five seasons from 2016–2017 to 2020–2021.

First, we clear existing variables and data from the workspace.

```
rm(list=ls())
```

Next, we download the historical match data for the five seasons from 2016–2017 to 2020–2021.

```
seasons = c(rep("1617",1), rep("1718",1), rep("1819",1), rep("1920",1), rep("2021",1))
division = c(rep(c("E0"),5)) # NB. "E0" is the EPL and 5 is the number of seasons.

urls = paste(seasons, division, sep="/")
urls = paste("https://www.football-data.co.uk/mmz4281", urls, sep="/")
```

Having specified the URL addresses, we use a 'for loop' to load the data into R, selecting just a few relevant variables for analysis.

```
download_data = NULL
for(i in 1:length(urls)){
  temp = read.csv(urls[i])
  temp = temp[,c("Div","Date","HomeTeam","AwayTeam","FTHG","FTAG",
               "FTR","HS","AS","HST","AST")]
  download_data = rbind(download_data, temp)
}
```

To check that the data has downloaded correctly, it is often helpful to inspect the data frame. This can easily be done using the 'head' and 'tail' functions in R, which visualise the top and bottom rows of the 'download_data' data frame.

```
head(download_data) # This displays the first six rows.
```

```
##
  Div     Date       HomeTeam   AwayTeam FTHG FTAG FTR HS AS HST AST
1  E0 13/08/16        Burnley    Swansea    0    1   A 10 17   3   9
2  E0 13/08/16 Crystal Palace West Brom    0    1   A 14 13   4   3
3  E0 13/08/16        Everton  Tottenham    1    1   D 12 13   6   4
4  E0 13/08/16           Hull  Leicester    2    1   H 14 18   5   5
5  E0 13/08/16       Man City Sunderland    2    1   H 16  7   4   3
6  E0 13/08/16  Middlesbrough      Stoke    1    1   D 12 12   2   1
```

```
tail(download_data) # This displays the last six rows.
```

```
##
     Div       Date         HomeTeam       AwayTeam FTHG FTAG FTR HS AS
1895  E0 23/05/2021        Leicester      Tottenham    2    4   A 10 11
1896  E0 23/05/2021        Liverpool Crystal Palace    2    0   H 19  5
1897  E0 23/05/2021         Man City        Everton    5    0   H 21  8
1898  E0 23/05/2021 Sheffield United        Burnley    1    0   H 12 10
1899  E0 23/05/2021         West Ham    Southampton    3    0   H 14 17
1900  E0 23/05/2021           Wolves     Man United    1    2   A 14  9
     HST AST
1895   6   4
1896   5   4
1897  11   3
1898   3   3
1899   7   5
1900   4   4
```

Now that the historical match data set has been safely loaded into R, we can proceed to search for our H2H matches, which in this example involve Arsenal and Chelsea.

```
teamH <- "Arsenal"  # This is the home team.
teamA <- "Chelsea"  # This is the away team.
```

The Arsenal versus Chelsea matches can be extracted from the data frame in various ways, but here we will use an indicator variable to do the job. Creating an indicator variable is a useful trick that can be helpful in many applications. In this case, we first create a new indicator vector, which is initially populated entirely with zeros.

```
n <- nrow(download_data)  # This determines the length of the indicator vector.
ind <- matrix(0,n,1)  # This created an [n,1] matrix (vector) populated with zeros.
```

Next, we bind this new vector onto the existing data frame and give it the name 'ndat'.

```
ndat <- cbind.data.frame(download_data, ind)
```

Now we populate the indicator variable using a 'for loop' and an 'if' statement. Here we will also introduce the AND operator '&' which takes two logical values and returns TRUE only if both values are TRUE themselves.

```
for (i in  1:n){
  if(ndat$HomeTeam[i] == teamH & ndat$AwayTeam[i] == teamA){ndat$ind[i] <- 1}
}
```

This inserts 1 in the indicator variable when the test criteria have been met (i.e., the home team is Arsenal and the away team is Chelsea). Having done this, we can then select only those matches for which the indicator variable is 1 using the following code:

```
H2H <- ndat[ndat$ind == 1,]
```

For completeness, we will also add the shots ratios for the respective matches.

```
H2H["HTSR"] <- NA # This creates the new column to store home team shots ratio.
H2H["ATSR"] <- NA # This creates the new column to store away team shots ratio.
H2H$HTSR <- round(H2H$HS/(H2H$HS+H2H$AS),3)  # Here we round to 3 dp
H2H$ATSR <- round(H2H$AS/(H2H$AS+H2H$HS),3)  # Here we round to 3 dp
```

Finally, we display the H2H match results.

```
print(H2H)
```

```
##
       Div       Date HomeTeam AwayTeam FTHG FTAG FTR HS AS HST AST ind
51      E0   24/09/16  Arsenal  Chelsea    3    0   H 14  9   5   2   1
599     E0 03/01/2018  Arsenal  Chelsea    2    2   D 14 19   6   6   1
981     E0 19/01/2019  Arsenal  Chelsea    2    0   H 13 13   5   1   1
1337    E0 29/12/2019  Arsenal  Chelsea    1    2   A  7 13   2   4   1
1661    E0 26/12/2020  Arsenal  Chelsea    3    1   H 15 19   7   3   1
        HTSR  ATSR
51     0.609 0.391
599    0.424 0.576
981    0.500 0.500
1337   0.350 0.650
1661   0.441 0.559
```

From this, we can see that Arsenal won three of the previous H2H matches, with Chelsea winning only one. One match was also drawn.

4.3 Producing PiT League Tables from Historical Match Data

Often, when analysing the performance of teams using historical data, it is useful to know the position in the league table of the various clubs at any given point in time. This can be very helpful when reviewing team performance over an entire season. Unfortunately, acquiring historical league position data can be a long and tedious task when done by hand. So, once again, it is much better to let R do all the hard work. Example 4.5 illustrates how point-in-time (PiT) league tables can be constructed retrospectively from historical match data using a function adapted from the original, which can be found at http://opisthokonta.net/?p=18 [2].

EXAMPLE 4.5: PRODUCING PiT LEAGUE TABLES FROM HISTORICAL MATCH DATA

In this example, we are going to use historical match data to construct the EPL league table after 98 matches in the 2020–2021 season, which was the table on 30 November 2020 after approximately ten rounds of competition. We shall do this using the two user-defined functions outlined below. (NB. For more information on user-defined functions, see Key Concept Box 4.1.)

First of all, we shall clear all existing variables and data from the workspace.

```
rm(list = ls())   # Clears all variables from workspace
```

Now we load the historical match data from https://www.football-data.co.uk/englandm.php. In this case, we will use just the first 98 EPL games and the first seven variables, but of course any league and number of matches can be selected to construct an appropriate PiT league table.

```
PiT_dat <- head(read.csv('https://www.football-data.co.uk/mmz4281/2021/E0.csv'),98)[,1:7]
```

NB. '98 matches' includes the matches on 30 November 2020 and equates to approximately 10 rounds of competition.

Let's inspect the downloaded data.

```
names(PiT_dat)
```

```
##
[1] "Div"       "Date"      "Time"      "HomeTeam" "AwayTeam" "FTHG"      "FTAG"
```

In order to utilise the user-defined functions described below, it is necessary to first create four vectors, 'HomeTeam', 'AwayTeam', 'HomeGoals', and 'AwayGoals', which will be loaded into the 'create.table' function.

```
# Define the four variables that are to be loaded into the user-defined function.
HomeTeam <- PiT_dat$HomeTeam
AwayTeam <- PiT_dat$AwayTeam
HomeGoals <- PiT_dat$FTHG
AwayGoals <- PiT_dat$FTAG
```

In order to produce a PiT league table, two user-defined functions need to be created: 'outcome' and 'create.table'. The first of these functions determines the match outcome (i.e., H, D, or A) based on goals scored by the home and away teams, while the second produces a PiT league table based on the match results. Importantly, the 'create.table' function utilises the 'outcome' function, so it is only necessary to apply the 'create.table' function when constructing a PiT league table.

```
# Function 1. (This creates a vector of match outcomes.)

outcome <- function(hGoals, aGoals){
  nMatches <- length(hGoals)
  results <- matrix(NA,nMatches,1) # Creates an empty vector to store match outcome results.

# This populates the results vector with match outcomes (i.e. H, A or D)
for(i in 1:nMatches){
    if(hGoals[i] > aGoals[i]){results[i] <- "H"}
```

```
    if(hGoals[i] < aGoals[i]){results[i] <- "A"}
    if(hGoals[i] == aGoals[i]){results[i] <- "D"}
  }
  return(results)
}


# Function 2. (This creates a current league table from the match results data.)

create.table <- function(hTeam, aTeam, hGoals, aGoals){

# Harvest team names and collate in to a vector
teams.temp <- unique(hTeam)
(teams <- sort(teams.temp)) # Arrange in alphabetical order.
nTeams = length(teams) # This identifies the number of teams in the league.

# Create a vector containing the match outcomes (i.e. H, A or D)
results <- outcome(hGoals, aGoals)

# Create empty vectors to store results.
x <- numeric(nTeams)
hWins <- x; hLoss <- x; hDraws <- x;
aWins <- x; aLoss <- x; aDraws <- x;
goals.for <- x; goals.against <- x; goal.diff <- x;
matches.played <- x; pts <- x;

# Populate vectors
for (i in 1:nTeams) {
   hResults <- results[hTeam == teams[i]]
   aResults <- results[aTeam == teams[i]]
   matches.played[i] <- length(hResults) + length(aResults)
   goals.H <- sum(hGoals[hTeam == teams[i]])
   goals.A <- sum(aGoals[aTeam == teams[i]])
   goals.for[i] <- goals.H + goals.A
   conceded.H <- sum(aGoals[hTeam == teams[i]])
   conceded.A <- sum(hGoals[aTeam == teams[i]])
   goals.against[i] <- conceded.H + conceded.A
   goal.diff[i] <- goals.for[i] - goals.against[i]
   hWins[i] <- sum(hResults == "H")
   hDraws[i] <- sum(hResults == "D")
   hLoss[i] <- sum(hResults == "A")
   aWins[i] <- sum(aResults == "A")
   aDraws[i] <- sum(aResults == "D")
   aLoss[i] <- sum(aResults == "H")

# Compute total points from the number of wins and draws for the respective teams.

# Points awarded for the match outcomes
   win.pts <- 3
   draw.pts <- 1
   pts[i] <- (win.pts*(hWins[i] + aWins[i])) + (draw.pts * (hDraws[i] + aDraws[i]))
  }

  table <- data.frame(cbind(matches.played, hWins, hDraws,hLoss, aWins, aDraws, aLoss,
                   goals.for, goals.against, goal.diff, pts), row.names=teams)

 names(table) <- c("PLD", "HW", "HD", "HL", "AW", "AD", "AL", "GF", "GA", "GD", "PTS")
 ord <- order(-table$PTS, -table$GD, -table$GF)
 table <- table[ord, ]
 return(table)
}
```

NB. The 'outcome' and 'create.table' functions listed above are adapted from code presented in opisthokonta.net [2].

Having written the user-defined functions, we can now apply them to produce the PiT league table, as follows.

```
League.table <- create.table(HomeTeam, AwayTeam, HomeGoals, AwayGoals)
print(League.table)
```

This produces:

##

	PLD	HW	HD	HL	AW	AD	AL	GF	GA	GD	PTS
Tottenham	10	2	2	1	4	1	0	21	9	12	21
Liverpool	10	5	0	0	1	3	1	22	17	5	21
Chelsea	10	2	2	1	3	2	0	22	10	12	19
Leicester	10	2	0	3	4	0	1	19	14	5	18
West Ham	10	3	1	1	2	1	2	17	11	6	17
Southampton	10	3	0	2	2	2	1	19	16	3	17
Wolves	10	2	2	1	3	0	2	11	11	0	17
Everton	10	2	1	2	3	0	2	19	17	2	16
Man United	9	1	1	3	4	0	0	16	16	0	16
Aston Villa	9	2	0	3	3	0	1	20	13	7	15
Man City	9	2	1	1	2	2	1	15	11	4	15
Leeds	10	1	2	2	3	0	2	15	17	-2	14
Newcastle	10	2	0	3	2	2	1	12	15	-3	14
Arsenal	10	2	0	3	2	1	2	10	12	-2	13
Crystal Palace	10	2	1	2	2	0	3	12	15	-3	13
Brighton	10	0	3	2	2	1	2	14	16	-2	10
Fulham	10	1	0	4	1	1	3	11	19	-8	7
West Brom	10	1	2	2	0	1	4	7	18	-11	6
Burnley	9	1	0	3	0	2	3	4	17	-13	5
Sheffield United	10	0	1	4	0	0	5	4	16	-12	1

From this, we can see that after approximately 10 rounds of competition, Tottenham were leading the EPL, with Liverpool second. However, this is not how the end-of-season (EoS) final table appeared because Manchester City actually won the EPL in season 2020–2021, with Liverpool finishing third and Tottenham finishing seventh.

KEY CONCEPT BOX 4.1: WRITING A USER-DEFINED FUNCTION IN R

In R, a user-defined function is a custom-built algorithm that is created by the user to perform a specific task. Generally, they are created to perform complex tasks or calculations for which there is not already a built-in function in R. An example of such a user-defined function in this chapter is 'create.table', which, as the name suggests, creates a point-in-time soccer league table from historical match data – a task that is highly specific and not generally called for by most people in everyday life.

When creating a user-defined function, we first need to define the function using the 'function' command, which is followed by all the arguments that the function will use and the operations that it will perform. For example, we could create a simple user-defined function to compute the area of a rectangle using the following code:

```
# Create function called 'calc_rectangle_area'
calc_rectangle_area <- function(length, width) {
  area <- length * width
  return(area)
}
```

This function, which is called 'calc_rectangle_area', uses two arguments, 'length' and 'width', to compute the area of a rectangle, which it returns as the output. With user-defined functions, all the input arguments (i.e., the input variables and values) must be declared in brackets directly after the 'function' term. The bit in the curly brackets {...} is where the user tells R what to do when the function is called. In this case, we have told R to compute the area of a rectangle by multiplying its length and width, and then to return the computed area as the output.

To use the 'calc-rectangle-area' function in practice, we simply type the following code, which inputs two parameters, 'length=5' and 'width=3'.

```
# Specify input values
length <- 5
width <- 3

# Apply function
calc_rectangle_area(length,width)
```

The function then returns the following output:

```
## [1] 15
```

Of course, user-defined functions are generally much more complex than the simple example shown here, but the general principle remains the same. We write an algorithm to do a specific task, and then we call it up every time we wish to use it. As a result, user-defined functions are particularly popular when performing complex tasks that must be repeated multiple times.

4.4 Compiling PiT Feature Tables from Historical Match Data

In machine learning, the term 'feature' is often used to refer to an individual measurable property or characteristic that can be used for predictive purposes. For example, home and away team total shots ratios (HTSR and ATSR) are features that can be used to both assess past performance and predict future performance. In Example 4.3, we have already seen how we can compile such features for a single club over an entire season. However, when making predictions, it is often useful to compile tables of features for all the teams in the league at specific points in time. To do this, a user-defined function is needed that can be applied to all the teams in the league, as illustrated in Example 4.6. This function (here called 'feature.Calc') is adapted from an algorithm by Tropiano [3] and sums the various performance metrics from the individual matches to produce a set of PiT feature scores that can be used to assess the relative performance of each of the teams in the league.

EXAMPLE 4.6: PRODUCING PiT FEATURE TABLES FROM HISTORICAL MATCH DATA

In this example, we are going to use historical match data to construct a table of features after 98 matches in season 2020–2021, which is approximately ten rounds of competition. We shall do this using the user-defined function outlined below.

Again, we will first clear the workspace.

```
rm(list = ls())   # Clears all variables from workspace
```

Now we load into R the match data for season 2020–2021, selecting only the first 20 variables and the first 98 matches.

```
fb_data <- head(read.csv('https://www.football-data.co.uk/mmz4281/2021/E0.csv'),98)[,1:20]
```

NB. This is the same data that we have used in the previous examples, but here we include the following additional variables:

- HC=Home Team Corners
- AC=Away Team Corners

Next, we select the variables that we want to keep in the analysis.

```
soc_dat <- fb_data[,c("Date","HomeTeam","AwayTeam","FTHG","FTAG",
                    "FTR","HS","AS","HST","AST","HC","AC")]
```

We then add some new derived variables:

```
soc_dat["HPts"] <- 0 # This creates the new column to store home team points per match
soc_dat["APts"] <- 0 # This creates the new column to store away team points per match
```

Which we populate as follows:

```
for(i in 1:nrow(soc_dat)){
  if(soc_dat$FTR[i] == "H"){soc_dat$HPts[i] <- 3}
  if(soc_dat$FTR[i] == "A"){soc_dat$APts[i] <- 3}
  if(soc_dat$FTR[i] == "D") {soc_dat$HPts[i] <- 1}
  if(soc_dat$FTR[i] == "D") {soc_dat$APts[i] <- 1}
}

# Rename column names
colnames(soc_dat)[colnames(soc_dat) == 'FTHG'] <- 'HG'
colnames(soc_dat)[colnames(soc_dat) == 'FTAG'] <- 'AG'
colnames(soc_dat)[colnames(soc_dat) == 'FTR'] <- 'Result'
```

To inspect the resultant data frame, we use the 'head' function.

```
head(soc_dat)
```

```
##
        Date       HomeTeam    AwayTeam HG AG Result HS AS HST AST HC AC
1 12/09/2020         Fulham     Arsenal  0  3      A  5 13   2   6  2  3
2 12/09/2020 Crystal Palace Southampton  1  0      H  5  9   3   5  7  3
3 12/09/2020      Liverpool       Leeds  4  3      H 22  6   6   3  9  0
4 12/09/2020       West Ham   Newcastle  0  2      A 15 15   3   2  8  7
5 13/09/2020      West Brom   Leicester  0  3      A  7 13   1   7  2  5
6 13/09/2020      Tottenham     Everton  0  1      A  9 15   5   4  5  3
  HPts APts
1    0    3
```

```
2    3    0
3    3    0
4    0    3
5    0    3
6    0    3
```

Having loaded the first 98 matches into R, we can now apply the following user-defined function, 'feature.Calc', to compute the PiT feature scores for any given team. Here we apply it to Tottenham, who after 10 rounds of competition were leading the EPL in 2020, and to Manchester City, who finished top at the end of the season.

```
# 'feature.Calc' function
feature.Calc <- function(df, team){
  Hmatches <- df[df$HomeTeam == team,] # This selects the target team's home matches.
  Amatches <- df[df$AwayTeam == team,] # This selects the target team's away matches.
  all <- rbind.data.frame(Hmatches,Amatches)
  n <- nrow(all) # Number of matches
  # Create empty vectors to store results.
  x <- numeric(n)
  GF <- x; GA <- x; SF <- x; SA <- x; STF <- x; STA <- x;
  CF <- x; CA <- x; Pts <- x;

  # Computes total goals for
  for(i in 1:n){
    if(all$HomeTeam[i] == team){GF[i] <- all$HG[i]}
    else {GF[i] <- all$AG[i]}
  }

  # Computes total goals against
  for(i in 1:n){
    if(all$HomeTeam[i] == team){GA[i] <- all$AG[i]}
    else {GA[i] <- all$HG[i]}
  }

  # Computes total shots for
  for(i in 1:n){
    if(all$HomeTeam[i] == team){SF[i] <- all$HS[i]}
    else {SF[i] <- all$AS[i]}
  }

 # Computes total shots against
  for(i in 1:n){
    if(all$HomeTeam[i] == team){SA[i] <- all$AS[i]}
    else {SA[i] <- all$HS[i]}
  }

 # Computes total shots on target for
  for(i in 1:n){
    if(all$HomeTeam[i] == team){STF[i] <- all$HST[i]}
    else {STF[i] <- all$AST[i]}
  }

# Computes total shots on target against
  for(i in 1:n){
    if(all$HomeTeam[i] == team){STA[i] <- all$AST[i]}
    else {STA[i] <- all$HST[i]}
  }
```

```
  # Computes total corners for
  for(i in 1:n){
    if(all$HomeTeam[i] == team){CF[i] <- all$HC[i]}
    else {CF[i] <- all$AC[i]}
  }

 # Computes total corners against
  for(i in 1:n){
    if(all$HomeTeam[i] == team){CA[i] <- all$AC[i]}
    else {CA[i] <- all$HC[i]}
  }

  # Computes total points awarded
  for(i in 1:n){
    if(all$HomeTeam[i] == team){Pts[i] <- all$HPts[i]}
    else {Pts[i] <- all$APts[i]}
  }
  Pld <- matrix(1,n,1) # Vector containing matches played
  GD <- GF-GA # Computes goal difference
  TG <- GF+GA # Computes total goals
  TSRF <- SF/(SF+SA) # Computes total shots ratio for
  TSRA <- SA/(SF+SA) # Computes total shots ratio against

  feats <- cbind.data.frame(Pld,GF,GA,GD,TG,SF,SA,STF,STA,TSRF,TSRA,CF,CA,Pts)
  featsSums <- colSums(feats)
  featsRes <- featsSums
  nOb <- nrow(feats)
  featsRes[10] <- featsSums[10]/nOb # This compute the average TSRF.
  featsRes[11] <- featsSums[11]/nOb # This computes the average TSRA.
  return(round(featsRes,2))
}
```

Now we can apply the 'feature.Calc' function to Tottenham and Manchester City.

```
Tot.features <- feature.Calc(soc_dat, "Tottenham")
print(Tot.features)
```

```
##
  Pld     GF     GA     GD     TG     SF     SA    STF    STA   TSRF
10.00  21.00   9.00  12.00  30.00 120.00 119.00  51.00  34.00   0.49
TSRA     CF     CA    Pts
0.51  42.00  50.00  21.00
```

```
MC.features <- feature.Calc(soc_dat, "Man City")
print(MC.features)
```

```
##
 Pld     GF     GA     GD     TG     SF     SA    STF    STA   TSRF
9.00  15.00  11.00   4.00  26.00 144.00  74.00  49.00  27.00   0.65
TSRA     CF     CA    Pts
0.35  57.00  31.00  15.00
```

NB. With the exception of TSRF and TSRA, which are average values, the feature scores reported here are the sums of the various metrics acquired from each match.

The feature scores for Tottenham, the league leaders at the time, are very interesting. Although they show that Tottenham exhibited a healthy positive goal difference of 12 after 10 games, the TSR results suggest that the team might not have been playing all that well. Unlike Manchester City and many of their other competitors (see below), Tottenham's average TSRA score of 0.51 was actually greater than their average TSRF score, which was 0.49, implying that in most matches they were conceding more shots than they were making. This suggests that they were possibly lucky to be leading the league after round 10. By comparison, Manchester City after nine games had an average TSRF score (i.e., 0.65) that was much greater than their TSRA score (i.e., 0.35), implying that perhaps they should have accrued more points at that stage of the competition. It is therefore not surprising that Tottenham's performance dropped away as the season progressed, with them eventually finishing in seventh place, while Manchester City steadily improved and finished top of the EPL in season 2020–2021.

The following code shows how the 'feature.Calc' function can be applied to all the teams in the EPL.

First, we identify the names of the teams in the league and collate them into a vector.

```
Teams <- unique(soc_dat$HomeTeam)
Teams <- sort(Teams) # Put them in alphabetical order
nTeams <- length(Teams) # Number of teams
print(Teams)
```

```
##
 [1] "Arsenal"          "Aston Villa"       "Brighton"
 [4] "Burnley"          "Chelsea"           "Crystal Palace"
 [7] "Everton"          "Fulham"            "Leeds"
[10] "Leicester"        "Liverpool"         "Man City"
[13] "Man United"       "Newcastle"         "Sheffield United"
[16] "Southampton"      "Tottenham"         "West Brom"
[19] "West Ham"         "Wolves"
```

Now we can apply the 'feature.Calc' function to all the teams in the league using a 'for loop'.

```
featureRes <- matrix(NA,nTeams,14)
for(i in 1:nTeams){
  featureRes[i,] <- feature.Calc(soc_dat, Teams[i])
}

# Compile feature table results
featureTab <- cbind.data.frame(Teams, featureRes)
colnames(featureTab) <- c("Team","Pld","GF","GA","GD","TG","SF","SA","STF",
                          "STA","TSRF","TSRA","CF","CA","Pts")
print(featureTab)
```

This produces:

```
##
               Team Pld GF GA  GD TG  SF  SA STF STA TSRF TSRA CF CA Pts
1           Arsenal  10 10 12  -2 22  95 124  32  39 0.46 0.54 47 50  13
2       Aston Villa   9 20 13   7 33 135 107  52  37 0.58 0.42 66 33  15
3          Brighton  10 14 16  -2 30 123  78  34  28 0.60 0.40 50 40  10
4           Burnley   9  4 17 -13 21  83 118  24  44 0.42 0.58 36 55   5
5           Chelsea  10 22 10  12 32 132  95  54  27 0.58 0.42 50 54  19
6    Crystal Palace  10 12 15  -3 27  94 133  36  45 0.40 0.60 51 55  13
7           Everton  10 19 17   2 36 116 129  49  49 0.48 0.52 48 43  16
8            Fulham  10 11 19  -8 30 126 124  42  57 0.50 0.50 38 45   7
9             Leeds  10 15 17  -2 32 153 136  55  49 0.52 0.48 55 49  14
10        Leicester  10 19 14   5 33 102 127  45  49 0.45 0.55 53 57  18
11        Liverpool  10 22 17   5 39 163  90  64  39 0.64 0.36 62 21  21
12         Man City   9 15 11   4 26 144  74  49  27 0.65 0.35 57 31  15
13       Man United   9 16 16   0 32 120  98  47  33 0.55 0.45 53 35  16
14        Newcastle  10 12 15  -3 27  86 158  28  59 0.36 0.64 42 55  14
15 Sheffield United  10  4 16 -12 20  97 152  31  57 0.38 0.62 50 67   1
16      Southampton  10 19 16   3 35 107 104  51  47 0.53 0.47 37 59  17
17        Tottenham  10 21  9  12 30 120 119  51  34 0.49 0.51 42 50  21
18        West Brom  10  7 18 -11 25  95 149  30  61 0.40 0.60 44 66   6
19         West Ham  10 17 11   6 28 118 109  40  39 0.51 0.49 47 51  17
20           Wolves  10 11 11   0 22 122 107  39  33 0.53 0.47 44 56  17
```

In Example 4.6, we have seen how R can be used to produce a range of feature scores for the teams in leagues. But how can these be utilised to evaluate the performance of teams? Well, one simple method is to produce a scatter plot of points earned and selected features, as illustrated in Example 4.7. This allows us to visualise the performance of the respective teams with respect to one attribute. One metric frequently used to evaluate team performance is TSR (i.e., TSRF), which combines both attacking and defensive qualities into a single score [4]. As such, it captures one of the key attributes (i.e., the ratio of shots made to those conceded) that can greatly influence the outcome of matches (see Chapter 5 for more details). Example 4.7, which utilises the 'featureTab' data frame for the 98 games produced in Example 4.6, illustrates how this can be done.

EXAMPLE 4.7: PRODUCING A TSR FEATURE SCATTER PLOT

Here we use the 'featureTab' data frame from Example 4.6 that was produced after 98 matches in season 2020–2021, to produce a scatter plot of TSRF and points awarded for all the teams in the EPL.

```
# Scatter plot of TSRF and points earned
x <- featureTab$TSRF
y <- featureTab$Pts
```

Before producing the actual scatter plot, we use the 'cor.test' function here to compute the Pearson correlation *r*-value for the relationship between TSRF and the points earned data.

```
# Correlation between TSRF and points earned
cor.test(x,y)
```

This produces the following output, from which we can see that r=0.513 and that the relationship is significant, with p=0.021.

```
##     Pearson's product-moment correlation

data:  x and y
t = 2.5353, df = 18, p-value = 0.02073
alternative hypothesis: true correlation is not equal to 0
95 percent confidence interval:
 0.09113012 0.77871828
sample estimates:
      cor
0.5129646
```

```
# Scatter plot
plot(x,y, pch=20, xlim=c(0.2,0.8), ylim=c(0,25), xlab="Average TSRF score", ylab="Points earned")
text(y~x, labels=Teams, cex=0.8, font=1, pos=4)  # This puts team names on the data points.
abline(lm(y~x), lty=2) # This draws a least squares best fit line through the data points.
```

The resulting scatter plot is shown in Figure 4.2.

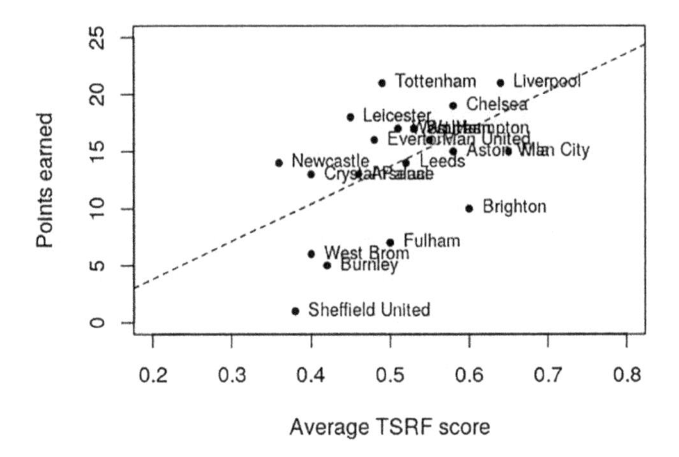

FIGURE 4.2
Scatter plot of the average total shots ratio for (TSRF) score and points earned for the teams in the EPL (season 2020–2021) after 98 games (i.e., approximately 10 rounds of competition).

From Figure 4.2, it can be seen that after 98 games, a reasonable correlation (r=0.513) exists between TSRF and points earned, with most teams clustering close to the best-fit line, despite the fact that some had only played 9 matches rather than 10. However, five teams (three of whom, Sheffield United, West Bromwich, and Fulham, were subsequently relegated) are positioned a long way below the best-fit line, indicating that for some reason these teams were failing to convert their TSRF score into league points. For example, consider Tottenham, who finished the season in 7th place with 62 points, and Fulham, who came 18th and was relegated with a total of just 28 points. After 10 matches, Tottenham had a TSRF score of 0.49 and a point total of 21, whereas Fulham had a marginally better TSRF score of 0.50 but had only accumulated 7 points. This suggests that the quality of the shots made by Tottenham (SoT=51) was generally better than those made by Fulham (SoT=42). Also, Tottenham conceded far fewer SoT (i.e., 34) compared with Fulham's 57. So although the TSRF scores for Tottenham and Fulham were similar, it is likely that the two teams differed greatly in the quality of their match play, hence the 14-point difference between the two teams after 10 games. Notwithstanding this, it is noticeable in Figure 4.2 that Tottenham is placed some distance above the best-fit line, perhaps suggesting that good luck might have played a role in some of their match performances early in the season since they were unable to maintain the same momentum as the season progressed.

If we compare Figure 4.2 with Figure 4.3, which represents the EoS position, it can be seen that the correlation between TSRF and points earned has strengthened (r=0.731) and that all but Brighton and Fulham are positioned relatively close to the best-fit line. This suggests that while these two teams were creating shooting opportunities, their shots were not being converted into goals. Also, it is noticeable that two of the relegated teams, Sheffield United and West Bromwich, had the lowest TSRF scores in the league, indicating that they were failing to create many chances to score.

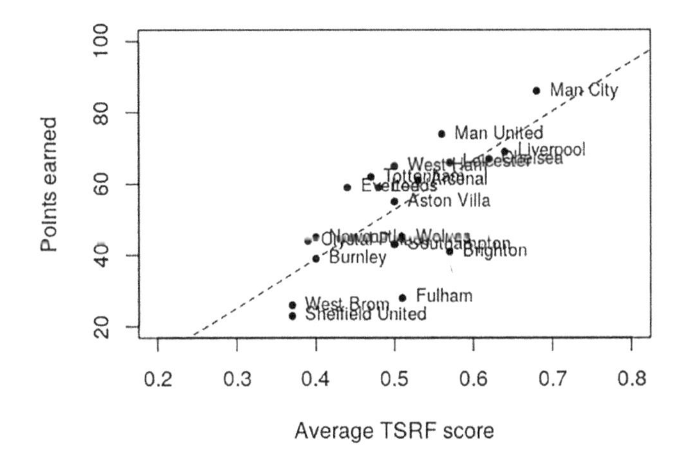

FIGURE 4.3
Scatter plot of the average total shots ratio for (TSRF) score and points earned for the teams in the EPL (season 2020–2021) after 380 games (i.e., at the end of the season).

References

1. Chamberlain M: Data-driven football predictions: Constructing the perfect over 2.5 goals betting strategy for the English Premier League. Independently published; *Footballdatalabs.com*; 2019.
2. opisthokonta.net: R functions for soccer league tables and result matrix. In: *opisthokontanet*. vol. https://opisthokonta.net/?p=18; 2012.
3. Tropiano A: Code a soccer betting model in a weekend. Independently published; 2020.
4. Joslyn LR, Joslyn NJ, Joslyn MR: What delivers an improved season in men's college soccer? The relative effects of shots, attacking and defending scoring efficiency on year-to-year change in season win percentage. *The Sport Journal* 29th June 2017:1–11.

5

Predicting End-of-Season League Position

In Chapter 4, we saw how historical match data can be used to construct point-in-time (PiT) league tables and feature tables, which can be very useful when assessing team performance. In this chapter, we will build on this and look at how PiT league performance data can be used to predict end-of-season (EoS) league position. Knowing where a team is likely to finish in its league at the end of any season is a key issue for fans, players, and managers alike, since promotion and relegation depend on league position. So, in this chapter, we will explore in depth how R can be used to gain insights into predicting likely EoS league position.

5.1 Why End-of-Season League Position Matters

For all the complexities of individual soccer matches, team performance is recorded and evaluated using a relatively simple process, with most teams belonging to a league that reflects the standings of the respective clubs after each round of competition. In a typical soccer league such as the English Premier League (EPL), teams are awarded three points for a win, one point for a draw, and no points for losing, with the respective partial league standings (i.e., the PiT positions in the league partway through the season) at any point in time based on the number of points accumulated.

The place in a league where a team finishes can have huge financial implications. For example, the top two or three teams (it varies from league to league) may be promoted to a higher division or selected to compete in an international competition, such as the *Union of European Football Associations* (UEFA) Champions League, while those in the bottom three or four places are generally relegated to a lower division. With relegation in particular, the financial implications may be immense, with, in some cases, relegated clubs forced into liquidation [1]. As such, it is the final league standing, rather than the points accumulated *per se*, that is the measure by which teams and their managers are judged. In fact, in the EPL, each additional placing in the final standings is worth £1.9 million (2017 data) [2], no matter the magnitude of the points difference between adjacent teams. So in short, EoS league position

DOI: 10.1201/9781003328568-5

matters a lot. Therefore, it is important to know at an early stage in the season if things are going wrong so that remedial action can be taken to avoid a catastrophic situation.

5.2 What Role Does Luck Play in League Position?

We all know the feeling of relief when the team that we support scores a winning goal against the run of play due to some unforeseen error made by the opposition goalkeeper. However, despite our joy, we know deep down that our team was extremely lucky and that it did not really deserve the victory. In other words, we are acknowledging that chance played an important role in the outcome of the match. But how much of a role does chance play in the outcome of matches? Can we quantify it? Well, the simple answer to this question is no, because every match is different and it is extremely difficult to quantify the contribution that luck plays in the outcome of specific matches.

To a greater or lesser extent, luck plays a contributory role in the outcome of all soccer games, but it is also true that in most matches, the stronger and more skilled team generally wins the encounter. This is reflected in the position of the teams in leagues – the stronger teams quickly rise to the top of the league, while the weaker ones tend to descend to the bottom [3]. Nevertheless, while it is difficult to quantify the contribution of luck to the outcome of individual matches, it is possible to use statistics to estimate how much chance contributes to the points distribution of teams in a league, as Spiegelhalter and Pearson demonstrated when they showed that for the 2006–2007 season, about 25% of the variance in the Premier League points was due to chance [4].

If we consider the data presented in Table 5.1 for the six seasons prior to the COVID pandemic, we can see that on average 46.0% of EPL matches resulted in a home win, 30.7% in an away win, and 23.4% in a draw. These values are consistent with the long-term probabilities of a home win, an away win, and a draw in the Premier League, which are approximately 0.46, 0.29, and 0.25, respectively. As Spiegelhalter and Pearson showed [4], it is possible to use these probabilities to compute the expected number of points that a team would earn in a season if the league was decided purely by chance, with all the teams of an equal standard and the match outcomes decided at random. In which case, if the EPL was completely random, the expected mean (μ_{exp}) and variance (v_{exp}) in the points awarded could be calculated using Equations 5.1 and 5.2.

$$\mu_{\text{exp}} = \frac{m}{2} \times (3 - p_{\text{D}}) \tag{5.1}$$

$$v_{\text{exp}} = \frac{m}{2} \times \left[9 - 7p_{\text{D}} - (3p_{\text{H}} + p_{\text{D}})^2 - (3p_{\text{A}} + p_{\text{D}})^2 \right] \tag{5.2}$$

TABLE 5.1

Overall League Statistics for the EPL from Season 2013–2014 to Season 2020–2021

Season	HW Fraction	AW Fraction	Draw Fraction	League Points Total	Team Points (Mean)	Team Points (Var)	Points Per Match (Mean)	Pandemic
2013–2014	0.471	0.324	0.205	1062	53.10	352.9	1.397	No
2014–2015	0.453	0.303	0.245	1047	52.35	253.9	1.378	No
2015–2016	0.413	0.305	0.282	1033	51.65	226.4	1.359	No
2016–2017	0.492	0.287	0.221	1056	52.80	374.4	1.390	No
2017–2018	0.455	0.284	0.261	1041	52.05	349.1	1.370	No
2018–2019	0.476	0.337	0.187	1069	53.45	419.2	1.407	No
2019–2020	0.453	0.305	0.242	1048	52.40	300.1	1.379	Yes
2020–2021	0.379	0.403	0.218	1057	52.85	270.9	1.391	Yes

where m is the number of matches played per season by each team (i.e., 380 matches); and p_H, p_A, and p_D are the probabilities of a home win, away win, and draw, respectively. (NB. For a detailed discussion of Equations 5.1 and 5.2, see https://plus.maths.org/content/understanding-uncertainty-premier-league [4].)

If we examine the points awarded over the eight seasons reported in Table 5.1, we see that the average points earned by the teams in each season are remarkably consistent, with the mean value being 52.6 points. However, by comparison, the variance in the points awarded to the teams is much greater, being on average 318.4 points. The observed variance (v_{obs}) in the points awarded to teams in each season can be calculated using Equation 5.3 for population variance.

$$v_{obs} = s^2 = \frac{1}{n} \times \sum_{i=1}^{n} \left(x_i - \bar{x}\right)^2 \tag{5.3}$$

where s^2 is the observed variance in the points awarded to the various teams, n is the number of teams in the league (i.e., 20 teams), x_i is the number of points awarded to the ith team, and $\bar{x}$ is the mean number of points awarded to teams in the league.

What Spiegelhalter and Pearson did was compute the expected variance in points (v_{exp}) that would otherwise occur if the league was decided purely by chance and then use this to compute the fraction of the observed variance in the EPL that could be attributed to chance using Equation 5.4, which is the quotient of the expected and observed variances.

$$\text{Fraction of variance due to chance}: fv_{ch} = \frac{v_{exp}}{v_{obs}} \tag{5.4}$$

If in Equations 5.1 and 5.2, we use the long-term probabilities values: p_H=0.46, p_A=0.29, and p_D=0.25, we find that the expected mean and variance values for the random Premier League are 52.25 and 63.44 points, respectively. While the expected mean of the random league is very close to the observed values shown in Table 5.1 (mean=52.6 points), it is noticeable that the observed variance in points awarded in the real league is much greater than the expected variance calculated using Equation 5.2. Indeed, Spiegelhalter and Pearson found that the expected variance in league points was only about a quarter of the observed variance for the 2006–2007 season, hence their conclusion that about 25% of the points awarded can be attributed to chance [4].

Using Spiegelhalter and Pearson's methodology, the computed fractions of variance due to chance (expressed as a percentage) for each of our eight example seasons are shown in Table 5.2, from which it can be seen that on average, about 20.7% of the variance exhibited in the points awarded to teams can be attributed to chance. Interestingly, before the COVID pandemic, which greatly disrupted seasons 2019–2020 and 2020–2021, the contribution of chance to the performance of teams in the EPL had been declining. However, with the disruption caused by the pandemic, the contribution of chance to league performance once again increased, possibly evidenced by the abnormally high number of away wins observed in season 2020–2021 (Table 5.1).

TABLE 5.2

Percentage of the Variance in Points Awarded That Is Due to Chance for the EPL from Season 2013–2014 to Season 2020–2021.

Season	Actual Team Points (Mean)	Actual Team Points (Var)	Expected Team Points	Expected Variance in Team Points	Variance Attributed to Chance (%)	Pandemic
2013–2014	53.1	352.9	52.25	63.44	18.0	No
2014–2015	52.35	253.9	52.25	63.44	25.0	No
2015–2016	51.65	226.4	52.25	63.44	28.0	No
2016–2017	52.8	374.4	52.25	63.44	16.9	No
2017–2018	52.05	349.1	52.25	63.44	18.2	No
2018–2019	53.45	419.2	52.25	63.44	15.1	No
2019–2020	52.4	300.1	52.25	63.44	21.1	Yes
2020–2021	52.85	270.9	52.25	63.44	23.4	Yes

The R code used to compute the values in Tables 5.1 and 5.2 is presented in Example 5.1, which evaluates the EPL for season 2020–2021.

EXAMPLE 5.1: QUANTIFYING THE CONTRIBUTION THAT CHANCE MAKES TO THE ENGLISH PREMIER LEAGUE

In this example, we shall use Spiegelhalter and Pearson's method to compute the percentage of the points variance that is attributable to chance for season 2020–2021, in the EPL. The source file for this example (EPL_final_table_2021.csv), which contains EoS data from the Premier League for season 2020–2021 can be found in the GitHub repository at: https://github.com/cbbeggs/SoccerAnalytics. To run the code in this example, the reader should download this file and save it in a directory called 'Datasets' on their computer.

First, we clear the workspace of all data and variables.

```
rm(list = ls())   # Clears all variables from the workspace
```

Next, we load the data file that contains the EoS table for the EPL.

```
# Here we assume that the file being loaded is stored in a directory called 'Datasets'.
tabdat <- read.csv("C:/Datasets/EPL_final_table_2021.csv")
```

Now we compute the average points per match for each team.

```
tabdat["avPPM"] <- NA # This creates the new column to store new data
tabdat$avPPM <- round(tabdat$PTS/tabdat$PLD, 4) # NB. Rounded to 3 dp

head(tabdat) # Display the first six rows of EoS league table
```

```
##
        Club PLD HW HD HL AW AD AL GF GA GD PTS  avPPM
1   Man City  38 13  2  4 14  3  2 83 32 51  86 2.2632
2 Man United  38  9  4  6 12  7  0 73 44 29  74 1.9474
3  Liverpool  38 10  3  6 10  6  3 68 42 26  69 1.8158
4    Chelsea  38  9  6  4 10  4  5 58 36 22  67 1.7632
5  Leicester  38  9  1  9 11  5  3 68 50 18  66 1.7368
6   West Ham  38 10  4  5  9  4  6 62 47 15  65 1.7105
```

Having loaded the data, we can now compute the number of matches and teams in the league.

```
m <- mean(tabdat$PLD) # Number of matches played per team
n <- nrow(tabdat) # Number of teams in league
```

Next, we compute the observed points average and variance.

```
LPPM.av <- mean(tabdat$avPPM) # Average number of points per match for the league
LPTS.av <- mean(tabdat$PTS) # Average team points total over entire season
```

We can compute the population variance of the observed league points using:

```
LPTS.error <- c() # Creates an empty vector to store results

for(i in 1:n){
  LPTS.error[i] <- tabdat$PTS[i] - LPTS.av
}

LPTS.var <- (1/n)*sum(LPTS.error^2) # Variance in team points
```

Now we specify the probabilities of a home win, an away win, and a draw.

```
pH <- 0.46 # Probability of a home win
pA <- 0.29 # Probability of an away win
pD <- 0.25 # Probability of a draws
```

Finally, we compile and display the league statistics.

```
league.stats <- round(c(m,pH,pA,pD,sum(tabdat$PTS),LPTS.av,LPTS.var,LPPM.av),3)
names(league.stats) <- c("Pld","HWprob","AWprob","Dprob","Pts.tot","Pts.av",
                         "Pts.var","Ptspm.av")
print(league.stats)
```

```
##
    Pld   HWprob   AWprob    Dprob  Pts.tot   Pts.av  Pts.var Ptspm.av
 38.000    0.460    0.290    0.250 1057.000   52.850  270.928    1.391
```

Having compiled the observed statistics for the actual league, we can now compute the expected mean and variance in the points total had the league been truly random.

```
# Expected average points total per team
exP <- (m/2)*(3-pD) # Expected average points total if league completely random

# Expected points variance for all the teams in the league
exVar <- (m/2)*(9-(7*pD)-(((3*pH)+pD)^2)-(((3*pA)+pD)^2)) # Expected variance in points
```

Finally, we can compute the fraction of the variance that is due to chance.

```
# But in reality the league points variance is:
Lvar <- LPTS.var

# Therefore, the proportion of the league points variance that is due to chance is:
fchance <- exVar/Lvar # NB. This is expressed as a fraction.

# Display results
print(exP)
```

```
## [1] 52.25
```

```
print(exVar)
```

```
## [1] 63.4353
```

```
print(Lvar)
```

```
## [1] 270.9275
```

```
print(fchance)
```

```
## [1] 0.2341412
```

So from this, we can see that during season 2020–2021, chance contributed about 23% towards the overall variance exhibited in the total points earned by the respective teams in the EPL.

5.3 The Dynamics of Soccer Leagues

The rigid structure of soccer leagues and the way that points are awarded mean that, despite the complexity of individual games, they all tend to behave in a similar manner. This was demonstrated by Beggs et al. [3], who showed that the top four soccer leagues in England all behave mathematically in a near identical manner, making their dynamics highly predictable. It has even been shown that the EPL and National (American) Football League (NFL) behave in a similar manner mathematically [5], despite the two leagues representing very different sports. The reason for the similarity in behaviour between apparently very different leagues is down to the mathematics associated with the points system. In most soccer leagues, the maximum number of points on offer to each team in every round of competition is just three points. So early in the competition, say in rounds 2 or 3, earning three points for a win can cause a team to leap perhaps five or six places up the league table. By contrast, however, when it comes to, say, round 36, gaining three points will generally only allow a team to move perhaps one or two places at best. So in effect, as the season progresses, the positions of the teams in all soccer leagues tend to become more fixed, with the ability for large movements becoming less and less [3]. This is illustrated in Example 5.2, which plots the Spearman correlations (see Key Concept Box 5.1 for information on correlations) between the partial standings and the EoS positions after every round of competition in the 2013–2014 EPL.

EXAMPLE 5.2: PLOTTING THE CORRELATIONS OF LEAGUE PARTIAL STANDINGS

In this example, the temporal dynamics of the EPL are illustrated using the partial standings after each round of competition. But before we start, we first need to clear the RStudio workspace of all existing data and variables.

```
rm(list = ls())   # Clears all variables from the workspace
```

Next, we load the data file (EPL_standings_2013_14.csv), which contains the EPL standings after every round of competition in the 2013–2014 season.

This file can be found in the GitHub repository at https://github.com/cbbeggs/SoccerAnalytics, along with the other data files for this book.

```
# Here we assume that the file being loaded is stored in a directory called 'Datasets'.
download_dat <- read.csv("C:/Datasets/EPL_standings_2013_14.csv")

standings <- download_dat[,2:39]
row.names(standings) <- download_dat[,1] # This names the rows of the data frame.
head(standings)
```

```
##
                R1 R2 R3 R4 R5 R6 R7 R8 R9 R10 R11 R12 R13 R14 R15 R16 R17
Arsenal         16  8  4  2  1  1  1  1  1   1   1   1   1   1   1   1   2
Aston Villa      3  9 13 17 13  9 10 13 13  14  11  12  11  10  11  11  13
Cardiff         17 11 11 12 16 11 14 17 16  12  14  15  17  15  16  15  15
Chelsea          4  1  2  6  4  5  3  2  2   2   4   3   2   2   3   3   4
Crystal Palace  12 19 14 18 19 19 19 19 20  20  20  19  20  19  19  18  18
Everton         10 14 15  9  6  4  7  7  6   7   6   7   5   5   5   5   5
                R18 R19 R20 R21 R22 R23 R24 R25 R26 R27 R28 R29 R30 R31 R32
Arsenal           1   1   1   1   1   2   1   2   1   2   3   4   4   4   4
Aston Villa      13  13  11  11  10  10  10  12  13  13  11  10  11  12  13
Cardiff          16  16  17  18  20  20  19  19  19  19  18  19  19  19  19
Chelsea           3   3   3   3   3   3   3   1   2   1   1   1   1   1   1
Crystal Palace   17  17  18  20  16  14  17  14  15  16  16  17  17  15  14
Everton           5   4   5   5   6   6   5   6   6   6   6   6   5   5   5
                R33 R34 R35 R36 R37 R38
Arsenal           4   4   4   4   4   4
Aston Villa      14  14  15  13  15  15
Cardiff          19  18  19  20  20  20
Chelsea           2   2   2   3   3   3
Crystal Palace   12  11  11  11  11  11
Everton           5   5   5   5   5   5
```

Now we can produce a Spearman correlation matrix with significance levels using the 'rcorr' function in the 'Hmisc' package. This method has the great advantage that it can be used to produce both *r* and *p*-values for the respective correlations – something that can often be extremely helpful when performing statistical analysis.

To use the 'rcorr' command, the data needs to first be in matrix form, which we can do using the 'as.matrix' function in R. The 'rcorr' function produces a data object officially called a 'list' in R, which contains three elements. Here we want elements 1 and 3, which correspond to the respective *r*-values and *p*-values, which we shall convert into a data frame using the 'as.data.frame' command (e.g., "as.data.frame(cor[1])").

```
# Perform Spearman correlation analysis using 'Hmisc' package

install.packages("Hmisc") # This installs the 'Hmisc' package.
# NB. This command only needs to be executed once to install the package.
# Thereafter, the 'Hmisc' library can be called using the 'library' command:
```

```
library(Hmisc)
cor <- rcorr(as.matrix(standings, type="spearman")) # type can be pearson or spearman
cor_r <- as.data.frame(cor[1]) # Convert to a data frame
cor_p <- as.data.frame(cor[3]) # Convert to a data frame
```

While these correlation data frames contain lots of interesting information, here we are only interested in the correlation results in the final column (i.e., number 38), which we round here to three decimal places.

```
round38_r <- round(as.data.frame(cor_r[,38]),3)
round38_p <- round(as.data.frame(cor_p[,38]),3)
```

For ease of display, we will transpose the *r*- and *p*-values results. This can be done using "t(...)". Transposition is a technique that is often used in linear algebra and during data analysis. It involves turning a data frame, matrix, or vector on its side and is often used to make data easier to display.

```
# r-values
print(t(round38_r)) # Transposed for ease of display.
```

```
##
               [,1]  [,2]  [,3]  [,4]  [,5]  [,6]  [,7]  [,8]  [,9]
cor_r[, 38] 0.37 0.544 0.708 0.797 0.838 0.606 0.728 0.777 0.735
            [,10] [,11] [,12] [,13] [,14] [,15] [,16] [,17] [,18]
cor_r[, 38] 0.728 0.701 0.794 0.768 0.805 0.823 0.826 0.854 0.848
            [,19] [,20] [,21] [,22] [,23] [,24] [,25] [,26] [,27]
cor_r[, 38] 0.833 0.821 0.824 0.823 0.863 0.902  0.92 0.905 0.883
            [,28] [,29] [,30] [,31] [,32] [,33] [,34] [,35] [,36]
cor_r[, 38] 0.901 0.911 0.898 0.926 0.929 0.946 0.947 0.949 0.986
            [,37] [,38]
cor_r[, 38] 0.997     1
```

From this, we see that the Spearman correlation between the partial standings of the teams at the end of round 1 and the EoS standings is $r=0.37$, whereas by round 10, this figure is $r=0.728$, and by round 37, it is $r=0.997$.

```
# p-values
print(t(round38_p)) # Transposed for ease of display.
```

```
##
              [,1]  [,2] [,3] [,4] [,5]  [,6] [,7] [,8] [,9] [,10]
cor_p[, 38] 0.108 0.013    0    0    0 0.005    0    0    0     0
            [,11] [,12] [,13] [,14] [,15] [,16] [,17] [,18] [,19]
cor_p[, 38] 0.001     0     0     0     0     0     0     0     0
            [,20] [,21] [,22] [,23] [,24] [,25] [,26] [,27] [,28]
cor_p[, 38]     0     0     0     0     0     0     0     0     0
            [,29] [,30] [,31] [,32] [,33] [,34] [,35] [,36] [,37]
cor_p[, 38]     0     0     0     0     0     0     0     0     0
            [,38]
cor_p[, 38]    NA
```

From this, we see that all the *r*-values except that for round 1 are significant (i.e., $p<0.05$).

Finally, we can plot the *r*-values for each round of the competition using:

```
plot(cor_r[,38],type = "l",col = "black", ylim=c(0,1), xlab="Round of competition",
     ylab = "Spearman correlation (r value)")
```

This produces the plot shown in Figure 5.1.

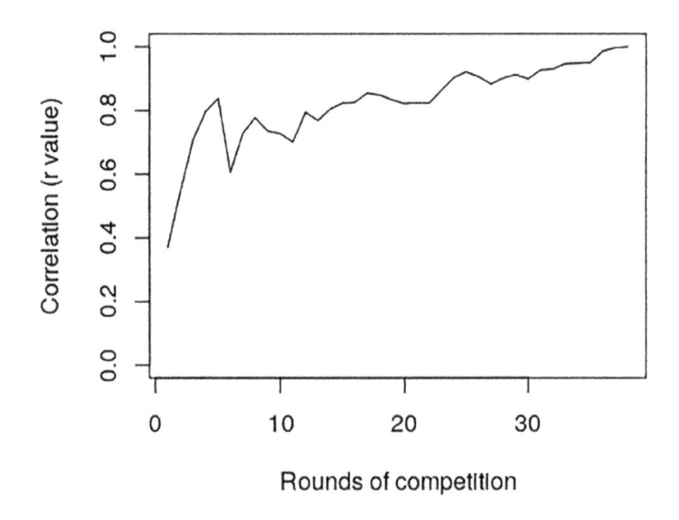

FIGURE 5.1
Change in Spearman correlation as the competition progressed for the EPL in the 2013–2014 season.

From Figure 5.1, it can be seen that the correlations at the beginning of the season are fairly weak with a lot of fluctuation, but by about round 10, the *r*-value is consistently greater than 0.7 and gradually increases towards unity as the season progresses. A similar picture is observed (Figure 5.2) for all the subsequent seasons up to 2018–2019, which was the last season not disrupted by the COVID pandemic. From this, a clear and consistent picture emerges, namely that by about rounds 8–10, most (i.e., 70%–80%) of the variation in the EPL standings has already taken place, with, in the main, only fine tuning taking place thereafter. This implies that the stronger teams quickly rise to the top of the league while the weak teams fall to the bottom, with the medium-strength teams arguing over the middle ground. Furthermore, it suggests that once this has happened, after about round 10, most teams tend to maintain their approximate position in the league for the rest of the season, with relatively few big changes occurring in league position after this.

Indeed, Beggs et al. [3] were able to compute that for the EPL, the team that was top after 10 rounds had a 77% chance of finishing the season in one of the top three places and a 41% chance of winning the league. Conversely, the team that was bottom of the league after round 10 had a 53% chance of being relegated.

Given that the mathematics of soccer leagues tend to hard-wire their behaviour, this makes it relatively easy to approximately predict, after say 10 rounds, where a team is likely to finish in its league. This can be very useful because, while we cannot be certain that the team that is bottom after 10 rounds will be relegated, we can say that there is a high likelihood of the team being involved in a relegation fight. Therefore, this should encourage the club's management team to take remedial action sooner rather than later in an attempt to avoid relegation.

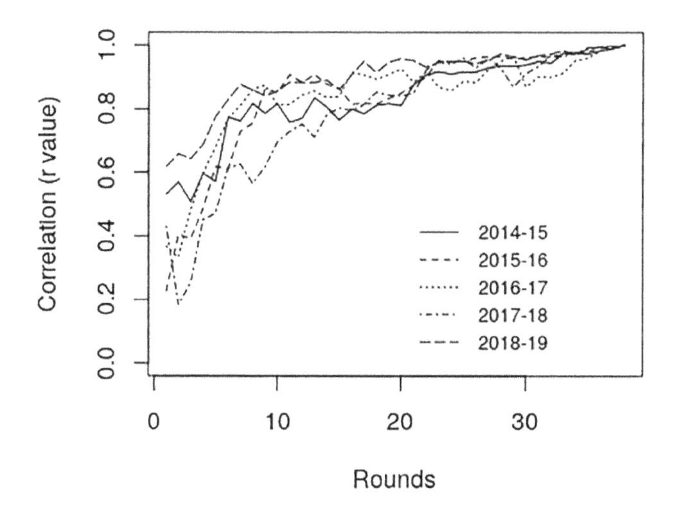

FIGURE 5.2
Change in Spearman correlation as the competition progressed for EPL seasons 2014–2015 to 2018–2019.

KEY CONCEPT BOX 5.1: CORRELATION

When analysing data, we often want to look at relationships between individual variables (vectors). This can be done by computing the covariance, which is a measure of how two variables, say x and y, change together (i.e., strength of their relationship). If the higher values of x mainly correspond with the higher values of y, and the same holds true for the lower values, then the covariance is said to be positive. On the contrary, if the higher values of x mainly correspond to the lower values of y, then the covariance is negative.

Covariance is, however, somewhat difficult to interpret because it is a non-standardised value. Therefore, correlation, which is simply normalised covariance, is generally used in statistics because it is easier to interpret. Correlation has no units and is characterised by the correlation coefficient r, which can be considered an effect size. With positive correlations, the r-value ranges from 0 (no correlation) to 1 (absolute correlation), whereas with negative correlations, the range is from 0 to –1. The r-value can be computed using the Pearson correlation method as follows:

$$r = \frac{\sum_{i=1}^{n}\left(x_i - \bar{x}\right)\left(y_i - \bar{y}\right)}{\sqrt{\sum_{i=1}^{n}\left(x_i - \bar{x}\right)^2 + \sum_{i=1}^{n}\left(y_i - \bar{y}\right)^2}} \tag{5.5}$$

where x_i and y_i are the individual values in the x and y vectors, and $\bar{x}$ and $\bar{y}$ are the respective mean values.

To illustrate how correlations work, we shall consider the four vectors below (here written in R code):

```
a <- c(2.1, 3.4, 2.8, 0.2, 8.8, 1.7, 6.6, 8.5, 3.5, 5.0)
b <- c(2.9, 2.7, 2.2, 0.9, 6.8, 2.2, 6.5, 7.2, 2.5, 4.1)
c <- c(9.1, 8.4, 4.8, 7.2, 6.8, 7.7, 5.6, 9.5, 6.5, 9.7)
d <- c(9.1, 8.4, 4.8, 8.2, 2.1, 7.7, 5.6, 1.5, 6.5, 4.7)
```

The correlations between these four vectors are as follows:

	a	*b*	*c*	*d*
a	1.000	0.968	0.085	–0.867
b	0.968	1.000	0.119	–0.789
c	0.085	0.119	1.000	0.031
d	–0.867	–0.789	0.031	1.000

From which we can see that vector a is strongly positively correlated with vector b (r=0.968) and strongly negatively correlated with vector d (r=–0.867). By contrast, vector c has little or no correlation with the other three vectors (r=0.085, 0.119, and 0.031), as illustrated in Figure 5.3.

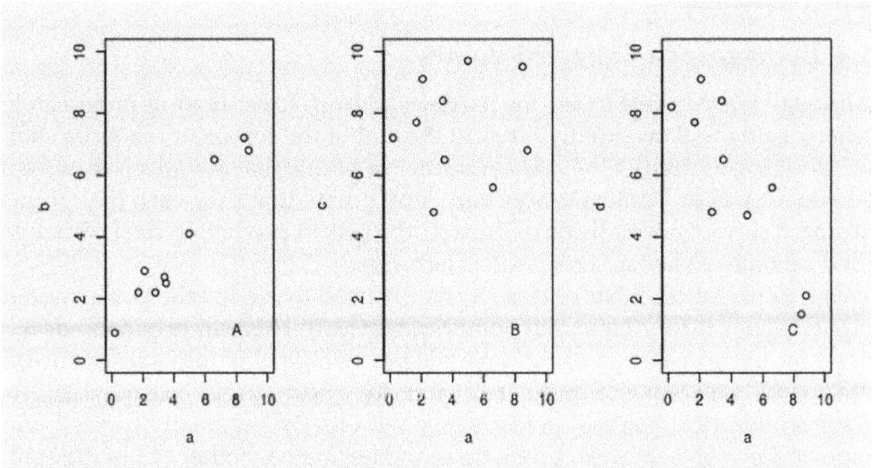

FIGURE 5.3
Scatter plots of: (A) vectors a and b (r = 0.968); (B) vectors a and c (r = 0.085); and (C) vectors a and d (r = −0.867).

Being a measure of the relatedness of one variable to another, the r-value is actually an effect size. As such, it can be compared with other effect sizes, as illustrated in Table 5.3. Another effect size that is often used in statistics is the coefficient of determination (r^2 or R^2), which is simply the r-value squared when only two variables are considered. The coefficient of determination is generally used to quantify the amount of variance explained when linear regression is performed.

TABLE 5.3
Table of Equivalent Effect Sizes for: Cohen's d, r value, and r^2 value

Effect	Cohen's d	Correlation Coefficient (r)	Coefficient of Determination (r^2 or R^2)
Small	0.2–0.5	0.1–0.3	0.01–0.09
Medium	0.5–0.8	0.3–0.5	0.09–0.25
Large	>0.8	>0.5	>0.25

While the Pearson correlation method produces the most widely used coefficient, other nonparametric alternatives exist, such as Spearman's rank correlation coefficient. With many data sets, the Pearson and Spearman methods produce similar results. However, Spearman correlation has the great advantage that it is less sensitive than Pearson correlation to outliers and therefore tends to yield more robust results when strong outliers are present.

5.4 Pythagorean Expected Points

Although it is possible to use the partial standings to estimate approximately where teams will eventually finish at the end of the season, it is a more challenging task to predict the actual number of points that any given team will accrue. However, because teams bank any points that they earn (i.e., points awarded cannot normally be deducted), the task of predicting the EoS points total becomes easier as the season progresses.

One technique that has been successfully used to predict the EoS expected points total is the Pythagorean points system [6,7]. This was originally developed for use in baseball to estimate the percentage of games a team should have won based on the number of runs they scored and conceded [6]. By comparing a team's actual performance with its Pythagorean expected win ratio, it is possible to make predictions and evaluate whether or not the team is over- or under-performing. Unfortunately, Pythagorean expectation in its original form cannot easily accommodate draws or the points system commonly used in soccer (i.e., 3 points for a win and 1 point for a draw), and so it has had to be modified to cope with the peculiarities of association football [8]. So, for soccer, the general form of the Pythagorean expected points formula is:

$$\text{Pythagorean expected points}: \text{pts}_{\text{exp}} = a \times \left[\frac{\text{GF}^b}{\text{GF}^c + \text{GA}^d} \right] \times m \qquad (5.6)$$

where pts_{exp} is the expected number of points awarded; GF is the total number of goals scored by the team; GA is the number of goals conceded by the team; m is the number of matches played; and a, b, c, and d are empirical coefficients.

There is some debate about the values that should be used for coefficients a, b, c, and d when applying Pythagorean expectation to soccer. For example, Eastwood [9] suggested $a=2.5$, $b=1.228$, $c=1.072$, and $d=1.127$, whereas Kingsman [10] preferred $a=2.28$, $b=1.17$, $c=1.06$, and $d=1.06$. However, Beggs, using data from 1995–2017, found that $a=2.78$, $b=1.24$, $c=1.24$, and $d=1.25$ produced superior predictions for the EPL. Example 5.3 uses the coefficient values developed by Beggs to illustrate the close relationship that exists between Pythagorean points and real points achieved in the EPL.

EXAMPLE 5.3: THE RELATIONSHIP BETWEEN PYTHAGOREAN POINTS AND ACTUAL POINTS AWARDED

In this example, we apply the Pythagorean points system, with Beggs' coefficients, to compute the expected points for the various clubs in the EPL for the 2020–2021 season. The source file for this example

(EPL_final_table_2021.csv) can be found in the GitHub repository at https://github.com/cbbeggs/SoccerAnalytics. The following code assumes that the data file is saved in a directory called 'Datasets'.

First, we clear the workspace of all data and variables.

```
rm(list = ls())   # Clears all variables from workspace
```

Next, we load the data, which consists of the EoS table for the EPL in season 2020–2021. We also specify the Pythagorean coefficients.

```
# Here we assume that the file being loaded is stored in a directory called 'Datasets'.
EoStab <- read.csv("C:/Datasets/EPL_final_table_2021.csv")
names(EoStab)
```

```
##
[1] "Club" "PLD" "HW" "HD" "HL" "AW" "AD" "AL" "GF" "GA" "GD" "PTS"
```

```
# Specify value of coefficients
a = 2.78
b = 1.24
c = 1.24
d = 1.25
```

Now we can compute the key metrics.

```
pygFrac <- ((EoStab$GF^b)/((EoStab$GF^c)+(EoStab$GA^d)))
pygPts <- a * pygFrac * EoStab$PLD
pygDiff <- EoStab$PTS - pygPts

# Round to specified dp
pygFrac <- round(pygFrac, 3)
pygPts <- round(pygPts, 2)
pygDiff <- round(pygDiff, 2)
```

Finally, we compile Pythagorean expectation results and concatenate these to the EoS league table, as follows:

```
pygtab <- cbind.data.frame(EoStab, pygFrac, pygPts, pygDiff)
print(pygtab)
```

```
##
           Club PLD HW HD HL AW AD AL GF GA  GD PTS pygFrac pygPts pygDiff
1      Man City  38 13  2  4 14  3  2 83 32  51  86   0.759  80.18    5.82
2    Man United  38  9  4  6 12  7  0 73 44  29  74   0.643  67.96    6.04
3     Liverpool  38 10  3  6 10  6  3 68 42  26  69   0.636  67.24    1.76
4       Chelsea  38  9  6  4 10  4  5 58 36  22  67   0.635  67.13   -0.13
5     Leicester  38  9  1  9 11  5  3 68 50  18  66   0.585  61.77    4.23
6      West Ham  38 10  4  5  9  4  6 62 47  15  65   0.576  60.81    4.19
7     Tottenham  38 10  3  6  8  5  6 68 45  23  62   0.616  65.11   -3.11
8       Arsenal  38  8  4  7 10  3  6 55 39  16  61   0.596  62.98   -1.98
9         Leeds  38  8  5  6 10  0  9 62 54   8  59   0.533  56.29    2.71
10      Everton  38  6  4  9 11  4  4 47 48  -1  59   0.484  51.11    7.89
11  Aston Villa  38  7  4  8  9  3  7 55 46   9  55   0.546  57.65   -2.65
12    Newcastle  38  6  5  8  6  4  9 46 62 -16  45   0.399  42.11    2.89
13       Wolves  38  7  4  8  5  5  9 36 52 -16  45   0.379  40.00    5.00
14  Crystal Pal  38  6  5  8  6  3 10 41 66 -25  44   0.347  36.66    7.34
15  Southampton  38  8  3  8  4  4 11 47 68 -21  43   0.377  39.88    3.12
16     Brighton  38  4  9  6  5  5  9 40 46  -6  41   0.447  47.25   -6.25
17      Burnley  38  4  6  9  6  3 10 33 55 -22  39   0.338  35.68    3.32
18       Fulham  38  2  4 13  3  9  7 27 53 -26  28   0.294  31.06   -3.06
19    West Brom  38  3  6 10  2  5 12 35 76 -41  26   0.268  28.31   -2.31
20 Sheffield Utd 38  5  1 13  2  1 16 20 63 -43  23   0.188  19.84    3.16
```

Here 'pygPts' is the expected EoS number of points using the Pythagorean method, with 'pygDiff' being the difference between the observed and expected points.

Having produced the table above, we can then determine the correlation between the Pythagorean expected points and the actual points achieved.

```
cor.test(pygtab$PTS, pygtab$pygPts) # This give the correlation r value.
```

```
##
Pearson's product-moment correlation

data:  pygtab$PTS and pygtab$pygPts
t = 17.608, df = 18, p-value = 8.574e-13
alternative hypothesis: true correlation is not equal to 0
95 percent confidence interval:
 0.9295660 0.9891556
sample estimates:
      cor
0.9721788
```

From this, we can see that there is a very high positive correlation (r=0.972) between the expected points and the actual points achieved by the clubs in the league, which we can visualise in a scatter plot.

```
# Scatter plot
plot(pygtab$PTS, pygtab$pygPts, col = "black", pch=21, xlab="EPL points",
ylab = "Pythagorean points")
abline(lm(pygtab$pygPts ~ pygtab$PTS), lty=1)
```

This produces the scatter plot displayed in Figure 5.4.

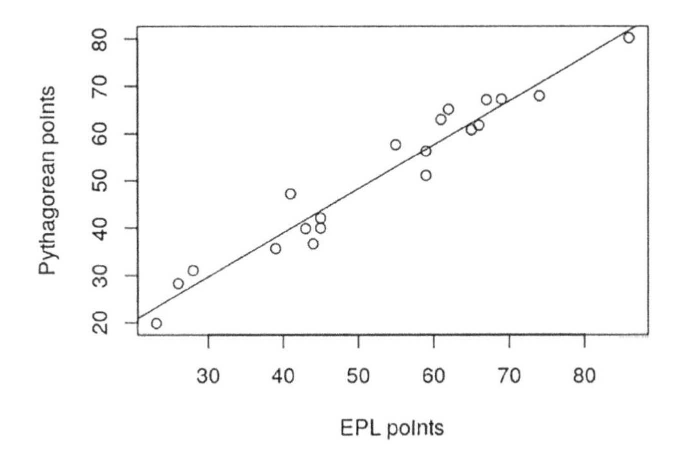

FIGURE 5.4
Scatter plot with best-fit line of Pythagorean expected points and actual points awarded after 38 rounds of competition for the EPL in season 2020–2021 ($r = 0.972$).

5.5 Pythagorean Points Prediction

While Example 5.3 demonstrates the close correlation between Pythagorean points and EoS points, it does not show how Pythagorean expectation can be used in practice to predict the total number of points that will be accumulated by a club in a given season. This, however, can easily be done part way through the season by applying the Pythagorean method to the games already played and then using this to predict how many points are likely to be earned from the remaining matches. In Example 5.4, we do exactly this by applying the user-defined function 'pythag_pred' to Liverpool after 10 rounds of competition in season 2020–2021. At that stage of the competition, Liverpool had played 10 matches, scored 22 goals, conceded 17 goals, and earned 21 points.

EXAMPLE 5.4: SINGLE PYTHAGOREAN EXPECTED POINTS PREDICTION

In this example, we apply the Pythagorean points prediction function, 'pythag_pred', which uses Beggs' coefficient values, to Liverpool after they had played just 10 of their league fixtures in season 2020–2021. At that stage of the competition, they had scored 22 goals, conceded 17 goals, and earned 21 points.

```
# Pythagorean expected points function.
pythag_pred <- function(PLD,GF,GA,PTS,nGames){

# Coefficients (NB. These can be fine tuned to suit the particular league.)
  a = 2.78
  b = 1.24
  c = 1.24
  d = 1.25

# Compute key metrics
  pythag_frac <- (GF^b)/((GF^c)+(GA^d))
  pythag_pts <- a * pythag_frac * PLD
  pythag_diff <- PTS - pythag_pts
  points_avail <- (nGames - PLD)*3
  pred_pts <- pythag_frac * a * (nGames - PLD)
  pred_total <- PTS + pred_pts
  pythag_total <- pythag_pts + pred_pts
  results <- round(c(PLD,GF,GA,PTS,pythag_frac,pythag_pts,pythag_diff, points_avail,
                      pred_pts,pred_total,pythag_total),1) # Creates results vector
  return(results)
}
```

Before applying the 'pythag_pred' function, we need to provide some information about the total number of matches played by each team in the league and also Liverpool's performance.

```
# Compute the total number of matches played.
nTeams <- 20 # Number of teams in EPL
nGames <- (nTeams-1)*2 # Computes the total number of games each team plays in a season.

# Specify the teams performance so far.
PLD <- 10 # Number of games played
GF <- 22 # Number of goals scored
GA <- 17 # Number of goals conceded
PTS <- 21 # Number of points achieved
```

Now we can apply the 'pythag_pred' function, as follows:

```
pred_res <- pythag_pred(PLD,GF,GA,PTS,nGames)
var_names <- c("PLD","GF","GA","PTS","PythagFrac","PythagPTS",
"PythagDiff","AvailPTS","PredPTS","PredTot","PythagTot")
Liv_pred10 <-cbind.data.frame(var_names, pred_res)
print(Liv_pred10)
```

```
##
      var_names pred_res
1           PLD     10.0
2            GF     22.0
3            GA     17.0
4           PTS     21.0
5    PythagFrac      0.6
6     PythagPTS     15.9
7    PythagDiff      5.1
8      AvailPTS     84.0
9       PredPTS     44.6
10      PredTot     65.6
11    PythagTot     60.5
```

From this, we see that the algorithm predicts that Liverpool would earn a further 44.6 points out of the 84 points that were available. When these were added to the 21 points that Liverpool had already earned after playing 10 matches, it gave a predicted total of 65.6 points for the whole season. In fact, Liverpool's EoS total was 69 points, which meant that the prediction had an error of just 3.4 points. However, the Pythagorean expected total for the whole season was predicted to be just 60.5 points, which reflects the fact that according to the algorithm, Liverpool overperformed in the first 10 games of the season.

What is so impressive about this prediction is that despite using just two predictors, 'goals for' and 'goals against', after 10 matches the algorithm was able to predict Liverpool's EoS points total very accurately. This is quite amazing when you think about it, because in the remaining 28 games until the end of the season, a lot of events would have occurred, with players getting injured or being sent off, penalties being awarded, and so on. Yet despite all these random events, the Pythagorean algorithm did a pretty good job of predicting Liverpool's EoS point total after only 10 rounds of competition. Having said this, as we will see in Example 5.5, the predictions for the other clubs in the league were not all as accurate as the prediction for Liverpool.

It is possible to perform league-wide Pythagorean point predictions on multiple teams using the 'pythag_pred' function using the approach taken in Example 5.5, where we initially apply the algorithm to make predictions after just 10 rounds of competition in the EPL for season 2020–2021.

EXAMPLE 5.5: PYTHAGOREAN POINTS PREDICTION APPLIED TO THE WHOLE LEAGUE

In this example, we will apply the 'pythag_pred' function to three time points in season 2020–2021: (i) after 98 matches (approximately 10 rounds); (ii) after 185 matches (approximately 19 rounds); and (iii) after 290 matches (approximately 29 rounds). The CSV data files used in this example (i.e., EPL_after_98_matches_2021.csv, EPL_after_185_matches_2021.csv, and EPL_after_290_matches_2021.csv) are stored on the GitHub repository, which can be accessed at: https://github.com/cbbeggs/SoccerAnalytics.

First, we shall load the league table after 98 matches have been played.

```
# Here we assume that the file being loaded is stored in a directory called 'Datasets'.
part_tab <- read.csv("C:/Datasets/EPL_after_98_matches_2021.csv") # Approx. 10 rounds
```

Then we input some league parameters and create an empty matrix to store the predicted results.

```
nTeams <- 20 # Number of teams in EPL
nGames <- (nTeams-1)*2 # Computes total number of games each team plays in a season.
predPTS <- matrix(NA, nTeams, 11) # Creates a [20 x 11] empty matrix populated with NAs.
```

Now we populate the empty matrix by applying the 'pythag_pred' function using a 'for loop'.

```
for(i in 1:nTeams){
  predPTS[i,] <- pythag_pred(part_tab$PLD[i], part_tab$GF[i],
                             part_tab$GA[i], part_tab$PTS[i], nGames)
}
```

Finally, we compile the results and display them.

```
Pred_EoSPTS <- cbind.data.frame(part_tab$Club, predPTS)
colnames(Pred_EoSPTS) <- c("Club", var_names)
head(Pred_EoSPTS) # This displays the top six rows.
```

```
##
         Club PLD GF GA PTS PythagFrac PythagPTS PythagDiff AvailPTS PredPTS
1   Tottenham  10 21  9  21        0.7      20.5        0.5       84    57.3
2   Liverpool  10 22 17  21        0.6      15.9        5.1       84    44.6
3     Chelsea  10 22 10  19        0.7      20.1       -1.1       84    56.2
4   Leicester  10 19 14  18        0.6      16.3        1.7       84    45.7
5    West Ham  10 17 11  17        0.6      17.4       -0.4       84    48.7
6 Southampton  10 19 16  17        0.5      15.2        1.8       84    42.5
  PredTot PythagTot
1    78.3      77.8
2    65.6      60.5
3    75.2      76.3
4    63.7      62.0
5    65.7      66.1
6    59.5      57.7
```

From this, we can see that the algorithm predicts an EoS point total of 78.3 for Tottenham and 65.6 for Liverpool. But how accurate were these predictions? In order to find this out, we need to download the EoS table for season 2020–2021 (EPL_final_table_2021.csv), as follows:

```
EoS_tab <- read.csv("C:/Datasets/EPL_final_table_2021.csv")
```

Having done this we sort the respective data frames alphabetically, so that we can make a comparison.

```
sortedEoS <- EoS_tab[order(EoS_tab$Club), ]
head(sortedEoS)
```

```
##
             Club PLD HW HD HL AW AD AL GF GA  GD PTS
8         Arsenal  38  8  4  7 10  3  6 55 39  16  61
11    Aston Villa  38  7  4  8  9  3  7 55 46   9  55
16       Brighton  38  4  9  6  5  5  9 40 46  -6  41
17        Burnley  38  4  6  9  6  3 10 33 55 -22  39
4         Chelsea  38  9  6  4 10  4  5 58 36  22  67
14 Crystal Palace  38  6  5  8  6  3 10 41 66 -25  44
```

```
sortedPred <- Pred_EoSPTS[order(Pred_EoSPTS$Club), ]
head(sortedPred)
```

```
##
             Club PLD GF GA PTS PythagFrac PythagPTS PythagDiff AvailPTS
14        Arsenal  10 10 12  13        0.4      12.2        0.8       84
10    Aston Villa   9 20 13  15        0.6      15.6       -0.6       87
16       Brighton  10 14 16  10        0.5      12.6       -2.6       84
19        Burnley   9  4 17   5        0.1       3.5        1.5       87
3         Chelsea  10 22 10  19        0.7      20.1       -1.1       84
15 Crystal Palace  10 12 15  13        0.4      11.8        1.2       84
   PredPTS PredTot PythagTot
14    34.1    47.1      46.2
10    50.3    65.3      66.0
16    35.2    45.2      47.7
19    11.2    16.2      14.7
3     56.2    75.2      76.3
15    33.1    46.1      44.9
```

Finally, we compile the EoS table with the Pythagorean predicted points added.

```
Pred <- sortedPred$PredTot
EoS_predtab <- cbind.data.frame(sortedEoS, Pred)
print(EoS_predtab)
```

```
##
                Club PLD HW HD HL AW AD AL GF GA  GD PTS Pred
8            Arsenal  38  8  4  7 10  3  6 55 39  16  61 47.1
11       Aston Villa  38  7  4  8  9  3  7 55 46   9  55 65.3
16          Brighton  38  4  9  6  5  5  9 40 46  -6  41 45.2
17           Burnley  38  4  6  9  6  3 10 33 55 -22  39 16.2
4            Chelsea  38  9  6  4 10  4  5 58 36  22  67 75.2
14    Crystal Palace  38  6  5  8  6  3 10 41 66 -25  44 46.1
10           Everton  38  6  4  9 11  4  4 47 48  -1  59 57.1
18            Fulham  38  2  4 13  3  9  7 27 53 -26  28 32.7
9              Leeds  38  8  5  6 10  0  9 62 54   8  59 49.4
5          Leicester  38  9  1  9 11  5  3 68 50  18  66 63.7
3          Liverpool  38 10  3  6 10  6  3 68 42  26  69 65.6
1           Man City  38 13  2  4 14  3  2 83 32  51  86 62.5
2         Man United  38  9  4  6 12  7  0 73 44  29  74 55.8
12         Newcastle  38  6  5  8  6  4  9 46 62 -16  45 47.1
20 Sheffield United  38  5  1 13  2  1 16 20 63 -43  23 12.6
15       Southampton  38  8  3  8  4  4 11 47 68 -21  43 59.5
7          Tottenham  38 10  3  6  8  5  6 68 45  23  62 78.3
19         West Brom  38  3  6 10  2  5 12 35 76 -41  26 24.0
6           West Ham  38 10  4  5  9  4  6 62 47  15  65 65.7
13            Wolves  38  7  4  8  5  5  9 36 52 -16  45 55.5
```

In order to evaluate the accuracy of the predictions made after 10 rounds we can compute the correlation *r*-value and the mean absolute error (MAE) as follows:

```
cor.test(sortedEoS$PTS,sortedPred$PredTot) # This computes the correlation r value.
```

```
##
	Pearson's product-moment correlation

data:  sortedEoS$PTS and sortedPred$PredTot
t = 5.216, df = 18, p-value = 5.831e-05
alternative hypothesis: true correlation is not equal to 0
95 percent confidence interval:
 0.5074660 0.9069449
sample estimates:
      cor
0.7757754
```

```
mae <- mean(abs(sortedEoS$PTS - sortedPred$PredTot)) # Mean absolute error
print(mae)
```

```
## [1] 9.18
```

This indicates that although the Pythagorean algorithm predicts the trend with reasonable accuracy (r=0.776), the MAE is not that great, being >9 points, which is evident when we produce a scatter plot of the predicted and actual EoS points.

```
# Scatter plot of EoS_PTS and PredTot
plot(sortedEoS$PTS, sortedPred$PredTot, col = "black", pch=21, xlab="Actual EoS points",
ylab = "Predicted EoS points")
abline(lm(sortedPred$PredTot ~ sortedEoS$PTS), lty=1)
```

This produces the scatter plot shown in Figure 5.5.

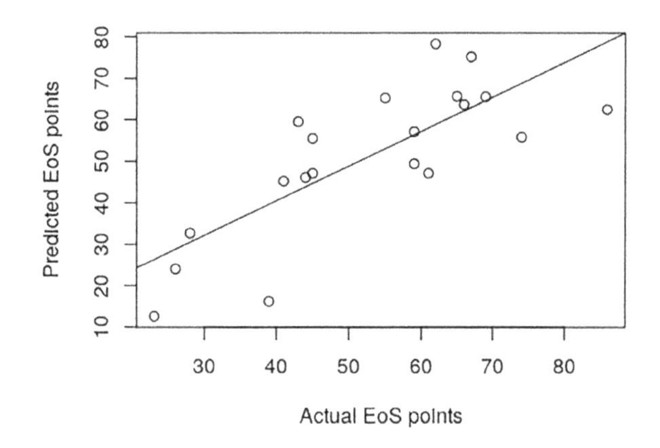

FIGURE 5.5
Scatter plot with best-fit line of predicted EoS points and actual EoS points, made after approximately 10 rounds of competition for the EPL in season 2020–2021 (r=0.778; MAE=9.18 points).

From this, we can see that the Pythagorean prediction after 10 rounds is less accurate at the extremes, with the errors being greatest for the clubs at the bottom and top of the league. For example, Manchester City and Sheffield United were predicted after 10 rounds of competition to achieve EoS totals of 62.5 and 12.6 points, respectively. In fact, they actually achieved 86 and 23 points, which is a long way from what was predicted.

The process above can be repeated for rounds 19 and 29 using the following league table files.

```
part_tab <- read.csv("C:/Datasets/EPL_after_185_matches_2021.csv") # Approx. 19 rounds
part_tab <- read.csv("C:/Datasets/EPL_after_290_matches_2021.csv") # Approx. 29 rounds
```

The results produced by this analysis are presented in Figures 5.6 and 5.7 and summarised in Table 5.4.

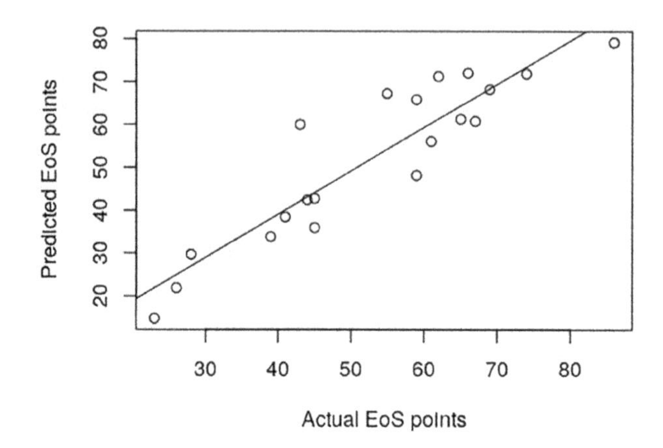

FIGURE 5.6
Scatter plot with best-fit line of predicted EoS points and actual EoS points, made after approximately 19 rounds of competition for the EPL in season 2020–2021 (r=0.917; MAE=6.06 points).

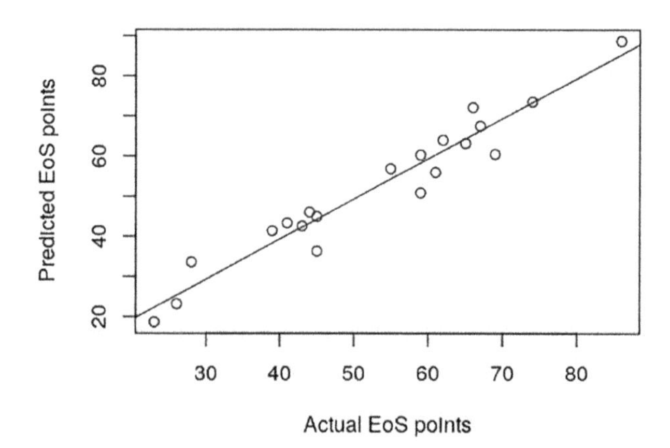

FIGURE 5.7
Scatter plot with best-fit line of predicted EoS points and actual EoS points, made after approximately 29 rounds of competition for the EPL in season 2020–2021 (r=0.968; MAE=3.36 points).

TABLE 5.4

Pythagorean Predictions Metrics Made after Rounds 10, 19, and 29 of Competition in the EPL for Season 2020–2021.

When Prediction Is Made	Correlation (r Value)	MAE (Points)
After approx. 10 rounds	0.776	9.18
After approx. 19 rounds	0.917	6.06
After approx. 29 rounds	0.968	3.36

From this, we can see that, unsurprisingly, the predictions made by the Pythagorean algorithm improve as the season progresses, with those made at the mid-point of the season being surprisingly accurate (i.e., MAE ≈ 6 points).

5.6 Performance Indicators and Predictor Metrics

In Section 5.5, we looked in depth at the Pythagorean expectation system, which is based on goals scored and conceded. However, this is just one of many performance indicators and metrics that can be used to assess and predict team performance. So, in this section, we will briefly look at several other key metrics that are often used and investigate how these relate to points earned.

There are three goal-based metrics that are frequently used to assess team performance. These are:

$$\text{Goal difference (GD): GF} - \text{GA} \tag{5.7}$$

$$\text{Goal ratio (GR): } \frac{\text{GF}}{\text{GF} + \text{GA}} \tag{5.8}$$

$$\text{Pythagorean goal ratio (PythGR) - here with Beggs' coefficients: } \frac{\text{GF}^{1.24}}{\text{GF}^{1.24} + \text{GA}^{1.25}} \tag{5.9}$$

Each of these reflects in different ways the number of goals scored and the goals conceded by any given team, which unsurprisingly strongly influences the number of points that the team will accrue. As such, they are all highly correlated with the points total achieved in any season (see Table 5.5), and therefore they can all in theory be used to predict EoS points. Having said this, they do not consider other aspects of the game, such as the number of goal chances created or the number of shots made and conceded by any given team. Consequently, from a coaching point of view, goal-based metrics are not necessarily that useful when assessing team performance. After all, a team can perform brilliantly on the pitch and still lose a match one-nil because, say, the opposition scored a lucky goal against the run of play. So when attempting to quantify overall team performance, it is often helpful to consider other non-goal-related metrics, such as total shots ratio (TSR) or total shots on target ratio (TSoTR) (Equations 5.10 and 5.11). These metrics do not rely on goal data and therefore cannot be acquired from league tables. However, these metrics can be computed on a match-by-match basis from historical match data, such as that found at https://www.football-data.co.uk/, as illustrated in Chapter 4.

TABLE 5.5

Correlation *r*-values between EoS Points Total and Three Goal Related Prediction Metrics for Eight Successive Seasons of the EPL

Season	Goal Difference (*r*-Value)	Goal Ratio (*r*-Value)	Pythagorean Goal Ratio (*r*-Value)
2013–2014	0.948	0.952	0.953
2014–2015	0.964	0.958	0.958
2015–2016	0.961	0.962	0.962
2016–2017	0.978	0.982	0.982
2017–2018	0.985	0.984	0.983
2018–2019	0.991	0.991	0.991
2019–2020	0.948	0.948	0.947
2020–2021	0.970	0.973	0.972

One of the most widely used match performance indicators is TSR, which attempts to capture in one metric the attacking and defensive shooting qualities of a team [11]. Like goal ratio, TSR is a ratio which accommodates both the shots attempted (SF) during a match and the shots conceded (SA) in a single metric. It is defined as follows:

$$\text{Total shots ratio (TSR)}: \frac{\text{SF}}{\text{SF} + \text{SA}} \tag{5.10}$$

The same approach can be taken to develop a similar metric for shots on target (SoT).

$$\text{Total shots on target ratio (TSoTR)}: \frac{\text{STF}}{\text{STF} + \text{STA}} \tag{5.11}$$

If we consider these various metrics, it becomes clear that a hierarchy exists. The number of goals scored is a subset of the number of shots on target, which in turn is a subset of the number of shots attempted. So, if a team makes only a few shots, it follows that they are unlikely to score many goals, if any at all. Conversely, if they attempt many shots, it is likely that they will score more goals. Therefore, it is clear that a relationship exists between shooting opportunities and goals scored. Consequently, shot-related metrics such as TSR and TSoTR can be used not only to assess team performance on a match-by-match basis but also to help predict EoS league point totals. This latter point is illustrated in Example 5.6, which uses EPL EoS data for season 2020-2021 to compute the correlation between points earned and the various metrics described above.

EXAMPLE 5.6: CORRELATION BETWEEN RATIO PERFORMANCE INDICATORS AND EOS POINTS TOTAL

In this example, we use data from season 2020–2021 to compute the EoS GR, TSR, and TSoTR metrics for each club in the EPL and look at how these are correlated with points earned.

First, we clear the workspace.

```
rm(list = ls())    # Clears all variables from workspace
```

Next, we load the data into R (NB. The original data file (EPL_EoS_indicators_2021.csv) can be found in the GitHub repository at: https://github.com/cbbeggs/SoccerAnalytics.).

```
# Here we assume that the file being loaded is stored in a directory called 'Datasets'.
perf_dat <- read.csv("C:/Datasets/EPL_EoS_indicators_2021.csv")

# Legend
# "Team" - club
# "Rank" - EoS standing in the league
# "Pld" - number of matches played
# "Pts" - points earned
# "GF" - goals for
# "GA" - goals against
# "GD" - goal difference
# "ShF" - number of shots made
# "SoTF" - number of shots on target made
# "ShA" - number of shots conceded
# "SoTA" - number of shots on target conceded
# "Poss" - percentage possession
# "PassComp" - percentage of passes completed
# "xG" - expected goals
```

Having done this, we inspect the data.

```
names(perf_dat)
```

```
##
[1] "Team" "Rank"  "Pld"  "Pts"  "GF"  "GA"  "GD"  "SF"  "SoTF"  "SA"
[11] "SoTA"   "Poss"   "PassComp"   "xG"
```

Next, we create a results data frame for the derived performance indicators (i.e., GR, TSR, and TSoTR) that we are going to compute.

```
ind1 <- perf_dat[,c(1:4)]

# Now we add some new derived variables
ind1["GR"] <- NA # Create empty vector to store goal ratio results
ind1["TSR"] <- NA # Create empty vector to store total shots ratio results
ind1["TSoTR"] <- NA # Create empty vector to store total shots on target ratio results

# Now we populate these new variables and round to 3 decimal places.
ind1$GR <- round(perf_dat$GF/(perf_dat$GF+perf_dat$GA),3)
ind1$TSR <- round(perf_dat$SF/(perf_dat$SF+perf_dat$SA),3)
ind1$TSoTR <- round(perf_dat$SoTF/(perf_dat$SoTF+perf_dat$SoTA),3)
```

Now we display the results.

```
print(ind1)
```

```
##
                Team Rank Pld Pts    GR   TSR TSoTR
1            Arsenal    8  38  61 0.585 0.521 0.524
2        Aston Villa   11  38  55 0.545 0.491 0.503
3           Brighton   16  38  41 0.465 0.575 0.524
4            Burnley   17  38  39 0.375 0.405 0.411
5            Chelsea    4  38  67 0.617 0.625 0.653
6     Crystal Palace   14  38  44 0.383 0.394 0.427
7            Everton   10  38  59 0.495 0.440 0.477
8             Fulham   18  38  28 0.338 0.511 0.429
9       Leeds United    9  38  59 0.534 0.488 0.509
10    Leicester City    5  38  66 0.576 0.561 0.562
11         Liverpool    3  38  69 0.618 0.649 0.595
12   Manchester City    1  38  86 0.722 0.682 0.705
13    Manchester Utd    2  38  74 0.624 0.549 0.593
14     Newcastle Utd   12  38  45 0.426 0.404 0.421
15     Sheffield Utd   20  38  23 0.241 0.371 0.310
16       Southampton   15  38  43 0.409 0.498 0.482
17         Tottenham    7  38  62 0.602 0.481 0.535
18         West Brom   19  38  26 0.315 0.362 0.313
19          West Ham    6  38  65 0.569 0.502 0.525
20            Wolves   13  38  45 0.409 0.514 0.509
```

From this, we can see that teams that generally have a high TSR also exhibit a high GR, and those with a low TSR tend to have a low GR. Consequently, teams with a low TSR tended to finish close to the bottom of the table, whereas those with a higher TSR tended to be near the top.

The relationship between the various indicator metrics and points earned can be evaluated by calculating the Pearson correlation *r*-values, as follows:

```
# Compute correlation r-values
ind1_cors <- round(cor(ind1[,c(4:7)]),3)
print(ind1_cors)
```

```
##
         Pts    GR   TSR TSOTR
Pts    1.000 0.972 0.727 0.906
GR     0.972 1.000 0.780 0.924
TSR    0.727 0.780 1.000 0.920
TSOTR  0.906 0.924 0.920 1.000
```

From this, we can clearly see that a hierarchy exists, with the correlation between GR and points earned (r=0.972) being the strongest and that between TSR and points (r=0.727) being the weakest. Having said this, the correlation between points and TSR is still very strong, suggesting that TSR is likely to be a good predictor of points earned. Noticeably, however, the relationship between TSoTR and points earned is even stronger (r=0.906), suggesting that shooting accuracy is a strong indicator as to whether or not a team will be successful.

5.7 Expected Goals

While it is clear from Example 5.6 that the metrics GR, TSR, and TSoTR are strongly correlated with each other and total points earned, they are somewhat difficult to interpret. Also, they fail to capture important information that might be helpful to coaches. For example, while the TSR tells us about the ratio between the number of shots made and those conceded, it tells us nothing about the quality of those shots or about how many are converted into goals. If we compare Manchester City (champions in season 2020–2021; GR=0.722; TSR=0.682) with Fulham (relegated in season 2020–2021; GR=0.338; TSR=0.511), we see not surprisingly that the GR for Manchester City was over twice that of Fulham's. Yet when we compare the TSRs of the two clubs, it can be seen that they are much closer. This may indicate that many more of Fulham's shots did not result in goals compared with those of Manchester City. However, it could also be that more of the shots conceded by Fulham were converted into goals. So, to rectify this ambiguity, it is sometimes helpful to compute the goals-to-shots ratio, which can be calculated as follows:

$$\text{Goals to shots ratio (GSR)}: \frac{GF}{SF} \tag{5.12}$$

If we do this for these two clubs, we find that the GSR for Manchester City is 0.13, while that for Fulham is only 0.05, indicating that indeed Manchester City were converting many more of their shots into goals compared with Fulham. Similarly, Manchester City exhibited a GSR against of 0.08, while the same figure for Fulham was 0.11. This suggests that, compared to Fulham, Manchester City were tending to restrict their opponents to longer-range shots, which were less likely to result in goals.

Based on the idea that both the quantity and the quality of chances contribute to the number of goals scored, the concept of *expected goals* (xG) was developed [12–14], which subsequently has become very popular as a method to evaluate the performance of both teams and players. The central idea underpinning the expected goals method is that not all goal attempts are equal and that the probability of any attempt resulting in a goal can be rated on a scale from 0 to 1, where 0 is a 0% chance of a goal being scored and 1 is a certain goal. So, for example, a tame long-range shot that has little chance of reaching the back of the net might be rated 0.01, whereas a simple tap-in in front of an empty goal could be given a rating as high as 0.9. By assigning a probability to each goal attempt (shot), it is possible to compute the expected goals (xG) for each team in any given match, as follows:

$$xG = p_{s1} + p_{s2} + p_{s3} + \ldots + p_{sn} \tag{5.13}$$

where p_{s1}, p_{s2}, p_{s3} … p_{sn} are the probabilities of each individual shot being converted into a goal and n is the total number of shots made by the team in the match.

For any given match, the expected goals for the respective teams are calculated by analysts working for sport analytics companies who use a database of comparative data to evaluate the probabilities that should be attributed to each individual shot. Once collated, this expected goal data can be used not only to evaluate individual matches but also player performance and the performance of teams over entire seasons [13].

If we consider the EPL results for 8 May 2022 (Table 5.6), we can see that the match results do not necessarily conform to the expected goals score. For example, the expected goal score of 2.04–1.73 in favour of Brighton against Manchester United suggests that the match was a tight affair, which slightly favoured Brighton. Yet the actual result was 4-0 to Brighton, despite Manchester United having most of the possession. The Brentwood versus Southampton and Leicester versus Everton matches also produced results that were at odds with the xG score. This has led some to question the validity of the expected goal concept. However, it should be remembered that the 'expected goals' method is derived from the statistical concept of 'expected value', which is a measure of the likelihood of a particular outcome occurring. So in effect, the computed expected goals should be viewed as an average value that would occur if the same match were repeated thousands of times. Sometimes the match outcome would be an extreme one, as happened in the Brighton versus Manchester United match on 8 May 2022, but more often the outcome would be much tighter, reflecting the computed *xG* score.

TABLE 5.6

EPL Results for 8 May 2022 Showing the Score, Expected Goals, and the Possession Breakdown for Each Match Played

Home Team	Away Team	Match Score	Expected Goals Score	Possession Breakdown (%)
Arsenal	Leeds United	2 – 1	2.75 – 0.58	64 – 36
Brentwood	Southampton	3 – 0	1.82 – 2.26	40 – 60
Brighton	Manchester United	4 – 0	2.04 – 1.73	43 – 57
Burnley	Aston Villa	1 – 3	1.77 – 1.57	49 – 51
Chelsea	Wolverhampton	2 – 2	2.28 – 1.49	59 – 41
Crystal Palace	Watford	1 – 0	2.04 – 0.82	68 – 32
Leicester City	Everton	1 – 2	2.23 – 1.20	66 – 34
Liverpool	Tottenham	1 – 1	2.45 – 1.05	66 – 34
Manchester City	Newcastle United	5 – 0	2.64 – 0.98	71 – 29
Norwich City	West Ham United	0 – 4	1.23 – 1.66	37 – 63

Source: https://footystats.org/ - accessed 12 May 2022.

It is also important to remember that *xG* scores can be misleading because the order in which goals are scored can not only influence the outcome of a match but also the expected goals. For example, if a team takes an early lead, they may then decide to sit back and defend their lead by packing the defence. In which case, two things are likely to happen: (i) the team will create fewer opportunities to score, which will be reflected in their *xG* tally; and (ii) for the remainder of the match, the opposition will try to generate more goal-scoring opportunities in pursuit of a comeback – something that is likely to result in an elevated *xG* score for the opposing team.

So why use expected goals? Well, the simple answer is that despite the obvious anomalies outlined above, overall there is a very strong correlation between *xG*'s and points earned, as illustrated in Example 5.7, which evaluates the correlations associated with GSR and *xG*'s for season 2020–2021.

EXAMPLE 5.7: CORRELATION BETWEEN GSR, *xG*'S, AND EOS POINTS TOTAL

In this example, we again utilise the 'perf_dat' data frame used in Example 5.6, but this time we focus on the 'GF', 'Poss', 'PassComp', and 'xG' variables.

First, we select the variables that we want to include in the analysis.

```
ind2 <- perf_dat[,c(1:5,12:14)]

# Now we add a new derived variable 'goal-to-shots ratio'.
ind2["GSR"] <- NA # Create empty vector to store goal ratio results

# Now we populate these new variables and round to 3 decimal places.
ind2$GSR <- round(perf_dat$GF/perf_dat$SF,3)
```

Now, we display the results.

```
print(ind2)
```

```
##
                Team Rank Pld Pts GF Poss PassComp   xG   GSR
1            Arsenal    8  38  61 55 53.8     83.6 53.5 0.121
2        Aston Villa   11  38  55 55 48.1     77.2 52.9 0.106
3           Brighton   16  38  41 40 51.3     80.2 51.6 0.084
4            Burnley   17  38  39 33 41.7     70.8 39.9 0.086
5            Chelsea    4  38  67 58 61.4     86.1 64.0 0.105
6     Crystal Palace   14  38  44 41 40.1     75.5 32.4 0.118
7            Everton   10  38  59 47 46.5     80.4 47.1 0.119
8             Fulham   18  38  28 27 49.9     80.5 41.3 0.061
9       Leeds United    9  38  59 62 57.6     79.8 57.5 0.118
10    Leicester City    5  38  66 68 54.6     81.3 56.0 0.144
11         Liverpool    3  38  69 68 62.4     84.3 72.6 0.113
12   Manchester City    1  38  86 83 63.9     88.3 73.3 0.141
13    Manchester Utd    2  38  74 73 55.8     84.2 60.2 0.141
14     Newcastle Utd   12  38  45 46 38.2     74.7 41.0 0.119
15     Sheffield Utd   20  38  23 20 41.5     76.1 31.4 0.063
16       Southampton   15  38  43 47 52.2     78.7 42.4 0.113
17         Tottenham    7  38  62 68 51.7     81.2 54.5 0.154
18         West Brom   19  38  26 35 37.6     71.4 33.8 0.104
19          West Ham    6  38  65 62 42.9     76.6 53.9 0.134
20            Wolves   13  38  45 36 49.3     81.6 39.9 0.078
```

And, we compute the Pearson correlation *r*-values.

```
cor.dat <- ind2[,c(4:9)]
ind2_cors <- round(cor(cor.dat),3)
print(ind2_cors)
```

```
##
               Pts    GF   Poss PassComp    xG   GSR
Pts       1.000 0.952 0.718    0.727 0.888 0.775
GF        0.952 1.000 0.692    0.645 0.875 0.861
Poss      0.718 0.692 1.000    0.903 0.873 0.307
PassComp  0.727 0.645 0.903    1.000 0.787 0.321
xG        0.888 0.875 0.873    0.787 1.000 0.538
GSR       0.775 0.861 0.307    0.321 0.538 1.000
```

From this, we see that although variation exists between individual matches, overall, *xG* is strongly correlated with both points earned (r=0.888) and goals scored (r=0.875). This indicates that the *xG* model has the potential to act as a good predictor of both league position and match outcome, as some commentators have suggested [13]. Also, it is noticeable that possession, pass completion, and GSR are all strongly positively correlated with points earned, which is unsurprising given that teams generally need good possession in order to make more attempts on goal and thus score goals. Having said this, it is noticeable that in season 2020–2021, GSR exhibited a stronger correlation with points earned (r=0.775) than either possession (r=0.718) or pass completion (r=0.727), suggesting that the quality of the teams' shooting was an influential factor in determining the league position of the respective teams.

An alternative way to produce a table of Pearson correlation values is to use the 'rcorr' function in the 'Hmisc' package, as follows. This produces both a table of the r-values and a table of the associated significance levels (i.e., p-values), which can be very useful when performing statistical analysis.

```
# Alternative method using the 'Hmisc' package
library("Hmisc")
cor.res <- rcorr(as.matrix(cor.dat))
cor.res
```

```
##
           Pts   GF Poss PassComp   xG  GSR
Pts       1.00 0.95 0.72     0.73 0.89 0.77
GF        0.95 1.00 0.69     0.64 0.87 0.86
Poss      0.72 0.69 1.00     0.90 0.87 0.31
PassComp  0.73 0.64 0.90     1.00 0.79 0.32
xG        0.89 0.87 0.87     0.79 1.00 0.54
GSR       0.77 0.86 0.31     0.32 0.54 1.00

n= 20

P
          Pts    GF     Poss   PassComp xG     GSR
Pts              0.0000 0.0004 0.0003   0.0000 0.0000
GF        0.0000        0.0007 0.0021   0.0000 0.0000
```

```
Poss     0.0004 0.0007        0.0000   0.0000 0.1876
PassComp 0.0003 0.0021 0.0000          0.0000 0.1669
xG       0.0000 0.0000 0.0000 0.0000          0.0144
GSR      0.0000 0.0000 0.1876 0.1669   0.0144
```

From this, we see that all the correlations, except those between 'Poss' and 'GSR', and 'PassComp' and 'GSR', are statistically significant (i.e., $p<0.05$), with most being strongly significant (i.e., $p<0.001$).

There are many ways to visualise correlation matrices in R, but here we will content ourselves with just two of the more useful techniques, involving the packages 'ggcorrplot' and 'PerformanceAnalytics'. These can be installed as follows:

```
install.packages("ggcorrplot") # This installs the 'ggcorrplot' package.
install.packages("PerformanceAnalytics") # This installs the 'PerformanceAnalytics' package.
# NB. These commands only needs to be executed once to install the package.
# Thereafter, the libraries can be called using the 'library' command:
```

The first of these visualisation methods uses the 'ggcorrplot' package, as follows, and produces the plot displayed in Figure 5.8, in which the respective r-values are shown in the boxes in the lower triangle.

```
# Produce correlation plot using 'ggcorrplot' package
library(ggcorrplot)
ggcorrplot(as.matrix(ind2_cors), type = "lower", ggtheme = theme_bw, lab = TRUE)
```

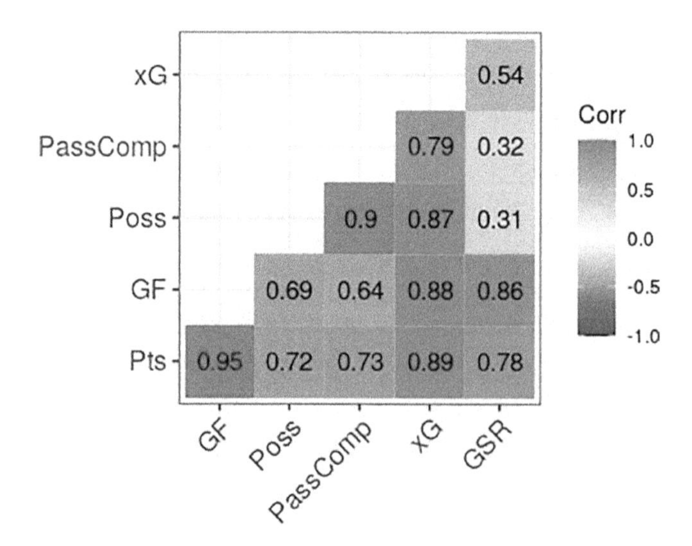

FIGURE 5.8
Multiple correlation plot displaying relevant r-values (produced using the 'ggcorrplot' package).

The second method uses the 'PerformanceAnalytics' package to produce the plot in Figure 5.9, which shows the scatter plots, histograms, and *r*-values associated with the various relationships in the data. This much more advanced plot can be very useful when analysing complex data sets, although the graphic produced may be overly complicated and rather confusing to some people.

```
# Produce advanced correlation plot using 'PerformanceAnalytics' package
library("PerformanceAnalytics")
chart.Correlation(cor.dat, histogram=TRUE, pch=19)
```

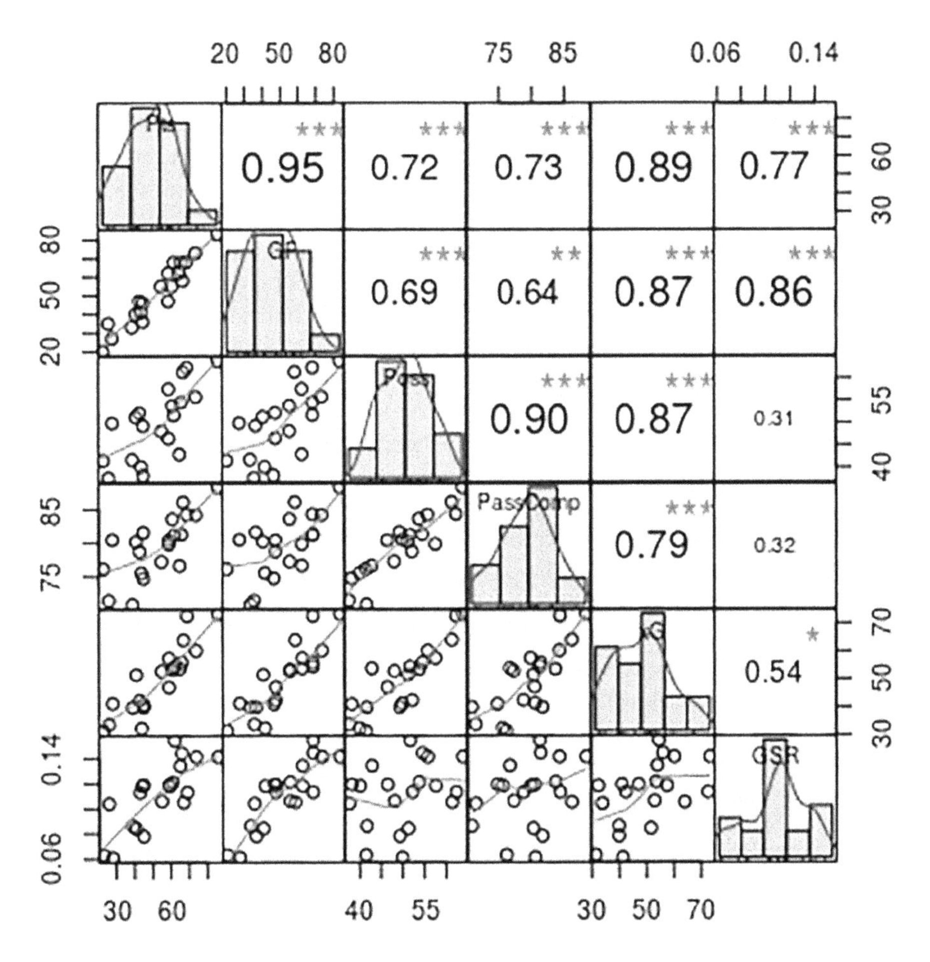

FIGURE 5.9
Multiple correlation plot displaying relevant scatter plots, histograms, and *r*-values (produced using the 'PerformanceAnalytics' package).

In Figure 5.9, the distribution of each variable is shown along the diagonal as a series of histograms, while the lower triangle shows the respective bivariate scatter plots with fitted lines. In the upper triangle, the *r*-values of the respective correlations are displayed, together with the significance levels expressed in stars (*p*-values (0, 0.001, 0.01, 0.05, 0.1, 1) are denoted by the symbols ("***", "**", "*", ".", " "), respectively).

5.8 Concluding Remarks

In this chapter, we have shown that relatively early in the season (i.e., after about 10 rounds), it is possible to predict with reasonable accuracy the likely EoS points total that will be achieved by most teams in the EPL. As such, the techniques presented in this chapter should be of great interest to managers, club owners, fans, pundits, and gamblers alike, all of whom want to know as early as possible the likely outcome of the season for their respective teams. Given the huge financial implications associated with promotion, relegation, and playing in Europe, it is not surprising that club owners, in particular, need to know how the season is likely to unfold for their team. They require accurate predictions so that, where applicable, appropriate remedial action (e.g., sacking managers, buying new players, etc.) can be taken as soon as possible. It is much better to act quickly and early rather than leave a problem until it is too late. With respect to this, the techniques presented in this chapter should be of great help. However, it is important to note that this chapter is only an introduction to the subject. Those readers who want to know more are advised to experiment and refine (tune) the models presented here so that more accurate predictions can be produced.

In addition to predicting EoS league position, in this chapter we have introduced various indicator metrics that can be used to predict league and match outcomes. While these can be very useful for analysing team performance, it is important to note that they are often strongly correlated with each other, which means that from a data science point of view, they generally contain similar information. For example, teams that make lots of shots tend to make more shots on target and, unsurprisingly, score more goals. So, indicators such as GSR, TSR, TSoTR, and *xG* that incorporate these metrics are likely to be strongly correlated, as illustrated in Examples 5.6 and 5.7. This means that we should be cautious about the merits of these and other performance indicators, no matter how much commentators on the Internet sing their praises. This is because pundits and commentators are often fond of promoting their favourite indicators as the best tools with which to predict match and league outcomes; however, because such indicators are often strongly correlated, the benefits of one indicator over another are likely to be marginal rather than transformational. The truth is that many of the performance indicators commonly used in soccer tell a similar story. Furthermore, they can only go so far because, as Spiegelhalter and Pearson demonstrated [4], chance still plays a major role in deciding the outcome of matches and leagues. Therefore, when using indicators to analyse match performance or make predictions, it is important to be aware of these issues.

References

1. McCourt I: Parma relegated to Serie D after failing to find a new owner. In: *The Guardian*; 22nd June 2015. https://www.theguardian.com/football/2015/jun/22/parma-relegated-serie-d-fail-new-owner
2. Davis C: Premier League 2017 prize money: How much your club is in line to earn this season? In: *The Telegraph*. London; 17th May 2017. https://www.telegraph.co.uk/football/2017/05/16/premier-league-2017-prize-money-much-club-line-earn-season/
3. Beggs CB, Bond AJ, Emmonds S, Jones B: Hidden dynamics of soccer leagues: The predictive 'power' of partial standings. *PLoS One* 2019, 14(12):e0225696.
4. Spiegelhalter D, Pearson M: Understanding uncertainty: The Premier League. In: *Plus Maths Magazine*. https://plus.maths.org/content/understanding-uncertainty-premier-league; 2008.
5. Shin S, Ahnert SE, Park J: Ranking competitors using degree-neutralized random walks. *PLoS One* 2014, 9(12):e113685.
6. Miller SJ: A derivation of the Pythagorean won-loss formula in baseball. *Chance* 2007, 20(1):40–48.
7. Heumann J: An improvement to the baseball statistic "Pythagorean Wins". *Journal of Sports Analytics* 2016, 2(1):49–59.
8. Hamilton HH: An extension of the Pythagorean expectation for association football. *Journal of Quantitative Analysis in Sports* 2011, 7(2):15.
9. Eastwood M: Applying the Pythagorean expectation to football: Part two. In: *Pena.lt/y*; 2012. https://pena.lt/y/2012/12/03/applying-the-pythagorean-expectation-to-football-part-two/
10. Kingsman T: Pythagorean expectation in football. In; 2016. https://tobykingsman.wordpress.com/2016/02/01/pythagorean-expectation-in-football/
11. Joslyn LR, Joslyn NJ, Joslyn MR: What delivers an improved season in men's college soccer? The relative effects of shots, attacking and defending scoring efficiency on year-to-year change in season win percentage. *The Sport Journal* 2017:1–12. https://www.statsperform.com/resource/assessing-the-performance-of-premier-league-goalscorers/
12. Green S: Assessing the performance of premier league goalscorers. In: *statsperform.com*; 2012. https://www.statsperform.com/resource/assessing-the-performance-of-premier-league-goalscorers/
13. Tippett J: The expected goals philosophy: A game-changing way of analysing football. Independently published 2019.
14. Rathke A: An examination of expected goals and shot efficiency in soccer. *Journal of Human Sport and Exercise* 2017, 12(2):514–529.

6

Predicting Soccer Match Outcomes

In Chapter 5, we saw how historical match data can be used to predict end-of-season (EoS) league positions with reasonable accuracy. In this chapter, we turn our attention to predicting the outcome of individual league matches using historical data. This is an issue that is of great interest to many, whether they be fans, players, or managers, with countless words devoted in newspapers and on websites to which team is most likely to win in forthcoming matches. Indeed, predicting the outcome of soccer matches and setting accurate odds are the lifeblood of the sports betting industry. However, many are not aware of the various data science techniques that can be used to make match predictions. So, in this chapter, we will explore some of the more popular techniques that can be used. As such, this chapter should be of interest to many readers.

6.1 Match Prediction and Betting

Although soccer match predictions are of great interest to the betting industry, with gamblers often using them to place bets, prediction and betting are actually two separate and distinct processes. Match prediction is simply the process of saying beforehand whether a particular soccer match will end either in a home win, a draw, or an away win, whereas gambling is a secondary activity that may or may not take place based on match predictions. So while wise gamblers will always be interested in accurate match predictions, they may only act on them if the odds offered by the bookmakers have enough value to make a bet worthwhile (see Chapter 7 for more details). Also, fans and managers who do not gamble may be interested in accurate predictions of match outcomes, as these will help to moderate their expectations. So in this chapter, we will focus purely on making match predictions and leave any discussion of betting to Chapter 7. Furthermore, because the techniques presented in this chapter rely on historical match data, we will focus solely on match prediction in a league context rather than the prediction of match outcomes in knockout cup competitions, which is covered in Chapter 9.

DOI: 10.1201/9781003328568-6

6.2 Soccer Match Prediction

Accurately predicting the outcome of individual soccer matches can be very difficult. Even the bookmakers, who are the experts in setting match odds, only get it 'right' about 50%–60% of the time. This is highlighted in Table 6.1, which shows the fraction of English Premier League (EPL) matches where the predicted favourite (i.e., the team with the shortest odds) actually won in the ten seasons from 2012–2013 to 2021–2022. From this, it can be seen that the favourite, as predicted by Pinnacle (a bookmaker with a good reputation for setting accurate odds), only won on average 55% of the matches played. While this might not appear to be a high percentage, it should be remembered that a purely random selection of teams would yield on average a prediction that was only 33% accurate, as illustrated in Table 6.1. So, overall, Pinnacle's forecasts were much better than random. Having said this, while accurately predicting the outcome of all soccer matches is extremely difficult, predicting the outcome of some matches can be extremely easy. For example, if we predicted that Manchester City (EPL champions in seasons 2011–2012, 2013–2014, 2017–2018, 2018–2019, 2020–2021, and 2021–2022) would win all their matches, we would on average be correct 76.3% of the time for home matches and 61.6% of the time for away matches (see Table 6.1).

TABLE 6.1

Fraction of Matches in Various Seasons Where the Outcome is Correctly Predicted by Random Selection, the Pinnacle Favourite, and the Selection of a Single Strong Team (i.e., Manchester City)

Season	Random Selection	Pinnacle Favourite	Manchester City (Home)	Manchester City (Away)
2012/2013	0.355	0.529	0.737	0.474
2013/2014	0.305	0.600	0.895	0.526
2014/2015	0.318	0.532	0.737	0.526
2015/2016	0.363	0.476	0.632	0.368
2016/2017	0.297	0.608	0.579	0.632
2017/2018	0.337	0.550	0.842	0.842
2018/2019	0.345	0.584	0.947	0.737
2019/2020	0.313	0.526	0.789	0.579
2020/2021	0.345	0.513	0.684	0.737
2021/2022	0.300	0.579	0.789	0.737
Mean	**0.328**	**0.550**	**0.763**	**0.616**

Unlike some other field sports, soccer has some peculiarities that make the accurate prediction of match outcomes difficult. Foremost amongst these is the fact that in soccer leagues many matches are drawn, with, for example, about 25% (on average) of EPL matches ending in a draw (see Chapter 5). This is in contrast to sports like rugby and basketball, where draws are very infrequent. This means that when making a prediction for a league soccer match, there are three possible outcomes to choose from instead of effectively two for sports such as rugby. So although home wins (probability $\approx$ 0.46) and away wins (probability $\approx$ 0.29) are more likely outcomes, approximately one in four matches are still drawn, which complicates things and reduces the likelihood of a correct prediction.

Another thing that makes prediction of matches difficult is that soccer is an inherently low-scoring sport, with a median of 3 goals scored per EPL match. This means that chance and luck play a much bigger role in deciding the outcome of a soccer match than, say, in a rugby game where more than 30 points are regularly scored by some teams. This means that even though a stronger team might dominate for most of a match, they can still be vulnerable to the weaker team scoring a late goal against the run of play and achieving a draw, or perhaps even inflicting a defeat. So in soccer, unlike some other sports, the stronger side does not always win, which makes the prediction of match outcomes difficult.

Finally, there are a plethora of other factors, such as home advantage, injuries to key players, league position, and even the weather, which can affect the outcome of soccer matches. However, although these factors often affect the outcome of matches, they can be difficult to quantify and assess. As such, this means that it is impossible to be 100% sure about the outcome of any match beforehand. Consequently, when predicting the outcome of matches, it is often helpful to take a probabilistic approach, estimating the likelihood of various outcomes occurring rather than trying to predict any clear-cut outcome. Indeed, as we will see, a probabilistic approach is favoured by some of the well-known techniques that are often employed to predict soccer matches.

Broadly speaking, match prediction techniques can be categorised as belonging to one of three categories [1]:

- Ranking systems, which attempt to rate the true strengths of the respective teams before any match, with the outcome predicted according to a prescribed set of rules.
- Poisson distribution models that predict the expected number of goals that will be scored by the home and away teams for any given match based on their historical performance.
- Machine learning systems that use advanced data science techniques such as random forests to make match predictions based on historical data.

Of these, ranking systems are often used to identify value in potential bets or in cup competitions or international fixtures where the traditional league structure does not apply. As such, these techniques will not be covered here but will be discussed in Chapter 9. The other two classes of technique will, however, be discussed in depth in this chapter. Notwithstanding this, it should be noted that many match prediction models exist (most of which are just variants on a basic theme), far too many to include in a single chapter. Therefore, we will content ourselves here with some basic models that are popular with sports analysts.

6.3 The Poisson Distribution of Goals Scored

For many years, it has been known that the goals scored by the home and away teams in soccer matches tend to conform to a Poisson distribution [1–3], with most games being low-scoring affairs and only about 5% involving more than five goals. In statistics, a Poisson distribution is a discrete probability distribution that expresses the probability of a given number of events occurring in a fixed interval of time. The use of the term 'discrete' here indicates that the probability distribution involves individually countable outcomes that occur at fixed intervals (i.e., 1, 2, 3, etc.), which of course is how goals are counted. In other words, a Poisson distribution does not allow us to have 1.5 goals. Importantly, the distribution assumes that each event occurs independently of other events. So in the case of a soccer match, this means that each goal can be considered an independent event, with the scoring of one goal not thought to change the probability of when the next one will arrive. While this assumption generally holds true, it has been shown that it is not strictly correct in all circumstances, with, in particular, marked deviations from Poisson expectation often observed during the last 5–10 minutes of matches when the away team is leading [4]. As a result, many attempts have been made to extend the basic Poisson model to better reflect goal-scoring behaviour during soccer matches [5–8]. However, discussion of these advanced extension techniques is beyond the scope of this chapter, which is primarily intended to be an introductory text.

Unlike a Gaussian (normal) distribution, which is bell-shaped, Poisson distributions tend to be skewed, with most events occurring towards the left-hand side (see Figure 6.1). However, their shape can vary, with the peak occurring in different places depending on the expected value, λ (lambda), which in the case of a soccer match is the expected number of goals that a team is likely to score based on its performance in previous matches. So, if in a given soccer match the home team had an expected goal score of 2.5 and the away team an expected score of 1.5, we would see that the Poisson distribution for the home team would be shifted more to the right compared with the away team, as illustrated in Figure 6.1. Effectively, this means that although we cannot be certain of the outcome of any specific match, it is more likely that the home team will score more goals than the away team.

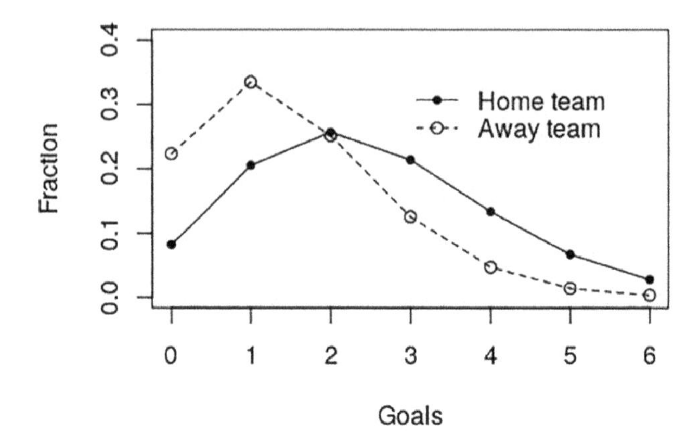

FIGURE 6.1
Poisson distributions of the goals scored per match by the home and away teams, assuming expected values λ, of 2.5 and 1.5, respectively.

The idea that the goals scored in soccer matches conform to a Poisson distribution can be extremely helpful and, as such, forms the basis of several match prediction models [1,3]. However, before we explore these models in detail, it is perhaps worth taking time to explore the extent to which goals scored in the EPL conform to a Poisson distribution. In Example 6.1, we do this using historical EPL match data from season 2018–19, which was the last uninterrupted season before the COVID pandemic.

EXAMPLE 6.1: POISSON DISTRIBUTION AND GOALS SCORED

In this example, we use match data from the 2018–2019 season to investigate the frequency distribution of goals scored in the EPL.

First, we clear any existing data from the workspace.

```
rm(list = ls())   # Clears all variables from the workspace
```

Now we import historical match data from the Internet.

```
mydata <- head(read.csv('https://www.football-data.co.uk/mmz4281/1819/E0.csv'),380)
```

Next, we inspect data.

```
names(mydata)
```

```
##
 [1] "Div"         "Date"        "HomeTeam"    "AwayTeam"    "FTHG"
 [6] "FTAG"        "FTR"         "HTHG"        "HTAG"        "HTR"
[11] "Referee"     "HS"          "AS"          "HST"         "AST"
[16] "HF"          "AF"          "HC"          "AC"          "HY"
[21] "AY"          "HR"          "AR"          "B365H"       "B365D"
[26] "B365A"       "BWH"         "BWD"         "BWA"         "IWH"
[31] "IWD"         "IWA"         "PSH"         "PSD"         "PSA"
[36] "WHH"         "WHD"         "WHA"         "VCH"         "VCD"
[41] "VCA"         "Bb1X2"       "BbMxH"       "BbAvH"       "BbMxD"
[46] "BbAvD"       "BbMxA"       "BbAvA"       "BbOU"        "BbMx.2.5"
[51] "BbAv.2.5"    "BbMx.2.5.1"  "BbAv.2.5.1"  "BbAH"        "BbAHh"
[56] "BbMxAHH"     "BbAvAHH"     "BbMxAHA"     "BbAvAHA"     "PSCH"
[61] "PSCD"        "PSCA"
```

This reveals that the data set contains lots of variables that are not of interest to us. So we shall produce a reduced data set that contains only a few selected variables that we will include in our analysis. In this case, we shall select the following variables for inclusion in a working data frame called 'dat', which we will use for all the examples in this chapter.

- Date = Match date (dd/mm/yy)
- HomeTeam = Home team
- AwayTeam = Away team
- FTHG = Full time home team goals
- FTAG = Full time away team goals
- FTR = Full time result (H = Home Win, D = Draw, A = Away Win)
- PSH = Decimal odds for a home win as computed by Pinnacle
- PSD = Decimal odds for a draw as computed by Pinnacle
- PSA = Decimal odds for a away win as computed by Pinnacle

```
dat <- mydata[,c("Date","HomeTeam","AwayTeam","FTHG","FTAG","FTR","PSH","PSD","PSA")]
names(dat)
```

```
##
[1] "Date" "HomeTeam" "AwayTeam" "FTHG" "FTAG" "FTR" "PSH" "PSD" "PSA"
```

For ease of reference, we will now rename some of the variables so that they are more easily recognisable.

```
# Rename column names
colnames(dat)[colnames(dat) == 'FTHG'] <- 'Hgoals'
colnames(dat)[colnames(dat) == 'FTAG'] <- 'Agoals'
colnames(dat)[colnames(dat) == 'FTR'] <- 'Result'
colnames(dat)[colnames(dat) == 'PSH'] <- 'HWodds'
colnames(dat)[colnames(dat) == 'PSD'] <- 'Dodds'
colnames(dat)[colnames(dat) == 'PSA'] <- 'AWodds'

# Display the first six rows of dat
head(dat)
```

```
##
        Date     HomeTeam       AwayTeam Hgoals Agoals Result HWodds Dodds
1 10/08/2018   Man United      Leicester      2      1      H   1.58  3.93
2 11/08/2018  Bournemouth        Cardiff      2      0      H   1.89  3.63
3 11/08/2018       Fulham Crystal Palace      0      2      A   2.50  3.46
4 11/08/2018 Huddersfield        Chelsea      0      3      A   6.41  4.02
5 11/08/2018    Newcastle      Tottenham      1      2      A   3.83  3.57
6 11/08/2018      Watford       Brighton      2      0      H   2.43  3.22
  AWodds
1   7.50
2   4.58
3   3.00
4   1.62
5   2.08
6   3.33
```

Now we can produce a table of home and away goal frequencies.

```
n <- nrow(dat)
location <- rep(c("Home", "Away"), each = n)
goals <- c(dat$Hgoals, dat$Agoals)
goal.dat <- cbind.data.frame(location, goals)
table(goal.dat)
```

```
##
        goals
location   0   1   2   3   4   5   6
    Away 119 122  87  36   9   6   1
    Home  88 116  95  48  22   8   3
```

From this, we see that when teams are playing away from home, they tend to score zero goals more often than when they are playing at home. Also, when playing at home, they are more likely to score two or more goals compared with when they are playing away. Clearly, for whatever reason, 'home advantage' is a very real phenomenon in the EPL.

We can produce a frequency plot of goals scored by the home and away teams using the following code:

```
# Plot home and away goal frequencies.
hg.freq <- table(dat$Hgoals)
ag.freq <- table(dat$Agoals)
hgoals.frac <- hg.freq/n
agoals.frac <- ag.freq/n

# Combined plot
yy <- rbind(hgoals.frac, agoals.frac)
barplot(yy, beside=TRUE,col=c("darkgray","lightgray"), names.arg=c(0:6),ylim=c(0,0.4),
axis.lty=1, xlab="Goals scored", ylab="Fraction")
legend(12,0.35, c("Home team","Away team"), col=c("darkgray","lightgray" ), pch=c(15,15),
bty="n")
```

This produces the plot in Figure 6.2.

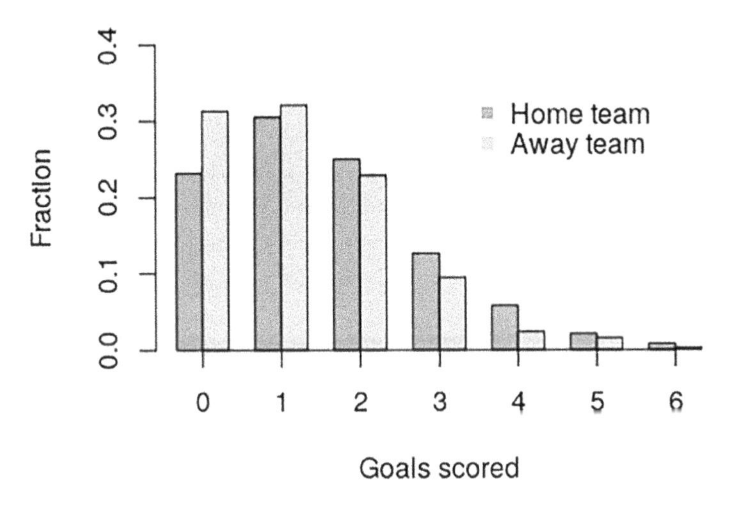

FIGURE 6.2
Frequency plot of the goals scored per match by the home and away teams in the English Premier League during the 2018–2019 season.

Having produced the frequency plot, we now produce the respective Poisson distributions based on the mean home and away goals scored, which we compute as follows:

```
# Compute and display the expected goals scored.
hgoals.mean <- mean(dat$Hgoals)
agoals.mean <- mean(dat$Agoals)

print(hgoals.mean)
```

```
## [1] 1.568421
```

```
print(agoals.mean)
```

```
## [1] 1.252632
```

Poisson distributions can be plotted using the Poisson probability mass function 'dpois' in R. This calculates the Poisson density distribution curve based on the mean lambda value for a specified series. In the following code, we make the series 0–6 goals and lambda equal 'hgoals.mean' for the home team and 'agoals.mean' for the away team.

```
# Plot Poisson distributions
scale <- c(0:6)
plot(scale, dpois(x=0:6, lambda=hgoals.mean), type="o", lty=1, pch=20, ylim=c(0,0.4),
xlab="Goals", ylab="Fraction")
lines(scale, dpois(x=0:6, lambda=agoals.mean), type="o", lty=2, pch=21)
legend(3,0.35, c("Home team","Away team" ), lty=c(1,2), pch=c(20,21), bty = "n")
```

This produces Figure 6.3.

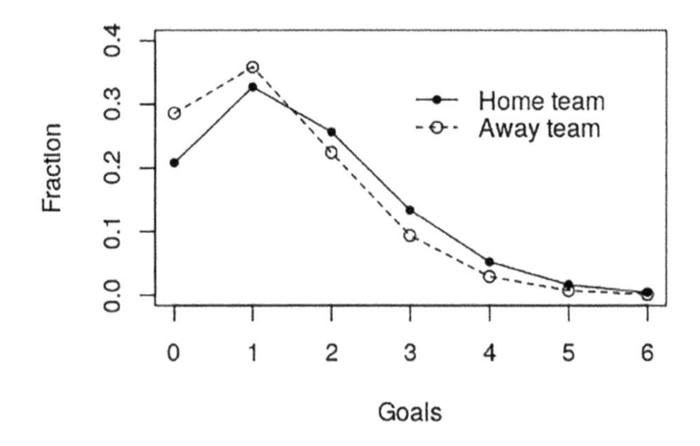

FIGURE 6.3
Poisson distributions of the goals scored per match by the home and away teams in the English Premier League during the 2018–2019 season.

From Figure 6.3, it can be seen that the Poisson distributions closely resemble the frequency distributions of the respective home and away goals scored. The closeness of the relationship can be quantified by computing the respective correlation *r*-values, as follows.

```
# Correlation with home team goals
cor(hgoals.frac, dpois(x=0:6, lambda=hgoals.mean))
```

```
## [1] 0.9945666  # r-value
```

```
# Correlation with away team goals
cor(agoals.frac, dpois(x=0:6, lambda=agoals.mean))
```

```
## [1] 0.9912843  # r-value
```

From this, we see that although the correlations are both very strong (i.e., >0.99), the one for away goals is slightly weaker than its home goals counterpart.

6.4 Poisson Regression Prediction Model

Because the goals scored during soccer matches are approximately Poisson distributed, it is possible to use a Poisson regression model, which is a type of generalised linear model (GLM), to compute expected home and away goals for each team, as illustrated in Example 6.2. Once the home and away teams expected goals have been calculated, the probabilities of the various match score lines can then be computed using the Poisson distributions for the respective teams.

GLMs are an extension of the linear regression model introduced in Chapter 1 (see Key Concept Box 1.2). They are characterised by a link function that links a set of linear predictors with a response variable. In the case of the Poisson regression model, it is assumed that the response variable, y, has a Poisson distribution and that the link function is the natural log (ln) of y. Using a log link function ensures that the response variable, y, must be positive, which is important when predicting goals in football matches. As such, the Poisson regression model has the general form:

$$\ln(y) = b_0 + b_1x_1 + b_2x_2 + \ldots \quad (6.1)$$

where y is the response variable; x_1, x_2, ... are the respective predictor variables; and b_0, b_1, b_2, ... are the linear coefficients applied to the respective predictor variables.

The predicted values of the response variable, y, can then be computed using the exponential of the linear expression, as follows:

$$y = e^{(b_0+b_1x_1+b_2x_2+\ldots)} = e^{b_0} + e^{b_1x_1} + e^{b_2x_2} + \ldots \quad (6.2)$$

In the context of soccer, we generally use Poisson regression models to predict the expected number of goals that the home and away teams will score in any given match based on their historical performance. This can then be used to predict the respective probabilities of a home win, an away win, or a draw occurring, which is why bookmakers often use such models when setting match odds.

Poisson regression match prediction models have been widely used to compute the likely outcome of individual soccer matches [1–3]. While not generally considered to be the most accurate of prediction models since they tend to underestimate the likelihood of low-scoring draws occurring, they are nonetheless popular and form the basis of many more sophisticated models. In Example 6.2, we use the first 370 matches from one EPL season to build a Poisson regression model, which we will then use to predict the outcomes of the remaining ten matches in that season.

EXAMPLE 6.2: POISSON REGRESSION MODEL EXAMPLE

In this example, we use EPL match data from the first 370 matches in season 2018–2019 to predict the outcome of the remaining ten matches in that season. To do this, we will use the 'dat' data frame that we created in Example 6.1, which contains the historical match data.

```
# Select data to build model.
names(dat)
```

```
##
[1] "Date"      "HomeTeam" "AwayTeam" "Hgoals"    "Agoals"    "Result"
[7] "HWodds"    "Dodds"     "AWodds"
```

Now we select only the first 370 rows and columns two to five and call this data frame 'build.dat'.

```
build.dat <- dat[1:370,2:5] # Select the first 370 matches and variables 2-5 for analysis.
```

In addition, for convenience we will rename some of the variables.

```
# Rename variables
names(build.dat)[names(build.dat) == 'HomeTeam'] <- 'Home'
names(build.dat)[names(build.dat) == 'AwayTeam'] <- 'Away'
names(build.dat)
```

```
##
[1] "Home"    "Away"    "Hgoals" "Agoals"
```

When building generalised linear models (GLMs), it is necessary to first transform the data from 'wide-form' to 'long-form' format, with all the home matches listed first on top of the away matches. This can be done as follows:

```
# Put data into long-form
long_dat <- rbind(
  data.frame(Home = 1,
             Team = build.dat$Home,
             Opponent = build.dat$Away,
             Goals = build.dat$Hgoals),
  data.frame(Home = 0,
             Team = build.dat$Away,
             Opponent = build.dat$Home,
             Goals = build.dat$Agoals))
```

This produces a date frame called 'long_dat', which has 740 rows and four columns (variables) called 'Home', 'Team', 'Opponent', and 'Goals'. For a home match, Home = 1, and for an away match, Home = 0. We can inspect 'long_dat' as follows:

```
head(long_dat)
```

```
##
  Home          Team        Opponent Goals
1    1    Man United       Leicester     2
2    1   Bournemouth         Cardiff     2
3    1        Fulham Crystal Palace     0
4    1 Huddersfield         Chelsea     0
5    1     Newcastle       Tottenham     1
6    1       Watford        Brighton     2
```

```
tail(long_dat)
```

```
##
    Home         Team     Opponent Goals
735    0 Southampton     West Ham     0
736    0      Fulham       Wolves     0
737    0    Brighton      Arsenal     1
738    0     Watford      Chelsea     0
739    0  Man United Huddersfield     1
740    0   Leicester     Man City     0
```

Now that we have the data in the correct format, we can build the Poisson regression model using the 'glm' function in R, as follows: Here the response variable is 'Goals' and the predictor variables are 'Home', 'Team', and 'Opponent'. The argument "data = long_dat" is used to specify that the model should be applied to the 'long_dat' data frame. The "family = poisson(link=log)" argument tells R that the GLM to be used is a Poisson model with a natural log link function.

```
# Build Poisson model
pois.mod <- glm(Goals ~ Home + Team + Opponent, family=poisson(link=log),
                     data=long_dat)
summary(pois.mod)
```

Which displays the following:

```
##
Call:
glm(formula=Goals~Home+Team+Opponent, family=poisson(link=log),
    data=long_dat)

Deviance Residuals:
     Min        1Q    Median        3Q       Max
-2.17511  -1.02684  -0.05616   0.48462   2.98913

Coefficients:
                       Estimate Std. Error z value Pr(>|z|)
(Intercept)             0.49255    0.19153   2.572 0.010120 *
Home                    0.25265    0.06268   4.031 5.56e-05 ***
TeamBournemouth        -0.27027    0.18251  -1.481 0.138637
TeamBrighton           -0.73344    0.20938  -3.503 0.000460 ***
TeamBurnley            -0.44926    0.19248  -2.334 0.019596 *
TeamCardiff            -0.77017    0.21376  -3.603 0.000315 ***
TeamChelsea            -0.12488    0.17402  -0.718 0.472976
TeamCrystal Palace     -0.41252    0.19021  -2.169 0.030099 *
TeamEverton            -0.31651    0.18342  -1.726 0.084416 .
TeamFulham             -0.69864    0.20946  -3.335 0.000852 ***
TeamHuddersfield       -1.18201    0.24914  -4.744 2.09e-06 ***
TeamLeicester          -0.32612    0.18448  -1.768 0.077101 .
TeamLiverpool           0.18490    0.16091   1.149 0.250511
TeamMan City            0.22907    0.15931   1.438 0.150462
TeamMan United         -0.06212    0.17271  -0.360 0.719084
TeamNewcastle          -0.61039    0.20184  -3.024 0.002493 **
TeamSouthampton        -0.43869    0.19284  -2.275 0.022912 *
TeamTottenham          -0.09173    0.17263  -0.531 0.595136
TeamWatford            -0.31059    0.18452  -1.683 0.092338 .
TeamWest Ham           -0.37878    0.18778  -2.017 0.043679 *
TeamWolves             -0.42390    0.18891  -2.244 0.024834 *
OpponentBournemouth     0.24578    0.18857   1.303 0.192441
OpponentBrighton        0.08898    0.19497   0.456 0.648101
OpponentBurnley         0.24133    0.18821   1.282 0.199748
OpponentCardiff         0.29496    0.18614   1.585 0.113057
OpponentChelsea        -0.25371    0.21407  -1.185 0.235938
OpponentCrystal Palace -0.02685    0.20040  -0.134 0.893410
OpponentEverton        -0.13655    0.20713  -0.659 0.509733
OpponentFulham          0.38609    0.18201   2.121 0.033901 *
OpponentHuddersfield    0.35513    0.18294   1.941 0.052224 .
OpponentLeicester      -0.05948    0.20249  -0.294 0.768947
OpponentLiverpool      -0.81119    0.25624  -3.166 0.001547 **
OpponentMan City       -0.80828    0.25625  -3.154 0.001609 **
OpponentMan United      0.02311    0.19852   0.116 0.907343
OpponentNewcastle      -0.08018    0.20244  -0.396 0.692067
OpponentSouthampton     0.20434    0.18914   1.080 0.279984
OpponentTottenham      -0.31066    0.21727  -1.430 0.152772
OpponentWatford         0.07114    0.19582   0.363 0.716397
OpponentWest Ham        0.05754    0.19669   0.293 0.769854
OpponentWolves         -0.12989    0.20714  -0.627 0.530607
```

```
---
Signif. codes:  0 '***' 0.001 '**' 0.01 '*' 0.05 '.' 0.1 ' ' 1

(Dispersion parameter for poisson family taken to be 1)

    Null deviance: 942.63  on 739  degrees of freedom
Residual deviance: 732.30  on 700  degrees of freedom
AIC: 2140.2

Number of Fisher Scoring iterations: 5
```

There is a lot of information here, but from the point of view of the model, the important metric that we want to use is the 'Estimate' value, which is the coefficient that should be applied to each team when either playing at home or away. The 'Estimate' is actually the logarithm of the expectation, and therefore we need to exponentiate the coefficients to get interpretable values. Therefore, we can use the intercept estimate value of 0.493 to compute the overall mean, which is $e^{0.493} = 1.637$, whereas the overall home advantage is $e^{0.253} = 1.288$.

Now, before going any further, it is perhaps worth reminding ourselves of the ten remaining matches to be played, the results of which we are trying to predict. This can be done as follows:

```
remain_matches <- dat[371:380,1:5]
print(remain_matches)
```

```
##
          Date        HomeTeam      AwayTeam Hgoals Agoals
371 12/05/2019        Brighton      Man City      1      4
372 12/05/2019         Burnley       Arsenal      1      3
373 12/05/2019 Crystal Palace   Bournemouth      5      3
374 12/05/2019          Fulham     Newcastle      0      4
375 12/05/2019       Leicester       Chelsea      0      0
376 12/05/2019       Liverpool        Wolves      2      0
377 12/05/2019      Man United       Cardiff      0      2
378 12/05/2019     Southampton  Huddersfield      1      1
379 12/05/2019       Tottenham       Everton      2      2
380 12/05/2019         Watford      West Ham      1      4
```

NB. Of course, if we were making predictions in real-time, we would not know the outcome scores of the matches beforehand. But for teaching purposes, we have included them here.

Now we can use "the model in anger", to try and predict the outcomes of the outstanding matches. So let's say we want to use the Poisson regression model to predict the result of the match between Burnley, at home, and Arsenal. We can do this by using the 'predict' function in R, as follows: This plugs the appropriate home and away estimate values for the respective clubs into the Poisson regression model to compute the expected goals for the match. However, we need to do this twice: once to predict the number of goals Burnley is expected to score and a second time for Arsenal.

```
# Apply Poisson model to predict match result
# Specify teams
HTeam <- "Burnley"
ATeam <- "Arsenal"

# Predict expected (average) goals.
# Home Team
Hgoals.exp <- predict(pois.mod, data.frame(Home=1, Team=HTeam, Opponent=ATeam),
type="response")
print(Hgoals.exp)
```

```
##
       1
1.344383
```

```
# Away Team
Agoals.exp <- predict(pois.mod, data.frame(Home=0, Team=ATeam, Opponent=HTeam),
type="response")
print(Agoals.exp)
```

```
##
       1
2.083151
```

This tells us that the model expects Burnley to score only 1.34 goals at home, while Arsenal is expected to score 2.08 goals away from home in this match. As such, the model predicts that Arsenal should win the match despite Burnley having the home advantage. This is perhaps not surprising since Arsenal were 5th in the league table going into this match, while Burnley were only in 15th place. In the real-life match, Arsenal actually won, scoring three goals to Burnley's one, so the prediction turned out to be correct.

While calculating the expected goals for the respective teams is useful, it has, of course, some limitations. Chief among these is the fact that teams can't score 1.34 or 2.08 goals. Goals must be integer numbers (i.e., 0, 1, 2, 3, …). So what do we do if the predicted expected goals turn out to be, say, 1.82 and 2.41? Would this be a 2–2 draw or some other result? Well, this is where the Poisson distribution introduced in Example 6.1 becomes very helpful, because we can use it to compute the likelihood of various match score lines occurring. However, to do this, we need to compute the probability matrix, which we can do as follows using the 'dpois' function to calculate the Poisson probability distributions for home and away teams. Here the away goals vector is transposed, and the matrix multiplication operator '%*%' used to compute the outer product matrix of the two vectors.

```
# Compute probability matrix
max_goals <- 8  # This specifies the maximum number of goals to be scored.
prob.mat <- round(dpois(0:max_goals, Hgoals.exp) %*% t(dpois(0:max_goals, Agoals.exp)),4)
print(prob.mat)
```

This produces the following probability matrix, in which the rows represent the goals scored by the home team and the columns represent the goals scored by the away team. Importantly, the first column and row represent zero goals scored by the respective teams. So, rows/columns 1–9 actually represent 0–8 goals.

```
##
        [,1]   [,2]   [,3]   [,4]   [,5]   [,6]   [,7]   [,8]  [,9]
 [1,] 0.0325 0.0676 0.0704 0.0489 0.0255 0.0106 0.0037 0.0011 3e-04
 [2,] 0.0436 0.0909 0.0947 0.0658 0.0342 0.0143 0.0050 0.0015 4e-04
 [3,] 0.0293 0.0611 0.0637 0.0442 0.0230 0.0096 0.0033 0.0010 3e-04
 [4,] 0.0131 0.0274 0.0285 0.0198 0.0103 0.0043 0.0015 0.0004 1e-04
 [5,] 0.0044 0.0092 0.0096 0.0067 0.0035 0.0014 0.0005 0.0001 0e+00
 [6,] 0.0012 0.0025 0.0026 0.0018 0.0009 0.0004 0.0001 0.0000 0e+00
 [7,] 0.0003 0.0006 0.0006 0.0004 0.0002 0.0001 0.0000 0.0000 0e+00
 [8,] 0.0001 0.0001 0.0001 0.0001 0.0000 0.0000 0.0000 0.0000 0e+00
 [9,] 0.0000 0.0000 0.0000 0.0000 0.0000 0.0000 0.0000 0.0000 0e+00
```

From this, we can see that the probability of a 0-0 draw is 0.0325, whereas the probability of Arsenal beating Burnley 1–3 is 0.0658.

We can compute the overall probabilities of a home win, a draw, or an away win by summing up the values in various parts of the probability matrix according to the following rules:

- The sum of the lower triangle of the matrix is the probability of a home win.
- The sum of the diagonal is the probability of a draw.
- The sum of the upper triangle of the matrix is the probability of an away win.

This can be executed in R using the following code:

```
# Compute probabilities
# Home win
sum(prob.mat[lower.tri(prob.mat)]) # Sum of lower triangle is the probability of a home win.
```

```
## [1] 0.2445
```

```
# Draw
sum(diag(prob.mat)) # Sum of diagonal is the probability of a draw.
```

```
## [1] 0.2108
```

```
# Away win
sum(prob.mat[upper.tri(prob.mat)]) # Sum of upper triangle is the probability of an away win.
```

```
## [1] 0.5441
```

From this, we see that the model computes that the probability of Burnley winning was only 0.245, whereas the probability of Arsenal winning was 0.544, with the probability of a draw being 0.211.

The predicted outcomes using the Poisson model for the final round of EPL matches in season 2018–2019 are presented in Table 6.2. These will be discussed in detail after we have first considered the Dixon–Coles model, which is an adaptation of the basic Poisson model.

6.5 Dixon–Coles Model

One major disadvantage of the Poisson regression model discussed in Section 6.4 is that it tends to underestimate the frequency with which low score lines (i.e., 0–0, 1–0, 0–1, and 1–1) occur in real life. Dixon and Coles [9] recognised this and proposed adding an extra parameter, ρ (rho), to adjust the probability matrix produced by the Poisson model. However, this is done in such a way that the marginal distribution remains Poisson-shaped. As such, the Dixon–Coles adaptation has been shown to produce predictions that are more accurate than the simple Poisson regression model [10].

While the Dixon–Coles model is popular, it is tricky to implement, and so here (in Example 6.3), we demonstrate a simplified version developed by opisthokonta.net [11], which produces similar results to more complex implementations of the model. This uses the basic Poisson regression model (as shown in Example 6.2) to predict the expected number of home goals, λ (lambda), and away goals, μ (mu), for the respective matches. These are then used to produce outcome probability matrices for the respective matches, as described in Example 6.2. Having done this, the Dixon–Coles adjustment is then applied to correct these matrices with respect to four low score lines (i.e., 0–0, 1–0, 0–1, and 1–1).

Implementation of the Dixon–Coles model is tricky because the value of ρ is not known, and therefore an iterative process is required in order to establish the optimum value of ρ – something that is computationally difficult. However, by utilising the user-defined functions developed by opisthokonta.net [11], as shown in Example 6.3, it is possible to successfully implement the Dixon–Coles model.

EXAMPLE 6.3: DIXON–COLES MODEL EXAMPLE

In this example, we use the Dixon–Coles extension to the Poisson regression model to predict the outcome of the Burnley versus Arsenal match in season 2018–2019, in a similar manner to Example 6.2. We do this by first applying the Poisson regression model 'pois.mod' that we created in Example 6.2 to the 'long_dat' data frame (containing the first 370 match results) to predict the expected home goals (lambda) and the expected away goals (mu), as follows:

```
expected <- fitted(pois.mod)
```

Note that here we have used the 'fitted' function in R to run the Poisson regression model. This is a generic function that extracts fitted values from objects that are returned by modelling functions. We can inspect the fitted value results as follows:

```
# Compile and display fitted results
exp.dat <- cbind.data.frame(long_dat, expected)

head(exp.dat)
```

```
##
  Home         Team       Opponent Goals  expected
1    1   Man United      Leicester     2 1.8656140
2    1  Bournemouth        Cardiff     2 2.1595038
3    1       Fulham Crystal Palace     0 1.0198948
4    1 Huddersfield        Chelsea     0 0.5013085
5    1    Newcastle      Tottenham     1 0.8387356
6    1      Watford       Brighton     2 1.6880772
```

```
tail(exp.dat)
```

```
##
    Home        Team     Opponent Goals  expected
735    0 Southampton     West Ham     0 1.1178492
736    0      Fulham       Wolves     0 0.7146329
737    0    Brighton      Arsenal     1 0.7859251
738    0     Watford      Chelsea     0 0.9307604
739    0  Man United Huddersfield     1 2.1936266
740    0   Leicester     Man City     0 0.5263179
```

We can then create vectors containing the expected home and away goals for the respective values as follows:

```
# Create home.exp and away.exp vectors
home.exp <- expected[1:nrow(build.dat)]
away.exp <- expected[(nrow(build.dat)+1):(nrow(build.dat)*2)]
```

To ensure that we have correctly created the respective expected goals vectors, we display their head and tail and compare these with the 'exp.dat' data frame above. This tells us that everything is OK.

```
# Inspect these vectors
head(home.exp)
```

```
##
        1         2         3         4         5         6
1.8656140 2.1595038 1.0198948 0.5013085 0.8387356 1.6880772
```

```
tail(away.exp)
```

```
##
      735       736       737       738       739       740
1.1178492 0.7146329 0.7859251 0.9307604 2.1936266 0.5263179
```

Here, the expected home goals are the 'lambda' values and the expected away goals are the 'mu' values. These will be used to compute the 'rho' value needed to make the Dixon–Coles adjustment to the probability matrix. However, in order to do this, we need to create three user-defined functions as described in opisthokonta.net [11,12].

The first of these user-defined functions is called 'tau'. This uses the input parameters lambda (expected home goals), mu (expected away goals), and rho to compute the adjustments to the probabilities for various low score lines. Here, *x* and *y* represent the various low-scoring options (i.e., 0–0, 0–1, 1–0, and 1–1).

```
tau <- Vectorize(function(x, y, lambda, mu, rho){
  if (x == 0 & y == 0){return(1 - (lambda*mu*rho))
  } else if (x == 0 & y == 1){return(1 + (lambda*rho))
  } else if (x == 1 & y == 0){return(1 + (mu*rho))
  } else if (x == 1 & y == 1){return(1 - rho)
  } else {return(1)}
})
```

The tau function is used by a second user-defined function called 'logLike', which computes the log-likelihood of the data in the 'build.dat' data frame. The log-likelihood (a logarithmic transformation of likelihood) is a measure of how well the observed data fits a particular statistical model. This function uses the vectors of lambda (expected home goals), mu (expected away goals), and the vectors of observed home ($y1$) and away ($y2$) goals to compute the log-likelihood when rho is initially set to zero (i.e., when the Dixon–Coles adjustment has not been applied).

```
logLike <- function(y1, y2, lambda, mu, rho=0){
  sum(log(tau(y1, y2, lambda, mu, rho)) + log(dpois(y1, lambda)) + log(dpois(y2, mu)))
} # Here, y1 is the home goals and y2 is the away goals.
```

Finally, a third user-defined function (optRho) is required. This computes the log-likelihood for various input values of rho and is utilised in the optimisation process when identifying the optimum value of rho.

```
optRho <- function(par){
  rho <- par[1]
  logLike(build.dat$Hgoals, build.dat$Agoals, home.exp, away.exp, rho)
}
```

NB. The codes for the 'tau', 'loglike', and 'optRho' functions listed above are as presented in opisthokonta.net [11,12].

Now the above functions can be put to work in the optimisation process using the Broyden–Fletcher–Goldfarb–Shanno (BFGS) algorithm in the 'optim' function as follows. The 'par' argument specifies the initial values for the parameters to be optimised over. Here the par parameter is initialised as 0.1.

```
res <- optim(par=c(0.1), fn=optRho, control=list(fnscale=-1), method='BFGS')
Rho <- res$par # This is the optimum value of rho.
print(Rho)
```

```
## [1] -0.02755523
```

Having computed the optimum value of rho, we can now use this to adjust the match probability matrices. But before we can do this, we need to use the Poisson regression model to compute the expected goals for the various matches. So, here we will compute the outcome probabilities for the Burnley versus Arsenal match.

```
HT <- "Burnley" # Home team
AT <- "Arsenal" # Away team
```

```
# Expected goals home
lambda <- predict(pois.mod, data.frame(Home=1, Team=HT, Opponent=AT), type='response')

# Expected goals away
mu <- predict(pois.mod, data.frame(Home=0, Team=AT, Opponent=HT), type='response')

print(lambda)
```

```
##
       1
1.344383
```

```
print(mu)
```

```
##
       1
2.083151
```

Having done this, we can now compute the raw probability matrix without the Dixon–Coles adjustment, as we did in Example 6.2.

```
maxgoal <- 8
prob_mat1 <- dpois(0:maxgoal, lambda) %*% t(dpois(0:maxgoal, mu))
print(round(prob_mat1,4)) # This is the unadjusted probability matrix
```

```
##
        [,1]   [,2]   [,3]   [,4]   [,5]   [,6]   [,7]   [,8]  [,9]
 [1,] 0.0325 0.0676 0.0704 0.0489 0.0255 0.0106 0.0037 0.0011 3e-04
 [2,] 0.0436 0.0909 0.0947 0.0658 0.0342 0.0143 0.0050 0.0015 4e-04
 [3,] 0.0293 0.0611 0.0637 0.0442 0.0230 0.0096 0.0033 0.0010 3e-04
 [4,] 0.0131 0.0274 0.0285 0.0198 0.0103 0.0043 0.0015 0.0004 1e-04
 [5,] 0.0044 0.0092 0.0096 0.0067 0.0035 0.0014 0.0005 0.0001 0e+00
 [6,] 0.0012 0.0025 0.0026 0.0018 0.0009 0.0004 0.0001 0.0000 0e+00
 [7,] 0.0003 0.0006 0.0006 0.0004 0.0002 0.0001 0.0000 0.0000 0e+00
 [8,] 0.0001 0.0001 0.0001 0.0001 0.0000 0.0000 0.0000 0.0000 0e+00
 [9,] 0.0000 0.0000 0.0000 0.0000 0.0000 0.0000 0.0000 0.0000 0e+00
```

Now we can apply the Dixon–Coles adjustment by computing a [2×2] scaling matrix using the optimised rho value that we computed above. The following code inserts the [2×2] scaling matrix in the top left corner of the probability matrix, replacing the original values.

```
scale_mat <- matrix(tau(c(0,1,0,1), c(0,0,1,1), lambda, mu, Rho), nrow=2)
prob_mat2 <- prob_mat1 # Makes copy of raw probability matrix
prob_mat2[1:2, 1:2] <- prob_mat1[1:2, 1:2] * scale_mat # Replaces original values
print(round(prob_mat2,4)) # This is the adjusted probability matrix
```

```
##
      [,1]   [,2]   [,3]   [,4]   [,5]   [,6]   [,7]   [,8]  [,9]
[1,] 0.0350 0.0651 0.0704 0.0489 0.0255 0.0106 0.0037 0.0011 3e-04
[2,] 0.0411 0.0934 0.0947 0.0658 0.0342 0.0143 0.0050 0.0015 4e-04
[3,] 0.0293 0.0611 0.0637 0.0442 0.0230 0.0096 0.0033 0.0010 3e-04
[4,] 0.0131 0.0274 0.0285 0.0198 0.0103 0.0043 0.0015 0.0004 1e-04
[5,] 0.0044 0.0092 0.0096 0.0067 0.0035 0.0014 0.0005 0.0001 0e+00
[6,] 0.0012 0.0025 0.0026 0.0018 0.0009 0.0004 0.0001 0.0000 0e+00
[7,] 0.0003 0.0006 0.0006 0.0004 0.0002 0.0001 0.0000 0.0000 0e+00
[8,] 0.0001 0.0001 0.0001 0.0001 0.0000 0.0000 0.0000 0.0000 0e+00
[9,] 0.0000 0.0000 0.0000 0.0000 0.0000 0.0000 0.0000 0.0000 0e+00
```

From this, we can see that the Dixon–Coles modification has adjusted the values in the first two rows of the first two columns.

Now we can compute the probabilities of the various match outcomes.

```
# Home win
sum(prob_mat2[lower.tri(prob_mat2)]) # Sum of lower triangle is probability of a home win.
```

```
## [1] 0.2420732
```

```
# Draw
sum(diag(prob_mat2)) # Sum of diagonal is probability of a draw.
```

```
## [1] 0.2157604
```

```
# Away win
sum(prob_mat2[upper.tri(prob_mat2)]) # Sum of upper triangle is probability of an away win.
```

```
## [1] 0.5418361
```

So we can see that the Dixon–Coles model computes the probability of Burnley winning to be 0.242 and Arsenal winning to be 0.542, with the probability of a draw being only 0.216 – values that are only slightly different from the probabilities predicted by the simple Poisson regression model. Comparing these with the match probabilities produced by Pinnacle, we find that although Pinnacle broadly agreed with our results, they gave Burnley a greater chance of winning the match, with a probability of 0.317, whereas they predicted the probability of an Arsenal win to be 0.444. The chance of a draw according to Pinnacle was only 0.263, which was slightly higher than we predicted.

A summary of the match outcome probabilities produced by the Poisson and Dixon–Coles models for the last ten EPL matches in season 2018–2019 is presented in Table 6.2, which also shows the probabilities computed by Pinnacle. From this, we can see the following:

- The Poisson and Dixon–Coles models produce very similar match probability predictions, with the probability of a draw being slightly higher with the Dixon–Coles model.
- For many of the matches, the probabilities produced by the Poisson and Dixon–Coles models are broadly similar to those produced by Pinnacle. Only for the Leicester versus Chelsea and Fulham versus Newcastle matches did the Poisson and Dixon–Coles models predict a different favourite than that identified by Pinnacle.
- In terms of the match favourites winning, the Poisson and Dixon–Coles models actually outperformed the predictions made by Pinnacle, with 50% of the match favourites winning with the former and only 40% winning with the latter.
- For none of the match predictions, a draw was the favourite outcome. Yet in this sample, 30% of the matches were drawn.

From this, we can conclude that both the Poisson and Dixon–Coles models appear to do a reasonable job, with the match outcome probabilities being generally in the same region as those produced by Pinnacle. Having said this, it is important to remember that our predictions were made using models built on the information contained in the previous 370 matches. It is therefore more likely that the Poisson and Dixon–Coles models would be less accurate at predicting match outcome probabilities early in the season, when the clubs have less of a track record.

The analysis presented in Table 6.2 also highlights the fallacy of thinking that the match favourite will always win, as exemplified by Cardiff beating Manchester United 0-2 at Old Trafford, even though Manchester United were the odds-on favourite to win. It is therefore preferable to view match outcomes in probabilistic terms, as this will give an indication of the closeness of any given match. For example, if we consider the Leicester versus Chelsea match, which ended in a 0-0 draw, we see that although Pinnacle did not make a drawn match the favourite outcome, they did predict the probabilities for a home win and an away win to be roughly of similar magnitude. This is a strong indicator that there wasn't really a clear match favourite and that there was a good chance that the match would be drawn.

Recent head-to-head comparisons between the Dixon–Coles model and several other Poisson-based models using EPL match data [10] showed that the Dixon–Coles model produced the best predictions, suggesting that it is superior to the basic Poisson regression model.

TABLE 6.2

Match Outcome Probabilities Produced by the Poisson and Dixon–Coles Models Compared with Those Computed by Pinnacle for the Last Ten English Premier League Matches in Season 2018–2019

						Pinnacle	Pinnacle	Pinnacle	Poisson	Poisson	Poisson	DC	DC	DC
Date	**Home Team**	**Away Team**	**Home Goals**	**Away Goals**	**Result**	**Home Win**	**Draw**	**Away Win**	**Home Win**	**Draw**	**Away Win**	**Home Win**	**Draw**	**Away Win**
12/05/2019	Brighton	Manchester City	1	4	A	**0.064**	**0.122**	**0.847**	**0.059**	**0.155**	**0.786**	**0.058**	**0.158**	0.783
12/05/2019	Burnley	Arsenal	1	3	A	**0.317**	**0.263**	**0.444**	**0.245**	**0.211**	**0.544**	**0.242**	**0.216**	0.542
12/05/2019	Crystal Palace	Bournemouth	5	3	H	**0.532**	**0.242**	**0.253**	**0.507**	**0.233**	**0.260**	**0.503**	**0.239**	0.257
12/05/2019	Fulham	Newcastle	0	4	A	0.392	0.280	0.353	**0.276**	**0.280**	**0.444**	**0.272**	**0.287**	**0.440**
12/05/2019	Leicester	Chelsea	0	0	D	0.410	0.274	0.341	0.324	0.266	0.410	0.321	0.273	0.406
12/05/2019	Liverpool	Wolves	2	0	H	**0.763**	**0.173**	**0.095**	**0.776**	**0.159**	**0.065**	**0.774**	**0.163**	**0.063**
12/05/2019	Man United	Cardiff	0	2	A	0.781	0.158	0.098	0.773	0.142	0.085	0.770	0.146	0.083
12/05/2019	Southampton	Huddersfield	1	1	D	0.694	0.207	0.131	0.689	0.202	0.109	0.686	0.207	0.106
12/05/2019	Tottenham	Everton	2	2	D	0.476	0.275	0.275	0.564	0.242	0.195	0.560	0.248	0.192
12/05/2019	Watford	West Ham	1	4	A	0.455	0.260	0.312	0.475	0.244	0.281	0.471	0.251	0.278

NB. Probabilities highlighted in bold are matches where the favourite won. Those highlighted with a shaded background are the matches where the favourite predicted by the Poisson and Dixon–Coles models differed from that predicted by Pinnacle.

6.6 Random Forest Model Using the Match Betting Odds

In addition to historical match data, another useful piece of information that can be utilised to help predict match outcomes are the odds published before the game by the betting companies. Match odds are often presented in terms of a fractional ratio (e.g., 5:3). However, it is usually more helpful to convert these to decimal odds (e.g., 2.67), which can then be used to compute the match outcome probability as follows:

$$\text{Computing decimal odds from fractional odds}: \text{Odds}_d = \text{Odds}_f + 1 \quad (6.3)$$

$$\text{Computing probability from decimal odds}: p = \frac{1}{\text{Odds}_d} \quad (6.4)$$

where Odds_d is the decimal odds, Odds_f is the fractional odds, and p is the probability.

As we have seen from Table 6.2, the match probabilities produced by the betting companies are often similar in magnitude to the probabilities produced by the Poisson and Dixon–Coles models. Therefore, without having to go to the trouble of building our own models, the published odds already give us a pretty good indication of the likely outcome in probabilistic terms for any given match. As such, the published betting odds represent a useful source of information that can potentially be utilised to predict the outcome of matches.

So, how can we harness the information contained in betting odds to make predictions about the outcome of matches? Well, one approach is to utilise machine learning algorithms to make predictions from the match odds. Machine learning is a branch of computer science that is primarily concerned with making predictions based on observed data [13,14]. As such, it could be argued that the Poisson and Dixon–Coles models fall into this category. However, classical machine learning uses data and algorithms in a way that tries to imitate the ways in which humans learn, gradually improving accuracy as more and more data is added. Often, this involves the use of pattern recognition techniques to identify clusters in the data or the identification of classification algorithms to accurately distinguish between separate and distinct groups within the data. One such classification technique that is widely used in machine learning is the random forest, which is made up of multiple decision trees, each of which is constructed using a random selection of predictor variables [15,16]. In this section, we are going to investigate how random forests can be utilised to predict match outcomes. Although a full discussion of random forests is beyond the scope of this chapter, a brief overview of random forest models is supplied in Key Concept Box 6.1 in order to familiarise readers with the technique.

Random forest models are widely used in the financial and technology sectors to make predictions about which class a particular event or person is likely to fall into. For example, when deciding whether or not to grant a loan, the bank will want to know into which class a potential customer is likely to fall – are they going to be someone who will repay the loan in full, or are they more likely to default? So, the bank asks lots of pertinent questions about the customer's work and financial status and then uses a machine learning algorithm to decide which class the customer is likely to belong to. If the algorithm predicts that the customer will not default, the loan will be granted; otherwise, it will be declined. Similarly, when it comes to predicting soccer matches, there are three potential outcome classes: a home win, an away win, or a draw. So random forests can potentially be useful when trying to predict match outcomes, as illustrated in Example 6.4, where we use a simple random forest model to predict the outcome of the last ten matches in the EPL for season 2018–2019. This opinion is supported by the pioneering work of Schauberger, Groll, and their associates, who found that random forests tended to outperform regression models when predicting the outcome of soccer competitions [17–19].

One central concept that is often used in machine learning is the idea of building a predictive model using some historical training data and then validating it against a known test data set. If the model performs well against the test data (i.e., it makes accurate predictions), then we keep it and use it as our prediction tool. However, if it does not produce accurate predictions, then we need to go back to the drawing board and build a better model. In Examples 6.2 and 6.3, although we did not use the terms 'train' and 'test', what we actually did was use the first 370 matches as a training data set to build the models, which we then tested against the validation data (i.e., the last ten matches) in Table 6.2. So, in Example 6.4, we will build on this and show how a random forest model built using a training data set can be used to predict outcomes using a testing data set.

EXAMPLE 6.4: RANDOM FOREST MODEL EXAMPLE

In this example, we use a simple random forest model to predict the outcome of the last ten matches in the EPL for season 2018–2019 in a similar manner to Examples 6.2 and 6.3. However, rather than using the distribution of the home and away goals, we will instead use the betting odds published by Pinnacle for these matches, which we incorporated into the 'dat' data frame in Example 6.1.

So, first, let us remind ourselves of the initial data.

```
head(dat) # Displays first six rows of dat
```

```
##
        Date      HomeTeam        AwayTeam Hgoals Agoals Result HWodds
1 10/08/2018   Man United       Leicester      2      1      H   1.58
2 11/08/2018  Bournemouth         Cardiff      2      0      H   1.89
3 11/08/2018       Fulham Crystal Palace      0      2      A   2.50
4 11/08/2018 Huddersfield         Chelsea      0      3      A   6.41
5 11/08/2018    Newcastle       Tottenham      1      2      A   3.83
6 11/08/2018      Watford        Brighton      2      0      H   2.43
  Dodds AWodds
1  3.93   7.50
2  3.63   4.58
3  3.46   3.00
4  4.02   1.62
5  3.57   2.08
6  3.22   3.33
```

Now we can select the variables that we want to include in our random forest analysis. Note that here we use the 'as.factor' function to ensure that the variable 'Result' is treated as a factor.

```
rf.dat <- dat[,c("Result","HWodds","Dodds","AWodds")]
rf.dat$Result <- as.factor(rf.dat$Result) # Ensure that the results are treated as a factor.
str(rf.dat)
```

```
##
'data.frame':    380 obs. of  4 variables:
 $ Result: Factor w/ 3 levels "A","D","H": 3 3 1 1 1 3 2 1 3 2 ...
 $ HWodds: num  1.58 1.89 2.5 6.41 3.83 2.43 2.36 4 1.27 1.86 ...
 $ Dodds : num  3.93 3.63 3.46 4.02 3.57 3.22 3.4 3.97 6.35 3.51 ...
 $ AWodds: num  7.5 4.58 3 1.62 2.08 ...
```

Having organised the data, the next step is to split it into the training and testing data sets, which we will use to perform our analysis.

```
rf_train.dat <- rf.dat[1:370,] # First 370 matches is the training data set
rf_test.dat <- rf.dat[371:380,] # Last 10 matches is the testing data set
```

In order to build a random forest model, we will need to install and load the 'randomForest' package.

```
install.packages("randomForest") # This installs the 'randomForest' package.
# NB. This command only needs to be executed once to install the package.
# Thereafter, the 'randomForest' library can be called using the 'library' command.
```

Because the random forest involves the generation of random numbers, we need to set a seed, which in this case is 123, although any arbitrary integer can be used. This initialises a pseudorandom number generator and ensures that every time the model is used, the same random numbers are selected, thus ensuring reproducibility. (NB. For reasons too complex to mention here, the 'set.seed' command may produce differing results on various computers, depending on the processor type and operating system used.)

```
# Load 'randomForest' library
library(randomForest)
set.seed(123) # Set a seed so that the results are repeatable.
rf.mod = randomForest(Result ~ HWodds + Dodds + AWodds, data=rf_train.dat)
```

Here the random forest model, which we have called 'rf.mod', uses the variables 'Hwodds', 'Dodds', and 'Awodds' (i.e., the odds for the respective possible match outcomes) to predict the class of the result (i.e., home win, away win, or draw) using the 'rf_train.dat' training data set.

We can view how the model performs against the training data as follows:

```
print(rf.mod)
```

```
##
Call:
 randomForest(formula=Result~HWodds+Dodds+AWodds, data=
rf_train.dat)
               Type of random forest: classification
                     Number of trees: 500
No. of variables tried at each split: 1

        OOB estimate of  error rate: 44.59%

Confusion matrix:
   A  D   H class.error
A 67 11  45   0.4552846
D 21 13  34   0.8088235
H 38 16 125   0.3016760
```

This reveals that with the training data set, the random forest model was best able to predict home wins, correctly identifying 69.8% of the matches in which the home team won. However, it could only correctly predict 54.5% of the matches where the away team won and only 19.1% of draws.

One major advantage of random forests is that they can be used to assess the relative importance of the various predictor variables. This can be done using the 'importance' function in the 'randomForest' package and visualised using the 'varImpPlot' function.

```
importance(rf.mod)
```

```
##
       MeanDecreaseGini
HWodds         78.48940
Dodds          66.94142
AWodds         80.57447
```

```
varImpPlot(rf.mod) # NB. This produces a variable importance plot.
```

This produces the plot in Figure 6.4, which indicates that 'AWodds' is actually the most influential indicator of match outcome, closely followed by 'Hwodds', with 'Dodds' being of less importance.

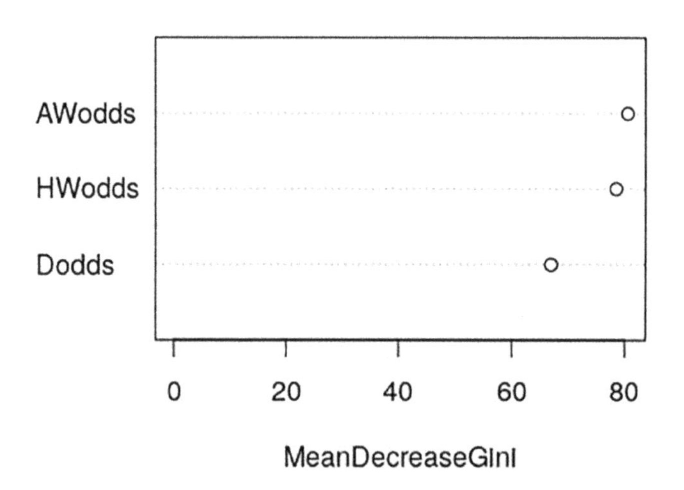

FIGURE 6.4
Variable importance plot of the random forest model.

Having built the random forest model, we can now validate it using the matches in the testing data set, as follows:

```
# Make predictions
rf.pred <- predict(rf.mod, newdata=rf_test.dat, type="response") # NB. Testing data set used.
rf.pred
```

```
##
371 372 373 374 375 376 377 378 379 380
  A   A   H   D   D   H   H   H   A   H
Levels: A D H
```

We can compare these with the real-life match results as follows:

```
# Compare predictions with actual results
pred.comp <- cbind.data.frame(rf_test.dat,rf.pred)
print(pred.comp)
```

```
##
    Result HWodds Dodds AWodds rf.pred
371      A  15.63  8.17   1.18       A
372      A   3.15  3.80   2.25       A
373      H   1.88  4.14   3.95       H
374      A   2.55  3.57   2.83       D
375      D   2.44  3.65   2.93       D
376      H   1.31  5.77  10.54       H
377      A   1.28  6.33  10.21       H
378      D   1.44  4.83   7.62       H
379      D   2.10  3.64   3.64       A
380      A   2.20  3.85   3.21       H
```

From this, we see that the random forest model correctly classified 50% of the matches, which is a similar performance to the Poisson and Dixon–Coles models.

Comprehensive evaluation of the performance of the random forest model can be done using the 'confusionMatrix' function in the 'caret' package, as follows:

```
# Calculate accuracy and kappa value for the predictions

install.packages("caret") # This installs the 'caret' package.
# NB. This command only needs to be executed once to install the package.
# Thereafter, the 'caret' library can be called using the 'library' command.

library(caret)
confusionMatrix(rf.pred, rf_test.dat$Result)
```

```
##
Confusion Matrix and Statistics

          Reference
Prediction A D H
         A 2 1 0
         D 1 1 0
         H 2 1 2

Overall Statistics
               Accuracy : 0.5
                 95% CI : (0.1871, 0.8129)
    No Information Rate : 0.5
```

```
    P-Value [Acc>NIR] : 0.6230

                Kappa : 0.2754

 Mcnemar's Test P-Value : 0.3916

Statistics by Class:

                     Class: A Class: D Class: H
Sensitivity            0.4000   0.3333   1.0000
Specificity            0.8000   0.8571   0.6250
Pos Pred Value         0.6667   0.5000   0.4000
Neg Pred Value         0.5714   0.7500   1.0000
Prevalence             0.5000   0.3000   0.2000
Detection Rate         0.2000   0.1000   0.2000
Detection Prevalence   0.3000   0.2000   0.5000
Balanced Accuracy      0.6000   0.5952   0.8125
```

This gives a complete overview of the predictive performance of the model, including the respective sensitivity and specificity scores.

KEY CONCEPT BOX 6.1: DECISION TREES AND RANDOM FORESTS

Back in the 1930s, the famous statistician Ronald Fisher created an iris data set [20], which contains biological data about species of flowers. This has subsequently been used for teaching purposes by data scientists around the world. It is even included in the base version of R and can be easily accessed by simply typing:

```
data(iris)
```

The iris data set contains 150 observations from three species of iris: Setosa, Versicolor, and Virginica (50 subjects in each group) and contains taxonomic data regarding the length and width of the sepals and petals. When viewed in a scatter plot, it is easy to spot clusters in the data that correspond to the iris species. However, the challenge that Fisher faced was to develop a mathematical algorithm that could reliably discriminate between the three species of iris using just the four taxonomic variables in the data set – something that, in the 1930s before computers were invented, was pretty difficult to do.

Without going into any detail about Fisher's analysis, it is possible to accurately discriminate between the three species very easily using just the petal length and the petal width, as illustrated in Figure 6.5, which shows sequential plots of the three iris species using: A, petal length; and B, petal width. If we first draw a horizontal line at $y = 2.45$ cm on plot A, we find that this completely separates all 50 of the Setosa plants from the other iris subjects. If we then perform a second split using the petal width data and draw a horizontal line at $y = 1.75$ cm on plot B, we find that this separates the remaining irises pretty well, with 49 of the subjects classified as being Versicolor with a petal width < 1.75 cm and 45 subjects classed as being Virginica with a petal width ≥ 1.75 cm. The optimum splits between the iris species can therefore be represented in the decision tree shown in Figure 6.6.

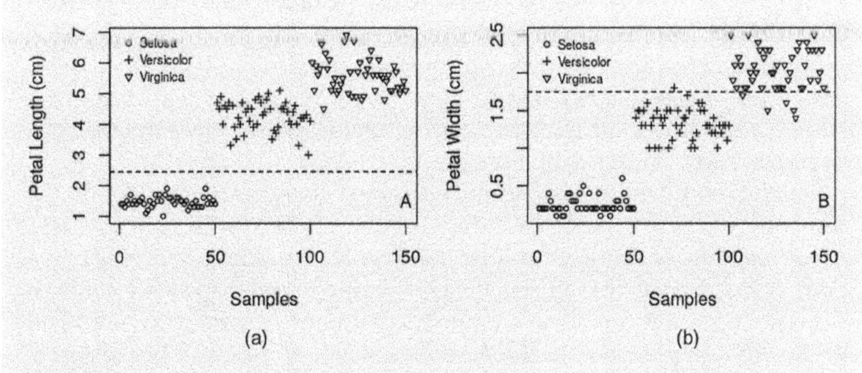

FIGURE 6.5
Sequential plots of the three iris species using (A) petal length and (B) petal width. The horizontal dotted lines represent the optimum split between (A) Setosa and Versicolor and (B) Versicolor and Virginica.

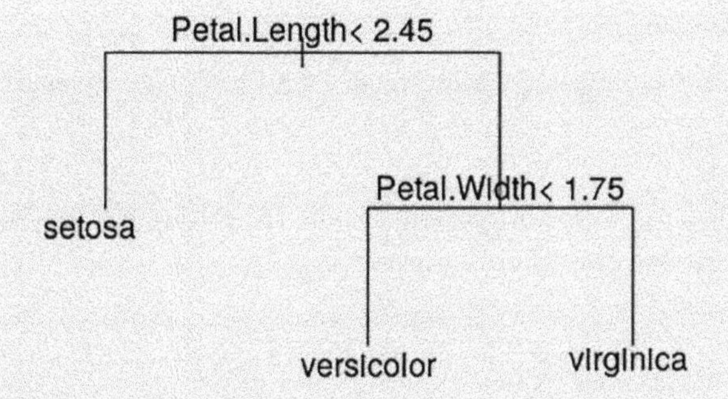

FIGURE 6.6
Decision tree to optimally discriminate between the three iris species using petal length and petal width, based on the Gini index.

Effectively, the decision tree shown in Figure 6.6 is a set of simple instructions that a user can execute in order to accurately classify iris species. If the petal length is < 2.45 cm, then the plant is probably a Setosa iris. Whereas if the petal length is ≥ 2.45 cm and the petal width is < 1.75 cm, then the iris is likely to be Versicolor – otherwise it is likely to be a Virginica iris.

Decision trees such as the one described above generally utilise something called Gini impurity (i.e., the Gini index), which is a measure of classification heterogeneity, to decide where the splits should go, thus ensuring that the classification groups are as homogenous as possible. However, decision trees that utilise the Gini index have the disadvantage that they are prone to bias and therefore may be unreliable. This problem, can however, be overcome by using conditional inference trees, which use a significance test to make splits instead of Gini impurity. As such, conditional inference trees tend to be more robust than simple decision trees and less prone to bias.

Another excellent way to produce a robust classification model that is resistant to bias is to build a random forest. Random forests are a widely used machine learning technique that involves building a forest of random decision trees that utilise the Gini index in order to make splits in the data. Random forests use a random number generator to build 500 (typically) unique trees, each of which is slightly different because it is constructed using a different random subset of variables and subjects sampled from the data set. As such, random forest analysis is an ensemble learning technique that is robust and generally produces superior results compared with individual classification trees.

In R, a random forest analysis can be performed on the iris dataset using the following code:

```
data(iris)
library(randomForest)
set.seed(123)
RF.model = randomForest(Species ~ Sepal.Length + Sepal.Width + Petal.Length + Petal.Width,
data = iris)
print(RF.model)
```

This produces the following results table, from which it can be seen that seven subjects are classified incorrectly.

```
## Call:
randomForest(formula = Species ~ Sepal.Length + Sepal.Width + Petal.Length
+ Petal.Width, data = iris)
               Type of random forest: classification
                     Number of trees: 500
No. of variables tried at each split: 2

        OOB estimate of  error rate: 4.67%
Confusion matrix:
```

```
           setosa  versicolor  virginica  class.error
setosa         50           0          0         0.00
versicolor      0          47          3         0.06
virginica       0           4         46         0.08
```

One great advantage of random forests is that the decrease in Gini impurity can be used to determine the relative importance of the respective variables when making a classification. In the 'randomForest' package, the relative importance of the various variables can be determined using the 'importance' command as follows:

```
importance(RF.model)
```

This produces the following table, from which it can be seen that petal length produces the greatest reduction in Gini impurity, followed closely by petal width. By comparison, sepal length and width are less useful as classification metrics.

```
##
              MeanDecreaseGini
Sepal.Length          9.659149
Sepal.Width           2.204130
Petal.Length         44.244822
Petal.Width          43.081313
```

6.7 Conditional Inference Tree Model

While random forest models can be useful, they are opaque because they are an ensemble technique that involves the creation of multiple random trees. As such, they are a black-box technique that produces answers when the appropriate data is inputted, making them difficult to interpret. However, there is a related technique called conditional inference trees that can be utilised to develop rules for predicting the outcome of matches.

A conditional inference tree is a kind of decision tree that uses recursive partitioning of predictor variables based on the value of correlations [21]. Conditional inference trees use a significance test to select the input variables and decide where splits should occur (see Key Concept Box 6.1 for more details). As such, they can be useful for creating rules that can be applied to decision trees, as illustrated in Example 6.5.

EXAMPLE 6.5: CONDITIONAL INFERENCE TREE MODEL EXAMPLE

In this example, we use a conditional inference tree to predict the outcome of the last ten matches in the EPL for season 2018–2019 in a similar manner to the other examples. As in Example 6.4, we will use the betting odds published by Pinnacle, which we will also use to develop some rules for predicting match outcomes.

In order to build the conditional inference tree model, we need to use the 'ctree' function in the 'party' library package in R [22], as follows:

```
install.packages("party") # This installs the 'party' package.
# NB. This command only needs to be executed once to install the package.
# Thereafter, the 'party' library can be called using the 'library' command.

# Load 'party' library
library(party)
ctree.mod = ctree(Result ~ HWodds + Dodds + AWodds, data = rf_train.dat)
print(ctree.mod)
```

```
##
Conditional inference tree with 5 terminal nodes

Response:  Result
Inputs:  HWodds, Dodds, AWodds
Number of observations:  370

1) HWodds <= 2.89; criterion=1, statistic=66.843
  2) HWodds <= 1.97; criterion=1, statistic=30.312
    3) Dodds <= 5.17; criterion=0.966, statistic=8.952
      4)*  weights=87
    3) Dodds>5.17
      5)*  weights=57
  2) HWodds>1.97
    6)*  weights=104
1) HWodds>2.89
  7) AWodds <= 1.76; criterion=0.971, statistic=9.231
    8)*  weights=57
  7) AWodds>1.76
    9)*  weights=65
```

```
# Plot the tree
plot(ctree.mod)
```

Which produces the following plot:

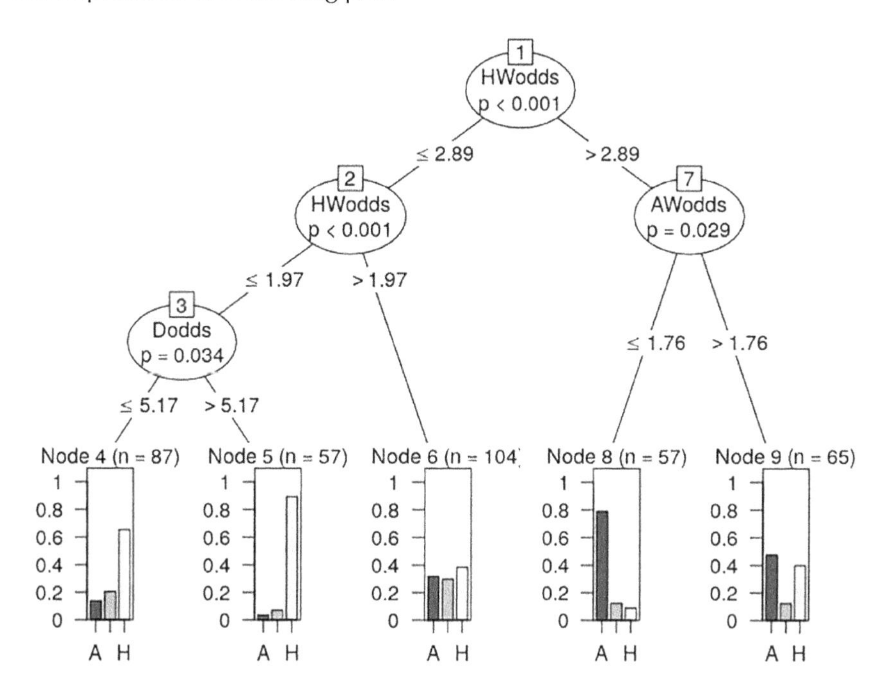

FIGURE 6.7
Conditional inference tree, showing significant splits.

From Figure 6.7, it can be seen that the conditional inference tree model parcels up the matches into five groups, which are classified according to some basic rules. For example, the model identified 57 matches in which the home win odds were >2.89 and the away win odds were ≤1.76, of which 78.9% resulted in an away win. Likewise, 89.5% of the matches that had home win odds of ≤1.97 and draw odds of >5.17 resulted in a home win.

A simple table of the results produced by the model using the build data can be constructed as follows:

```
table(predict(ctree.mod), rf_train.dat$Result)
```

```
##
      A   D   H
  A  76  15  31
  D   0   0   0
  H  47  53 148
```

Now we can apply the conditional inference tree model to make predictions for the remaining ten matches based on the rules established above.

```
# Produce predictions
cif.pred <- predict(ctree.mod, newdata=rf_test.dat, type="response")
print(cif.pred)
```

```
##
[1] A A H H H H H H H H
Levels: A D H
```

From this, we see that the conditional inference tree model does not perform all that well, only correctly classifying 40% of the matches. In particular, the model is poor at predicting draws. Having said this, if we are only interested in predicting the outcome of some matches rather than all the matches, then the conditional inference model performs much better. For example, if we set the following rules:

- If HWodds <1.97, assume a home win.
- If HWodds >2.89 and AWodds <1.76, assume an away win.
- Otherwise, do not make a prediction.

Then the percentage of match outcomes correctly predicted is likely to increase.

```
n <- nrow(rf_test.dat)
res.vec <- matrix(0,n,1) # This creates an empty vector of length n to store the results.

for(i in 1:n){
  if(rf_test.dat$HWodds[i] <=1.97){res.vec[i] <- "H"}
  else if(rf_test.dat$HWodds[i] >2.89 & rf_test.dat$AWodds[i] <=1.76){res.vec[i] <- "A"}
  else {res.vec[i] <- "NA"}
}

print(cbind(rf_test.dat, res.vec))
```

```
##
    Result HWodds Dodds AWodds res.vec
371      A  15.63  8.17   1.18       A
372      A   3.15  3.80   2.25      NA
373      H   1.88  4.14   3.95       H
374      A   2.55  3.57   2.83      NA
375      D   2.44  3.65   2.93      NA
376      H   1.31  5.77  10.54       H
377      A   1.28  6.33  10.21       H
378      D   1.44  4.83   7.62       H
379      D   2.10  3.64   3.64      NA
380      A   2.20  3.85   3.21      NA
```

From this, we see that of the matches for which a prediction was made, 60% of the predictions were correct, which is actually better than the performance of the random forest model.

6.8 Concluding Remarks

In this chapter, we have explored several alternative methodologies that can be employed to predict the outcome of soccer matches. While the methods illustrated are by no means exhaustive, they are indicative of the various types of strategies that can be employed. However, it is important to remember that the methods introduced here are relatively basic and that there are plenty of refinements that could be undertaken to improve their performance. For example, shots could be used in the Poisson model instead of goals scored. Also, a time limit could be set on the historical data used so that match results from the distant past are treated with less importance compared with recent results. Nevertheless, if you can master the techniques presented in this chapter, you will be well on your way. Then you should be in a position to tackle more sophisticated prediction techniques as you grow in your experience and knowledge.

No matter which prediction technique you may wish to employ, it is important to remember that no method is foolproof. By nature, soccer is a low-scoring sport in which chance plays a major contributory role, so it is inevitable that draws and unexpected results will frequently occur. Remember, from Table 6.1, we see that the bookmaker's favourite only wins about 55% of the time. Therefore, it is generally better to consider match predictions in probabilistic terms rather than purely binary outcomes, as this will help you moderate your expectations.

References

1. Karlis D, Ntzoufras I: On modelling soccer data. *Student* 2000, 3(4):229–244.
2. Reep C, Pollard R, Benjamin B: Skill and chance in ball games. *Journal of the Royal Statistical Society: Series A (General)* 1971, 134(4):623–629.
3. Maher MJ: Modelling association football scores. *Statistica Neerlandica* 1982, 36(3):109–118.
4. Heuer A, Rubner O: How does the past of a soccer match influence its future? Concepts and statistical analysis. *PLoS One* 2012, 7(11):e47678.
5. Karlis D, Ntzoufras I: Analysis of sports data by using bivariate Poisson models. *Journal of the Royal Statistical Society: Series D (The Statistician)* 2003, 52(3):381–393.

6. Rue H, Salvesen O: Prediction and retrospective analysis of soccer matches in a league. *Journal of the Royal Statistical Society: Series D (The Statistician)* 2000, 49(3):399–418.
7. Koopman SJ, Lit R: A dynamic bivariate Poisson model for analysing and forecasting match results in the English Premier League. *Journal of the Royal Statistical Society Series A (Statistics in Society)* 2015, 52(3):167–186.
8. Egidi L, Pauli F, Torelli N: Combining historical data and bookmakers' odds in modelling football scores. *Statistical Modelling* 2018, 18(5–6):436–459.
9. Dixon MJ, Coles SG: Modelling association football scores and inefficiencies in the football betting market. *Journal of the Royal Statistical Society: Series C (Applied Statistics)* 1997, 46(2):265–280.
10. Maozad SN, Razali SNAM, Mustapha A, Nanthaamornphong A, Wahab MHA, Razali N: Comparative Analysis for Predicting Football Match Outcomes based on Poisson Models. In: *2022 19th International Conference on Electrical Engineering/Electronics, Computer, Telecommunications and Information Technology (ECTI-CON): 2022*: IEEE, Prachuap Khiri Khan; 2022: 1–4.
11. opisthokonta.net: A simple re-implementation of the Dixon–Coles model. In: *opisthokontanet.* vol. https://opisthokonta.net/?p=1685; 2018.
12. opisthokonta.net: The Dixon-Coles model for predicting football matches in R (part 1). In: *opisthokontanet.* https://opisthokonta.net/?p=890; 2014.
13. Olden JD, Lawler JJ, Poff NL: Machine learning methods without tears: A primer for ecologists. *The Quarterly Review of Biology* 2008, 83(2):171–193.
14. Larranaga P, Calvo B, Santana R, Bielza C, Galdiano J, Inza I, Lozano JA, Armananzas R, Santafe G, Perez A: Machine learning in bioinformatics. *Briefings in Bioinformatics* 2006, 7(1):86–112.
15. Breiman L: Random forests. *Machine Learning* 2001, 45(1):5–32.
16. Liaw A, Wiener M: Classification and regression by randomForest. *R news* 2002, 2(3):18–22.
17. Schauberger G, Groll A: Predicting matches in international football tournaments with random forests. *Statistical Modelling* 2018, 18(5–6):460–482.
18. Groll A, Ley C, Schauberger G, Van Eetvelde H: A hybrid random forest to predict soccer matches in international tournaments. *Journal of Quantitative Analysis in Sports* 2019, 15(4):271–287.
19. Groll A, Ley C, Schauberger G, Van Eetvelde H, Zeileis A: Hybrid Machine Learning Forecasts for the FIFA Women's World Cup 2019. *arXiv preprint arXiv:190601131*, 2019.
20. Fisher RA: The use of multiple measurements in taxonomic problems. *Annals of Eugenics* 1936, 7(2):179–188.
21. Hothorn T, Hornik K, Zeileis A: Unbiased recursive partitioning: A conditional inference framework. *Journal of Computational and Graphical Statistics* 2006, 15(3):651–674.
22. Hothorn T, Hornik K, Strobl C, Zeileis A, Hothorn MT: Package 'party'. *Package Reference Manual for Party Version 09-998* 2015, 16:37.

7

Betting Strategies

In this chapter, we shall investigate how R can be used to develop strategies for betting on the outcome of Premier League matches. This chapter is an introduction to the subject and aims to make the reader aware of how the sports betting industry operates as well as provide some useful R code that might help when developing a successful betting strategy.

7.1 The Sports Betting Industry

This book is primarily a data science text aimed at introducing the reader to R and its potential in soccer analytics. However, it is impossible to ignore the subject of sports betting because it drives much of the data traffic associated with soccer worldwide. The global sports betting market is huge. According to a 2021 study by Statistica [1], the total amount wagered in sports betting worldwide was approximately 1.5 trillion euros in 2020. Of this, the global amount wagered on the English Premier League (EPL) alone in season 2019–2020 was over 68 billion euros worldwide, while that wagered on La Liga reached roughly 41 billion euros. Given this, anyone considering betting on the outcome of soccer matches needs to be mindful of a few simple facts of which most people are totally unaware – facts that the betting industry is generally not keen to promote. These can be briefly summarised as follows:

- The sports betting industry is very good at what it does. It makes money, and lots of it, mainly from punters (gamblers), who generally lose money. It loves naive gamblers who make uninformed bets, and it spends vast sums of money advertising its services to potential punters. For example, in the UK alone, the industry spent £1.5 billion on marketing in 2018 [2]. Indeed, the old adage that the bookmaker always wins appears to be basically true, with the US National Bureau of Economic Research concluding that there is "little evidence that there exist bettors who are systematically able to beat the bookmaker" [3]. Furthermore, statistical simulations indicate that, irrespective of the bettors' skill level, sports betting does not provide long-term opportunities for profit for the vast majority of bettors [4].

DOI: 10.1201/9781003328568-7

- Because the sports betting industry is such a big business, they can afford to employ the best statisticians and PhD data scientists to make sure that it remains profitable. Most gamblers are unaware of this and are, in effect, enthusiastic amateurs taking on hardened professionals. So, if you want to stand any chance of making a profit from sports betting, it is advisable to learn some data science and be able to code in a statistical programming language like R or Python.
- Predicting the outcome of individual soccer matches is difficult. As we saw in Chapter 6, prediction models are generally only able to get it right about 50%–60% of the time. Indeed, in soccer, the bookies' favourite (i.e., the outcome that the bookmaker predicts is most likely to occur) generally only wins about 55% of the time. So rather than trying to pick individual winners, it is often better to play the bookmakers at their own game and take a probabilistic approach. That is what the bookmakers do – they generally don't care who wins individual matches; rather, they care about setting the match odds correctly so that they make money.
- Bookmakers employ sophisticated machine learning algorithms to ensure that their match odds are set as accurately as possible, and generally, they are very good at setting match odds despite not knowing in advance the 'true' probabilities of match outcomes (i.e., probabilities of a home win, a draw, or an away win). Indeed, analysis of 52,411 Betfair odds from worldwide soccer league matches during the period from 29 October 2004 to 31 October 2005, found an almost perfect correlation ($r=0.995$) between the probabilities implied *a priori* by volume-weighted average betting prices and the probabilities calculated a *post priori* by the actual results [5]. So in short, the bookmaker's odds are generally pretty accurate. Therefore, the average punter stands little chance of beating the bookie [3], unless they employ sophisticated machine learning techniques to spot opportunities that the bookmakers might have missed. This involves the application of subject knowledge, data science, and a lot of hard work.
- Much of what is published on betting, especially on the Internet, is either wrong or misleading. The sports betting industry has a strong vested interest in promoting gambling and is only too happy to create and support 'helpful' websites giving useful tips and insights on how to bet 'successfully' – things that are likely to entice potential punters to place bets [2,6]. Similarly, while many gambling websites provide useful tips concerning betting strategies, they are generally very light on technical content (e.g., useful machine learning code, etc.) and shy away from difficult statistical and data science concepts. When code or examples are provided, the reader needs to be cautious because websites will often cherry-pick examples where

the technique works and ignore situations where it fails. Also, if a technique works and people are making money, they generally don't want to share it for free. So, would-be sports bettors should always be cautious of anything that they read on the Internet.

- It is possible to beat the bookies under certain circumstances if enough hard work, discipline, and rigour are applied. Unfortunately, when this happens, bookmakers have a habit of closing the accounts of successful sports bettors, with staff advised to restrict the activities of customers who "look like bad business" [7].

Collectively, this means that anyone seeking to become successful at sports betting needs to: be on top of their game; be prepared to work hard; and have a good knowledge of data science and statistics. Having said this, there are opportunities where sports betting can be profitable, and in this chapter, we shall explore these in the context of the EPL. Example R codes will also be presented, which will illustrate how these opportunities might be exploited.

It is also worth pointing out that this book is an educational text designed to introduce R in the context of soccer analytics, not a comprehensive account of how to get rich through sports betting. Therefore, the examples and codes presented should be treated with caution, as they are for illustrative purposes only. Therefore, before applying them in real life, the reader is advised to test out the various techniques presented in other contexts (e.g., in other leagues and for different seasons) to validate their robustness.

7.2 How Soccer Betting Works

When they place a bet on the outcome of a soccer match, many gamblers think that they are betting against the bookmaker; hence, terms like 'beat the bookie' are often used. However, while it is true that the bet is placed with the bookmaker, this is not the whole story because bookmakers actually pay winnings out of the money that they receive from those who lose when betting on matches. So alternatively, we can look at a match as being a kind of betting exchange in which the bookmaker makes the rules (i.e., sets the odds) and extracts a profit.

For every match, the bookmaker will make a 'book' (i.e., a record of the bets on all the possible match outcomes), and this will of necessity involve them taking into account the bets placed by all those gambling on the match. When creating a book, the bookmaker strives to offer odds that will ensure that they always make a profit regardless of the outcome of the match. Importantly, this will involve accepting bets on the outcome in the right proportions to ensure that they profit no matter which outcome prevails. So, if in the lead-up to a match there is heavy betting on one particular outcome, say a home win, then this might expose the bookmaker to the risk of making a loss. Therefore, they will adjust (shorten) the match odds accordingly to reduce their exposure and ensure that they still make a profit. So in theory, the bookmaker should not care who wins a particular match because the gamblers who lose will be the ones who pay the winners, not the bookmaker. Indeed, if the bookmaker has set the match odds correctly, they cannot fail to make a profit.

Making a successful book is a skilled task that involves a lot of mathematics in order to ensure that the bookmaker always makes a profit regardless of the match outcome. Importantly, this means that the bookmaker does not gamble. Rather, bookmakers skillfully set and adjust the odds so that they are slightly in their favour, which means that no matter the outcome of any given match, the bookmaker should always return a profit. To illustrate this, let us look, for example, at the match between Manchester United and Tottenham at Old Trafford, which occurred on 27 August 2018. For this match, Pinnacle offered the following decimal odds: 2.61 for a home win, 2.93 for an away win, and 3.36 for a draw. The implied probabilities can be computed from these odds using the following equation:

$$\text{Computing implied probability from decimal odds: } p = \frac{1}{\text{Odds}_d} \quad (7.1)$$

where Odds_d is the decimal odds and p is the implied probability.

This produces the following implied probabilities for the match: $p=0.383$ for a home win; $p=0.341$ for an away win; and $p=0.298$ for a draw, which when totalled equals $p=1.022$.

Those of you who know a little bit about statistics will instantly spot that the overall implied probability is greater than 100%, which of course is impossible because the total probability cannot be greater than $p=1$. The bookmakers know this, and the additional 2.2% is, in this case, Pinnacle's profit, known as the 'over-round'. So in theory, no matter the match outcome (on the day Tottenham won 0–3), the bookmaker should have made a profit. However, the over-round alone will not necessarily guarantee a profit for the bookmaker. This is because at the last minute, there could have been a rush of people putting money on Tottenham to win because they fancied the odds. This would potentially leave Pinnacle exposed to a large payout if the away team won, in which case they would have to shorten (adjust) the odds of Tottenham winning in order to ensure that they still made a profit. Of course, we will never know what actually happened behind the scenes at Pinnacle in the days before the Manchester United versus Tottenham match, but at least the example illustrates how the process operates in theory.

From this, we can see that the bookmaker does not gamble; rather, they simply do three things:

- Set accurate odds that ensure, no matter the match outcome, they make a profit.
- Adjust the odds to ensure that they are not financially exposed should a lot of money suddenly be placed on one team.
- Sit back and let chance take its course.

One additional point that is worth remembering is that bookmakers only make a profit if people place bets with them. Furthermore, the more people who bet with them, the greater the profit that they make. Therefore, it is very much in a bookmaker's interest to offer match odds that encourage people to bet with them, especially given that there are lots of bookmakers competing with each other for business. So bookmakers generally like to offer keen odds that will entice punters to place bets with them while still ensuring that they make a profit. Paradoxically, if they are too greedy, they actually make less money because fewer people will place bets with them. Therefore, the whole exercise is very much a balancing act in which mathematics plays a central role.

7.3 Roulette

Perhaps the best way to demonstrate how R can be used to assist in formulating a betting strategy is to simulate a roulette wheel in a typical European casino, as this is one of the purest examples of betting and illustrates how chance and the odds set ensure that the house (i.e., the casino) will always make a profit. In Example 7.1, we simulate a European roulette wheel using R and assume that ten consecutive players each bet £10 on ten spins.

For those not familiar with the game of roulette, a European roulette wheel has 18 black segments, 18 red segments, and one green segment. In the purest form of the game, players place bets as to whether the ball will land on a red or black segment when the wheel is spun. If a player places a bet of £10 that the ball will land on, say, red, and they win, then the house will give them £10; otherwise, if the ball lands on black or green, they lose their £10 stake. If we assume that the roulette wheel is completely fair and unbiased, then the probability of a player winning is 0.4865 or 48.65% (i.e., 18 out of 37 segments), which is pretty good. Players in the USA are not so lucky because casinos there have roulette wheels that have two green segments. So the overall win probability in the USA is only 47.37% (i.e., 18 out of 38 segments),

EXAMPLE 7.1: EUROPEAN ROULETTE WHEEL EXAMPLE

In this example, we simulate the action of a European roulette wheel by creating a vector called 'wheel', which contains 18 ones (representing wins) and 19 minus ones (representing loses). Then all we do is use the 'sample' function in R to select one random value from the 'wheel' vector, which represents the outcome each time the wheel is spun. If the randomly selected number is 1, then the player wins; otherwise, if the value selected is –1, then the player loses. As we will see, no matter how much each player wins or loses, the house always wins in the end because the odds are loaded in its favour.

First, we clear the workspace and specify the simulation parameters.

```
rm(list = ls())  # Clears all variables from the workspace

# Specify number of players
nplayers <- 10  # Number of players playing roulette
nspins <- 10 # Number of spins per player
wager <- 10 # Wager per spin (pounds)
```

We next create the 37-element vector, which we use to simulate the roulette wheel. Here we use the 'rep' function to create 18 consecutive 1s and 19 consecutive –1s, which we combine into a single vector.

```
# Create empty matrix to store results
outcome <- matrix(0,nplayers,nspins)
wheel <- c(rep(1,18),rep(-1,19)) # European roulette wheel (i.e. 18 black, 18 red & 1 green)
print(wheel) # This displays the roulette wheel vector.
```

```
##
[1]  1  1  1  1  1  1  1  1  1  1  1  1  1  1  1  1  1  1 -1 -1 -1 -1
[23] -1 -1 -1 -1 -1 -1 -1 -1 -1 -1 -1 -1 -1 -1 -1
```

Having done this, we now simulate the outcomes of the random spins of the wheel for each of the ten players. Note that here we have set a seed in the code to ensure that we get the same results every time we run the simulation. (NB. If different random results are required from each simulation, then simply disable (rem-out) the set.seed command line by adding a # in front of the code.) Here the 'sample' function randomly selects a single value from the 'wheel' vector.

```
# Simulate random spins of the roulette wheel
set.seed(234) # This sets the seed so that the results are reproducible.
for (i in 1:nplayers){
  for (j in 1:nspins){
    outcome[i,j] <- sample(wheel,1) # This randomly samples one value from the wheel vector.
  }

  }

# Display the results for each player
results <- outcome*wager # This computes the value of the wins and losses
print(results) # Here negative values represent losses and positive values represent wins
```

This produces the following matrix, which presents the outcome of each spin for all ten players:

```
##
      [,1] [,2] [,3] [,4] [,5] [,6] [,7] [,8] [,9] [,10]
 [1,]  -10  -10  -10  -10   10   10   10   10   10    10
 [2,]   10  -10  -10  -10   10   10  -10  -10  -10    10
 [3,]  -10   10  -10   10  -10  -10   10  -10  -10   -10
 [4,]   10   10   10  -10   10   10   10  -10  -10    10
 [5,]  -10   10   10   10   10   10   10  -10   10   -10
 [6,]   10  -10   10   10  -10   10  -10  -10  -10    10
 [7,]  -10  -10  -10  -10   10   10  -10  -10  -10   -10
 [8,]  -10   10  -10   10  -10   10  -10  -10  -10   -10
 [9,]  -10   10  -10  -10   10  -10   10   10  -10    10
[10,]   10  -10   10   10   10   10   10  -10  -10    10
```

Finally, we compute the profit and loss for each player as well as those for the casino. We do this by using the 'colSums' function, which produces the sum of each column in the data frame.

```
# This computes the total profit for each player.
profit <- colSums(results)
print(profit)
```

```
##
[1] -20   0 -20   0  40  60  20 -60 -60  20
```

From this, we see that four players made a profit and four players made losses, with two breaking even.

```
# This computes the profit for the casino.
house.profit <- -1*sum(results) # NB. This converts the negative values into positive ones.
print(house.profit)
```

```
## [1] 20
```

By comparison, the casino made a modest profit of £20.

We can compute the overall probability of winning using the following code:

```
# Now we can compute the overall number of wins and losses
win <- matrix(0,nplayers,nspins)
lose <- matrix(0,nplayers,nspins)

for (i in 1:nplayers){
  for (j in 1:nspins){
    if(outcome[i,j] <0){lose[i,j] <- outcome[i,j]}
    if(outcome[i,j] >0){win[i,j] <- outcome[i,j]}
  }
}

# Display overall wins and loses
print(sum(win)) # Total number of winning spins.
```

```
## [1] 49
```

```
print(sum(lose)) # Total number of losing spins.
```

```
## [1] -51
```

(NB. This means 51 losing spins)

```
# Compute win probability
winprob <- sum(win)/(sum(win)-sum(lose))
print(winprob)
```

```
## [1] 0.49
```

(i.e., 49% chance of winning)

From this, we can see that although the selection of winning and losing players is a completely random process, the casino always wins in the end because it sets odds that are marginally in the house's favour.

Example 7.1 perfectly illustrates the roles that the various parties involved in the sports betting industry perform when they gamble. Here, the casino is the bookmaker who sets the odds and facilitates the market in which the various players partake. Note that at no point does the casino gamble because there is no uncertainty as to whether or not the house will win – the house always wins in the end. Rather, the gambling is left up to the players, who either win or lose depending on the random behaviour of the roulette wheel.

The example also highlights 'the gambler's fallacy', which occurs when players think that they can recoup their losses by placing more bets on the roulette wheel. This comes about because of the misguided belief that in games of chance like roulette, if a certain outcome has not happened in a while, then it is more likely to occur in the future. This is wrong, as many gamblers have found out to their cost. In reality, because the casino has set the odds in their favour, the longer you play, the greater the chance of making a loss.

Soccer is not a game of pure chance, but because chance makes a significant contribution to the outcome of matches, the gambler's fallacy still applies to a certain extent. It is pointless to place bets on matches where the odds are stacked in the bookmaker's favour, because even if you can predict the winner most of the time, so can the bookmaker, with the result that they will not offer very favourable odds. In the end, if gamblers try to pick winners, the bookmaker will generally win. That is why professional gamblers often concentrate more on the match odds offered by the bookmakers rather than trying to predict the outcome of individual matches, which is where the concept of 'value' comes in.

7.4 Value Betting

The central concept behind value betting is to turn the tables on the bookmakers by finding bets where the quoted odds give the bettor the 'edge' over the bookmaker. If there is a better chance of an event occurring than that implied by the bookie's odds, then we say that the bet has 'value'. So value betting involves spotting opportunities where the bookmaker has underestimated the chance of an event occurring. In other words, bets where the bookmaker has got the odds wrong, i.e., when the implied probability for a particular outcome is actually greater than that suggested by the bookie's odds.

For any given match, the implied probability can be computed from the decimal odds using Equation 7.1. If fractional odds are stated, then these can be converted into decimal odds as follows:

$$\text{Computing decimal odds from fractional odds: } \text{Odds}_d = \text{Odds}_f + 1 \quad (7.2)$$

where Odds_d is the decimal odds and Odds_f is the fractional odds.

Should the bet be successful, then the profit that will be accrued can be calculated as follows:

$$\text{Computing the winnings from decimal odds: } w = \text{Odds}_d - 1 \quad (7.3)$$

where w is the winnings or profit.

If we consider again the match that occurred between Manchester United and Tottenham at Old Trafford on 27 August 2018, we can see that the bookmaker, Pinnacle, offered the following decimal odds: 2.61 for a home win;

2.93 for an away win, and 3.36 for a draw, which equate to implied probabilities of p=0.383 for a home win, p=0.341 for an away win and p=0.298 for a draw. From this, we can see that Pinnacle thought that the match would be a close run affair, with Manchester United having a slight advantage. In reality, however, the match was an upset, and Tottenham won 0–3, much to the annoyance of the Manchester United fans. Given the magnitude of Tottenham's away win, it is reasonable to suggest that Pinnacle might have underestimated the chance of Tottenham winning and that the 'true' probability of this outcome was actually much greater than that implied by the bookmaker's odds. Of course, we will never know if this is the case because it is not possible to work out the 'true' probability of any match outcome – all we can do is estimate it and hope that we have got our sums right! Nevertheless, if we had a crystal ball and could see the 'true' probability of Tottenham winning, we might have found that it was actually 0.4 (40%), rather than just 0.341 (34.1%). In which case, it would be worth taking a bet on Tottenham to win because the odds would be in our favour.

Taking the analogy of the roulette wheel, value betting is equivalent to replacing the casino's wheel with your own wheel in which the win probability is set in your favour – say with 19 segments where you win and only 18 where the casino wins. Now, the gambler's fallacy is reversed – the more you play, the more you win, as illustrated in Example 7.2.

EXAMPLE 7.2: VALUE BETTING EXAMPLE

In this example, we extend the roulette wheel model to simulate the outcome of the Manchester United versus Tottenham match using the odds produced by Pinnacle for the match on 27 August 2018. Here we assume that because Tottenham won 0–3, the *a priori* odds were probably a little long and that the 'true' odds should have been 2.50 for a Tottenham win. For simplicity, instead of creating a wheel with just 37 segments, here we create a match 'wheel' that has 1000 segments and apportion the ones and minus ones according to the 'true' probability (i.e., 0.4) that we have assumed for a Tottenham win.

The code to do this is presented below:

```
rm(list = ls())  # Clears all variables from workspace

b.odds <- 2.93 # Odds offered by bookmaker
t.odds <- 2.50 # Assumed 'true' odds

# Compute probabilities

# Bookmaker's estimate
b.prob <- 1/b.odds
print(b.prob)
```

```
## [1] 0.3412969
```

```
# Assumed true probability
t.prob <- 1/t.odds
print(t.prob)
```

```
## [1] 0.4
```

Now we can create the virtual match wheel and perform ten simulations of the match. Remember, just like on the real roulette wheel, all that matters is whether the ball landed on an 'away win' or not. If it did, then we attribute 1 to the outcome of the simulated match; otherwise, –1 is logged.

```
# Create a virtual roulette wheel to simulate the match outcomes
n <- round(t.prob*1000) # This is the total number of segments required on the virtual wheel.
sims <- 10 # Number of simulations (i.e. spins of the wheel).
res <- matrix(0,sims,1)
vir_wheel <- c(rep(1,n),rep(-1,(1000-n))) # Virtual roulette wheel
bet <- 10 # Amount wagered (i.e. £10)

set.seed(234)
for (i in 1:sims){
  res[i] <- sample(vir_wheel,1)
}

# Display random match outcomes
print(res)
```

```
##
       [,1]
 [1,]    -1
 [2,]     1
 [3,]     1
 [4,]    -1
 [5,]     1
 [6,]     1
 [7,]    -1
 [8,]    -1
 [9,]     1
[10,]     1
```

Assuming that £10 is bet on each simulated match, we can then compute the potential winnings as follows:

```
# Compute winnings
winnings <- matrix(0,sims,1)
for (i in 1:sims){
  if(res[i] <0){winnings[i] <- -1*bet}
  if(res[i] >0){winnings[i] <- bet*(b.odds-1)}
}

# Display winnings
print(winnings)
```

```
##
        [,1]
 [1,] -10.0
 [2,]  19.3
 [3,]  19.3
 [4,] -10.0
 [5,]  19.3
 [6,]  19.3
 [7,] -10.0
 [8,] -10.0
 [9,]  19.3
[10,]  19.3
```

From this, we can see that of the ten simulated bets made, six were successful and four were not, which, if we use the odds offered by Pinnacle, would have yielded a profit of £75.8 over the ten simulated matches.

```
# Compute profit over ten matches
prof <- sum(winnings)
print(prof)
```

```
## [1] 75.8
```

However, in reality, only one Manchester United versus Tottenham match was actually played. Therefore, the expected winnings for that match are the average of the 'prof' vector, which in this case is £7.58.

```
# Compute expected winnings per match
exp.winnings <- mean(winnings)
print(exp.winnings)
```

```
## [1] 7.58
```

Of course, Example 7.2 is just a hypothetical example based on the result of a real match, which yielded a rather unexpected result. Although we have assumed that there was value in the bet, we cannot be certain of this because the result might just have been a fluke. If, however, we go with the theory that the match odds do not actually reflect the true odds, then the more bets that we place, the more likely it is that we will yield a profit – simply because the odds are stacked in our favour. In fact, when the match simulation was repeated 100 times, the expected profit increased to £494.30. So it is well worth looking for value in bets because it greatly increases the chances of achieving a profit.

7.5 Spotting Value

Value betting all sounds very easy and logical, but the trouble comes when we try to decide which bets have value. This is because nobody, including the bookmakers, actually knows beforehand the 'true' probability of a given event occurring. Therefore, how can we decide which bets might have value? This is a major problem that has exercised the thoughts of many commentators [8–10], with many complex solutions suggested which are beyond the scope of this book. One solution is to compare the match odds offered by the bookmakers with those set on betting exchanges such as Betfair, which is generally considered a more efficient market place and thus more capable of delivering the true match odds. Another method that we will use here (Example 7.3) for simplicity is to use the odds produced by Pinnacle, as these are considered to be some of the keenest (i.e., the highest) offered, with the over-round generally much lower than that offered by other bookmakers.

In Example 7.3, which illustrates how value bets can be identified, we use the Pinnacle odds as baseline values. This is because for most matches they generally offer the keenest odds, which means that if any other bookmaker is offering higher odds for a given match, there might well be value in that bet. In which case, we should take the bet with the bookmaker offering the best odds. So, in Example 7.3, the betting strategy is simply to identify all the times when the other bookmakers offer better odds than Pinnacle and then take a bet with the bookmaker offering the highest odds. For example, for the match between Manchester United and Chelsea on 11 August 2019, which Manchester United won 4-0, Victor Chandler (VC) offered odds of 2.25 on a home win, whereas Pinnacle only offered odds of 2.21. Therefore, with hindsight, there appears to have been value in the bet offered by VC, and if we had taken that bet, we would have earned a profit of £12.50 for an original stake of £10. Of course, if there is no value in the odds being offered, we should simply decline to place a bet.

EXAMPLE 7.3: VALUE BETTING STRATEGY

In this example, we apply a value betting strategy using the odds offered by Pinnacle as benchmark values, which we then use to identify value in the odds offered by the other bookmakers. If we find that the odds offered by another bookmaker are higher than those offered by Pinnacle, we flag it as having value and then place a bet of £10 with the bookmaker offering the best odds. Otherwise, if no value is identified, we decline to place a bet.

In Example 7.3, we apply this strategy to the EPL for the entirety of season 2018–2019, which was the last uninterrupted season before the COVID-19 pandemic. Also, in order to keep things simple, we restrict ourselves here to identifying value in the home win bets only, although the code for betting on draws and away wins is included but is disabled (i.e., remmed out). In order to run the code for draws or away wins, all you need to do is select the appropriate lines of code by removing the # at the front to enable the code.

```
rm(list = ls())   # Clears all variables from workspace

# Load data
mydata <- head(read.csv('https://www.football-data.co.uk/mmz4281/1819/E0.csv'),380)

# For ease of use we will select only the variables in which we are interested.
dat <- mydata[,c("Date","HomeTeam","AwayTeam","FTHG","FTAG","FTR",
                 "B365H","B365D","B365A","BWH","BWD","BWA","IWH","IWD","IWA",
                 "PSH","PSD","PSA","WHH","WHD","WHA","VCH","VCD","VCA")]

# Inspect data
names(dat)
```

```
##
 [1] "Date"     "HomeTeam" "AwayTeam" "FTHG"     "FTAG"     "FTR"
 [7] "B365H"    "B365D"    "B365A"    "BWH"      "BWD"      "BWA"
[13] "IWH"      "IWD"      "IWA"      "PSH"      "PSD"      "PSA"
[19] "WHH"      "WHD"      "WHA"      "VCH"      "VCD"      "VCA"
```

Having loaded the data into R, we next divide the home win, draw, and away win odds offered by the various bookmakers into sub-groups, which we save as data frames called 'hwin', 'draw', and 'awin'.

```
# Group odds
hwin <- dat[,c("B365H","BWH","IWH","PSH","WHH","VCH")]
draw <- dat[,c("B365D","BWD","IWD","PSD","WHD","VCD")]
awin <- dat[,c("B365A","BWA","IWA","PSA","WHA","VCA")]
```

Next, we create several 'empty' vectors in which we are going to store our results.

```
# Create value vectors
val <- matrix(0,nrow(dat),1) # Value indicator
odds <- matrix(0,nrow(dat),1) # Bet odds
OC <- matrix(-1,nrow(dat),1) # Bet outcome indicator
W <- matrix(0,nrow(dat),1) # Wins
L <- matrix(0,nrow(dat),1) # Losses
```

Now, we select the type of bet in which we are interested. In this case, we will select only the home win bets. (NB. In order to analyse for draws and away wins, select either 'draw' or 'awin' instead of 'hwin'.)

```
# Select the appropriate data set (i.e. hwin, draw, or awin)
data <- hwin   # Select when evaluating home win bets.
#data <- draw # Select when evaluating draw bets.
#data <- awin  # Select when evaluating away win bets.
```

Here we have selected the home win data frame 'hwin', which we have given the generic name 'data'. We can inspect this using the 'head', which here we use to view the first 20 rows.

```
# Inspect data
head(data, 20)
```

This displays the decimal odds on a home win offered by Bet365 (B365H), Bet&Win (BWH), Interwetten (IWH), Pinnacle (PSH), William Hill (WHH), and VC Bet (VCH), respectively.

```
##
    B365H  BWH  IWH  PSH  WHH  VCH
1    1.57 1.53 1.55 1.58 1.57 1.57
2    1.90 1.90 1.90 1.89 1.91 1.87
3    2.50 2.45 2.40 2.50 2.45 2.50
4    6.50 6.25 6.20 6.41 5.80 6.50
5    3.90 3.80 3.70 3.83 3.80 3.90
6    2.37 2.35 2.20 2.43 2.38 2.40
7    2.37 2.35 2.25 2.36 2.30 2.38
8    4.00 3.70 3.60 4.00 3.80 3.90
9    1.25 1.20 1.25 1.27 1.25 1.25
10   1.85 1.80 1.80 1.86 1.83 1.85
```

```
11   3.25 3.10 2.90 3.25 3.10 3.25
12   1.80 1.78 1.80 1.83 1.83 1.80
13   1.90 1.91 2.00 1.89 1.91 1.91
14   2.04 2.00 2.10 2.05 2.05 2.05
15   1.28 1.28 1.27 1.28 1.25 1.29
16   2.10 2.10 2.05 2.09 2.05 2.10
17   5.50 5.50 5.10 5.37 4.80 5.50
18   2.45 2.40 2.40 2.47 2.38 2.40
19   1.10 1.07 1.08 1.10 1.06 1.06
20   7.00 6.75 7.00 7.34 6.00 7.00
```

Now, we can identify the matches where there is value in the potential bets. We identify these by placing a 1 in the 'val' indicator vector.

```
for(i in 1:nrow(dat)){
  odds[i] <- max(data[i,])
  if(data$PSH[i] < odds[i]){val[i] <- 1} # Select for home win.
  #if(data$PSD[i] < odds[i]){val[i] <- 1} # Select for draw.
  #if(data$PSA[i] < odds[i]){val[i] <- 1} # Select for away win.
}
```

Having identified the value bets, we need to specify how much we are going to wager on each match, which in this case is £10 for every match identified.

```
# Specify wager value on each bet.
stake <- 10 # £10 wagered on the bet
```

Now, we compute the winnings and losses achieved on the bets made and compile them into a results table.

```
for(i in 1:nrow(dat)){
  if(dat$FTR[i] == "H" & val[i] >0){OC[i] <- 1} # Select for home win.
  #if(dat$FTR[i] == "D" & val[i] >0){OC[i] <- 1} # Select for draw.
  #if(dat$FTR[i] == "A" & val[i] >0){OC[i] <- 1} # Select for away win.
  if(OC[i] >0){W[i] <- stake*(odds[i]-1)}
  else if(val[i] >0 & OC[i] <0){L[i] <- stake}
}

results <- cbind(dat[,c(1:6)],data,val,odds,OC,W,L)
head(results,20)
```

This displays the results for the first 20 matches, as follows:

```
##
         Date        HomeTeam       AwayTeam FTHG FTAG FTR B365H  BWH  IWH
1  10/08/2018      Man United      Leicester    2    1   H  1.57 1.53 1.55
2  11/08/2018     Bournemouth        Cardiff    2    0   H  1.90 1.90 1.90
3  11/08/2018          Fulham Crystal Palace    0    2   A  2.50 2.45 2.40
4  11/08/2018    Huddersfield        Chelsea    0    3   A  6.50 6.25 6.20
5  11/08/2018       Newcastle      Tottenham    1    2   A  3.90 3.80 3.70
6  11/08/2018         Watford       Brighton    2    0   H  2.37 2.35 2.20
7  11/08/2018          Wolves        Everton    2    2   D  2.37 2.35 2.25
8  12/08/2018         Arsenal       Man City    0    2   A  4.00 3.70 3.60
9  12/08/2018       Liverpool       West Ham    4    0   H  1.25 1.20 1.25
10 12/08/2018     Southampton        Burnley    0    0   D  1.85 1.80 1.80
11 18/08/2018         Cardiff      Newcastle    0    0   D  3.25 3.10 2.90
12 18/08/2018         Chelsea        Arsenal    3    2   H  1.80 1.78 1.80
13 18/08/2018         Everton    Southampton    2    1   H  1.90 1.91 2.00
14 18/08/2018       Leicester         Wolves    2    0   H  2.04 2.00 2.10
15 18/08/2018       Tottenham         Fulham    3    1   H  1.28 1.28 1.27
16 18/08/2018        West Ham    Bournemouth    1    2   A  2.10 2.10 2.05
17 19/08/2018        Brighton     Man United    3    2   H  5.50 5.50 5.10
18 19/08/2018         Burnley        Watford    1    3   A  2.45 2.40 2.40
19 19/08/2018        Man City   Huddersfield    6    1   H  1.10 1.07 1.08
20 20/08/2018 Crystal Palace      Liverpool    0    2   A  7.00 6.75 7.00
    PSH  WHH  VCH val odds OC    W  L
1  1.58 1.57 1.57   0 1.58 -1  0.0  0
2  1.89 1.91 1.87   1 1.91  1  9.1  0
3  2.50 2.45 2.50   0 2.50 -1  0.0  0
4  6.41 5.80 6.50   1 6.50 -1  0.0 10
5  3.83 3.80 3.90   1 3.90 -1  0.0 10
6  2.43 2.38 2.40   0 2.43 -1  0.0  0
7  2.36 2.30 2.38   1 2.38 -1  0.0 10
8  4.00 3.80 3.90   0 4.00 -1  0.0  0
9  1.27 1.25 1.25   0 1.27 -1  0.0  0
10 1.86 1.83 1.85   0 1.86 -1  0.0  0
11 3.25 3.10 3.25   0 3.25 -1  0.0  0
12 1.83 1.83 1.80   0 1.83 -1  0.0  0
13 1.89 1.91 1.91   1 2.00  1 10.0  0
14 2.05 2.05 2.05   1 2.10  1 11.0  0
15 1.28 1.25 1.29   1 1.29  1  2.9  0
16 2.09 2.05 2.10   1 2.10 -1  0.0 10
17 5.37 4.80 5.50   1 5.50  1 45.0  0
18 2.47 2.38 2.40   0 2.47 -1  0.0  0
19 1.10 1.06 1.06   0 1.10 -1  0.0  0
20 7.34 6.00 7.00   0 7.34 -1  0.0  0
```

Here, when val=1, it indicates that there is value in the bet, and when OC=1, it indicates that the bet was successful and that the home team won. The win and lose columns indicate how much was won or lost on any bets made.

Finally, we compute how much was won and lost and the total profit made over the entire season.

```
sum(W)
```

```
## [1] 1644.9
```

```
sum(L)
```

```
## [1] 1240
```

```
Profit <- sum(W)-sum(L)
print(Profit)
```

```
## [1] 404.9
```

From this, we see that for season 2018–2019, by adopting a value betting strategy based on home wins only, we achieved a tidy profit of £404.90 for a £10 stake on each match.

While in Example 7.3 a reasonable profit was achieved for season 2018–2019, it is important not to cherry-pick. The big question is therefore: will the same value betting strategy be successful when applied to other seasons, as well as to draws and away wins? To answer this question, the code in Example 7.3 was applied to eight EPL seasons from 2013 to 2021. The results of this analysis are presented in Table 7.1, which reveals that the home win and away win value betting strategies yielded total profits of £895.30 and £1195.33, respectively, while the draw strategy was a dismal failure, resulting in a £541.90 loss. So we can say with reasonable confidence that the value betting strategy, based on the Pinnacle odds, works pretty well when identifying home and away win bets that have value but is no good when applied to draws.

TABLE 7.1

Profits Achieved by Various Value Betting Strategies for the English Premier League from Season 2013–2014 to Season 2020–2021

Season	HW Strategy Profit (£)	Draw Strategy Profit (£)	AW Strategy Profit (£)	HW Prob.	Draw Prob.	AW Prob.	Pandemic
2013–2014	214.50	–354.90	63.23	0.471	0.205	0.324	No
2014–2015	115.90	–28.50	–178.80	0.453	0.245	0.303	No
2015–2016	–65.00	159.00	529.60	0.413	0.282	0.305	No
2016–2017	95.20	–66.90	–79.60	0.492	0.221	0.287	No
2017–2018	177.40	231.00	–374.50	0.455	0.261	0.284	No
2018–2019	404.90	–444.70	5.00	0.476	0.187	0.337	No
2019–2020	331.10	–4.00	243.3	0.453	0.242	0.305	Yes
2020–2021	–378.70	–32.90	987.1	0.379	0.218	0.403	Yes
Mean	**111.91**	**–67.74**	**149.42**	**0.449**	**0.233**	**0.319**	**n.a.**
Total	**895.30**	**–541.90**	**1195.33**	**n.a.**	**n.a.**	**n.a.**	**n.a.**

The results in Table 7.1 are particularly interesting because they reveal that the home win (HW) strategy was more consistent than the away win (AW) strategy. In particular, the performance of the two strategies appears to have been heavily influenced by the disruption to the EPL caused by the COVID-19 pandemic. For example, in season 2020–2021, at the height of the pandemic, there was a noticeable drop in the probability of the home team winning, which was matched by a corresponding rise in the probability of away wins occurring. This might explain why, for season 2020–2021, the home win strategy was very unsuccessful, achieving a £378.70 loss, while the away win strategy was unusually successful, achieving a £987.10 profit.

7.6 Arbitrage

The words 'gambling' and 'betting' are often used interchangeably, with most people thinking that they are the same thing. However, there is a subtle difference between the two. Gambling implies uncertainty – hence "taking a gamble", because the outcome is not known. Betting is a deliberate, calculated action, which does not necessarily involve uncertainty – although often it does. So when formulating a betting strategy, we can take a high-risk, but potentially high-reward approach that involves a lot of uncertainty (i.e., gambling), or alternatively, we can minimise the risk (i.e., reduce the uncertainty) at the cost of a much lower return on our investment. The word 'investment' is used deliberately here because there are a lot of similarities between successful betting and investment strategies. If we consider things from this point of view, then arbitrage betting is the ultimate low-risk strategy because, in theory, it involves no uncertainty at all and therefore is not, strictly speaking, gambling. It involves exploiting inefficiencies in the sports betting market [11–13], and if executed correctly, arbitrage betting will yield a guaranteed profit.

In soccer, arbitrage betting opportunities come about when competing bookmakers set odds that have pricing discrepancies that leave room for a bet that cannot lose, provided, of course, that the appropriate wager is placed on each of the three possible match outcomes (i.e., home win, draw, and away win). The sports betting industry is fiercely competitive, and so companies compete with each other to try and provide gamblers with the best odds while still making a profit. Although every individual bookmaker will always over-round and offer match odds that add up to an overall probability greater than one (i.e., greater than 100%), for various technical reasons, pricing discrepancies sometimes occur between the odds offered by different bookmakers, and this gives rise to arbitrage opportunities. For example, take the match that occurred on 8 August 2015 in the EPL between Bournemouth and Aston Villa, which Aston Villa won 0-1. The best odds that were offered by the various bookmakers on this match are summarised in Table 7.2.

TABLE 7.2

Potential Profits Achieved Using an Arbitrage Betting Strategy on the Match between Bournemouth and Aston Villa – 8 August 2015

Nominal Wager (£) (A)	Bookmaker	Outcome	Odds Offered (B)	Outcome Probability ($C = 1 \div B$)	Wager per Outcome (£) ($D = A \times C$)	Individual Bet Profit (£) ($E = D \times (B-1)$)	Potential Arbitrage Profit (£)
1000	Interwetten	Home win	2.10	0.476	476	523.60	15.60
	Pinnacle	Draw	3.65	0.274	274	726.10	16.10
	Pinnacle	Away win	4.27	0.234	234	765.18	15.18
	Total	**n.a.**	**n.a.**	**0.984**	**984**	**n.a.**	**n.a.**

A nominal wager of £1000 is assumed.

From this, we can see that Interwetten offered decimal odds of 2.10 for a home win, while Pinnacle offered odds of 3.65 and 4.27 for a draw and an away win, respectively. The important point is that when these three match outcome odds are converted to implied probabilities using equation 7.1 and are then summed up, the total of the implied probabilities comes to just 0.984, which is less than one – implying that an arbitrage opportunity exists. By contrast, if the total of the implied probabilities had been greater than one, then no opportunity would have existed because the over-round would have guaranteed that the bookmakers made a profit.

From Table 7.2, we can see that for this match, the implied probabilities of a home win, a draw, and an away win were 0.476, 0.274, and 0.234, respectively. As well as telling us that we have an arbitrage opportunity here, these probabilities can also be used to compute how much money should be wagered on each bet, as follows:

$$\text{Stake to be wagered on each outcome bet: } s = p \times m \tag{7.4}$$

where s is the stake to be wagered on the individual outcome bet; p is the implied probability of the outcome; and m is the nominal amount of money to be wagered on the match (£).

When equation 7.4 is applied, assuming a nominal wager of £1000 on the whole match, it tells us that we need to bet: £476 on a home win with Interwetten; and £274 and £234 on a draw and an away win, respectively, with Pinnacle. If Bournemouth won, then the profit would be £15.60, and if Aston Villa won, it would be £15.18. For a draw, the profit would be £16.10. In reality, however, Aston Villa won the match, and so the arbitrage bet would have yielded a total profit of £15.18 (i.e., £765.18 – £476 – £274), which is a yield of about 1.6% on the money 'invested'.

So if arbitrage betting guarantees a profit, why isn't everyone doing it? Well, the answer to this is complex and comes down to several factors, which include:

- Arbitrage betting is somewhat boring because it involves no risk and is essentially not gambling. This means that it doesn't produce the adrenaline rush that some people find exciting when gambling. Arbitrage betting feels more like accounting because there is almost no uncertainty, with all the match outcomes producing a similar profit.
- The yields produced by each arbitrage bet can be very low – typically 1%–3% of the overall amount wagered, depending on the league. Consequently, in order to achieve a decent return for any given match, relatively large sums of money need to be wagered in the first place.

- Arbitrage opportunities are relatively rare and difficult to spot, particularly in high-profile leagues like the EPL. They do, however, crop up more frequently in obscure minor soccer leagues, which may be unfamiliar to bookmakers. Therefore, the challenge becomes one of spotting arbitrage opportunities rather than actually placing the bets. For this reason, many arbitrage bettors (arbers) employ commercial software packages to spot opportunities. However, this can be expensive and adds overhead costs to the arbitrage betting process, which must be recuperated from the winnings.
- Arbitrage betting can go wrong. Despite being foolproof in theory, in practice, things can go wrong procedurally when placing arbitrage bets. One of the major problems lies in the fact that for any given match, multiple bets have to be placed with different bookmakers, which takes time, and requires multiple bookmaker accounts. Unfortunately, with the online sports markets, arbitrage bets generally have a very short lifespan, with many opportunities only existing for a few minutes [14]. So, if one leg of the arbitrage bet vanishes midway through the process, say after we have placed the other two bets, then the whole arbitrage opportunity disappears and is transformed instead into a conventional bet with the usual risks involved. The same problem occurs if one of the bookmakers decides to cancel your bet for any reason, in which case the whole arbitrage falls apart.
- Many bookmakers discourage arbitrage betting and will close or restrict the accounts of customers whom they suspect of being 'arbers'. This is very much an occupational hazard with arbitrage betting. For this reason, many arbitrage bettors wager rounded sums of money (e.g., £90) on any bets made, as placing, say, £89.31 on Liverpool to win at home is a dead giveaway that the bet is one leg of an arbitrage.

Notwithstanding this, arbitrage betting can be profitable if opportunities are spotted and the corresponding bets are executed successfully. However, spotting opportunities can be difficult. Example 7.4 presents some R code that can be used to retrospectively spot and assess arbitrage opportunities in the EPL. This example illustrates how arbitrage opportunities can be spotted and exploited despite the familiarity of the EPL to bookmakers.

EXAMPLE 7.4: ARBITRAGE BETTING EXAMPLE

In this example, we will use the data frames 'dat', 'hwin', 'draw', and 'awin' that we created in Example 7.3 to illustrate how an arbitrage betting strategy can be applied to the EPL. Here we apply the R code to the EPL for season 2018–2019.

First, we generate some empty vectors and matrices in which to store the results.

```
# Create empty vectors to store results
hmax <- matrix(0,nrow(dat),2) # Best home win odds offered
dmax <- matrix(0,nrow(dat),2) # Best draw odds offered
amax <- matrix(0,nrow(dat),2) # Best away win odds offered
prob <- matrix(0,nrow(dat),1) # Sum of implied probabilities
arb <- matrix(0,nrow(dat),1) # Arbitrage opportunity indicator
```

Next, we compute the arbitrage opportunities and compile them into a data frame. This is done by identifying the best (maximum) odds offered on each match by the various bookmakers and then computing their inverse, which yields the implied probability. Here we also use the 'which.max' function to identify the bookmaker offering the best odds.

```
for(i in 1:nrow(dat)){
  hmax[i,1] <- round(1/(max(hwin[i,])),3) # Identifies best odds and computes the probability
  hmax[i,2] <- which.max(hwin[i,]) # Identifies location of best odds in 'hwin' data frame
  dmax[i,1] <- round(1/(max(draw[i,])),3)
  dmax[i,2] <- which.max(draw[i,])
  amax[i,1] <- round(1/(max(awin[i,])),3)
  amax[i,2] <- which.max(awin[i,])
  prob[i] <- hmax[i,1]+dmax[i,1]+amax[i,1]
  if(prob[i] <1){arb[i] <- 1}
}

# Compile arbitrage opportunity data frame
arb.temp <- cbind.data.frame(hmax,dmax,amax,prob,arb)
colnames(arb.temp) <- c("Hprob","Hbm","Dprob","Dbm","Aprob","Abm","Prob","Arb")
arb.ops <- cbind.data.frame(dat[,c(1:3,6)],arb.temp)
head(arb.ops,20)  # NB. Hbm, Dbm & Abm are the best odds offered.
```

This displays the first 20 matches, as follows:

```
##
         Date      HomeTeam        AwayTeam FTR Hprob Hbm Dprob Dbm Aprob
1  10/08/2018    Man United       Leicester   H 0.633   4 0.250   2 0.133
2  11/08/2018   Bournemouth         Cardiff   H 0.524   5 0.275   4 0.211
3  11/08/2018        Fulham Crystal Palace   A 0.400   1 0.289   4 0.333
4  11/08/2018  Huddersfield         Chelsea   A 0.154   1 0.249   4 0.617
5  11/08/2018     Newcastle       Tottenham   A 0.256   1 0.280   4 0.476
6  11/08/2018       Watford        Brighton   H 0.412   4 0.303   3 0.294
7  11/08/2018        Wolves         Everton   D 0.420   6 0.294   4 0.303
8  12/08/2018       Arsenal        Man City   A 0.250   1 0.250   6 0.500
9  12/08/2018     Liverpool        West Ham   H 0.787   4 0.148   2 0.071
10 12/08/2018   Southampton         Burnley   D 0.538   4 0.278   3 0.192
11 18/08/2018       Cardiff       Newcastle   D 0.308   1 0.312   3 0.392
12 18/08/2018       Chelsea         Arsenal   H 0.546   4 0.250   1 0.222
13 18/08/2018       Everton     Southampton   H 0.500   3 0.281   4 0.211
14 18/08/2018     Leicester          Wolves   H 0.476   3 0.281   4 0.250
15 18/08/2018     Tottenham          Fulham   H 0.775   6 0.159   4 0.080
```

```
16 18/08/2018        West Ham    Bournemouth  A 0.476   1 0.274   4 0.267
17 19/08/2018        Brighton     Man United  H 0.182   1 0.278   1 0.552
18 19/08/2018         Burnley        Watford  A 0.405   4 0.323   1 0.286
19 19/08/2018        Man City   Huddersfield  H 0.909   1 0.077   1 0.029
20 20/08/2018 Crystal Palace       Liverpool  A 0.136   4 0.196   4 0.690
   Abm  Prob Arb
1    1 1.016   0
2    6 1.010   0
3    1 1.022   0
4    4 1.020   0
5    6 1.012   0
6    1 1.009   0
7    1 1.017   0
8    3 1.000   0
9    1 1.006   0
10   6 1.008   0
11   6 1.012   0
12   1 1.018   0
13   1 0.992   1
14   6 1.007   0
15   4 1.014   0
16   2 1.017   0
17   4 1.012   0
18   6 1.014   0
19   6 1.015   0
20   3 1.022   0
```

From this, we see that in the first 20 matches of the 2018–2019 season, only one arbitrage opportunity arose. This is the match (number 13) between Everton and Southampton, indicated by a 1 in the 'Arb' column.

In order to compute the winnings, we first need to convert the probabilities for the selected bets back into decimal odds, as follows:

```
# Convert probabilities to odds
arb.bets <- arb.ops
arb.bets[,5] <- round(1/arb.ops[,5],3)
arb.bets[,7] <- round(1/arb.ops[,7],3)
arb.bets[,9] <- round(1/arb.ops[,9],3)
colnames(arb.bets) <- c("Date","HomeTeam","AwayTeam","FTR",
                    "Hodds","Hbm","Dodds","Dbm","Aodds","Abm","Prob","Arb")
head(arb.bets,20)
```

```
##
         Date       HomeTeam       AwayTeam FTR Hodds Hbm  Dodds Dbm
1  10/08/2018     Man United      Leicester   H 1.580   4  4.000   2
2  11/08/2018    Bournemouth        Cardiff   H 1.908   5  3.636   4
3  11/08/2018         Fulham Crystal Palace   A 2.500   1  3.460   4
4  11/08/2018   Huddersfield        Chelsea   A 6.494   1  4.016   4
5  11/08/2018      Newcastle      Tottenham   A 3.906   1  3.571   4
6  11/08/2018        Watford       Brighton   H 2.427   4  3.300   3
7  11/08/2018         Wolves        Everton   D 2.381   6  3.401   4
8  12/08/2018        Arsenal       Man City   A 4.000   1  4.000   6
9  12/08/2018      Liverpool       West Ham   H 1.271   4  6.757   2
10 12/08/2018    Southampton        Burnley   D 1.859   4  3.597   3
11 18/08/2018        Cardiff      Newcastle   D 3.247   1  3.205   3
```

```
12 18/08/2018         Chelsea         Arsenal  H 1.832  4  4.000  1
13 18/08/2018         Everton     Southampton  H 2.000  3  3.559  4
14 18/08/2018       Leicester          Wolves  H 2.101  3  3.559  4
15 18/08/2018       Tottenham          Fulham  H 1.290  6  6.289  4
16 18/08/2018        West Ham     Bournemouth  A 2.101  1  3.650  4
17 19/08/2018        Brighton      Man United  H 5.495  1  3.597  1
18 19/08/2018         Burnley         Watford  A 2.469  4  3.096  1
19 19/08/2018        Man City    Huddersfield  H 1.100  1 12.987  1
20 20/08/2018  Crystal Palace       Liverpool  A 7.353  4  5.102  4
    Aodds Abm  Prob Arb
1   7.519   1 1.016   0
2   4.739   6 1.010   0
3   3.003   1 1.022   0
4   1.621   4 1.020   0
5   2.101   6 1.012   0
6   3.401   1 1.009   0
7   3.300   1 1.017   0
8   2.000   3 1.000   0
9  14.085   1 1.006   0
10  5.208   6 1.008   0
11  2.551   6 1.012   0
12  4.505   1 1.018   0
13  4.739   1 0.992   1
14  4.000   6 1.007   0
15 12.500   4 1.014   0
16  3.745   2 1.017   0
17  1.812   4 1.012   0
18  3.497   6 1.014   0
19 34.483   6 1.015   0
20  1.449   3 1.022   0
```

Next, we calculate how much should be wagered on each of the possible match outcomes using the computed probabilities and a nominal total wager of £1000.

```
# Specify nominal wager on each bet
nom.wager <- 1000 # i.e. £1000

# Compute stakes to be placed on each bet
Hstake <- matrix(0,nrow(dat),1)
Dstake <- matrix(0,nrow(dat),1)
Astake <- matrix(0,nrow(dat),1)
wager <- matrix(0,nrow(dat),1)

for(i in 1:nrow(dat)){
  Hstake[i] <- round((arb.ops$Hprob[i]*nom.wager),2)
  Dstake[i] <- round((arb.ops$Dprob[i]*nom.wager),2)
  Astake[i] <- round((arb.ops$Aprob[i]*nom.wager),2)
  wager[i] <- round((Hstake[i]+Dstake[i]+Astake[i]),2)
}

# Compile data frame of monies wagered.
money.bet <- cbind.data.frame(Hstake,Dstake,Astake,wager)

# Display 'money.bet' data frame
head(money.bet, 20)
```

```
##
   Hstake Dstake Astake wager
1     633    250    133  1016
2     524    275    211  1010
3     400    289    333  1022
4     154    249    617  1020
5     256    280    476  1012
6     412    303    294  1009
7     420    294    303  1017
8     250    250    500  1000
9     787    148     71  1006
10    538    278    192  1008
11    308    312    392  1012
12    546    250    222  1018
13    500    281    211   992
14    476    281    250  1007
15    775    159     80  1014
16    476    274    267  1017
17    182    278    552  1012
18    405    323    286  1014
19    909     77     29  1015
20    136    196    690  1022
```

From this, we can see that for match number 13, the one with the arbitrage opportunity, the algorithm tells us to place £500 on a home win, £281 on a draw, and £211 on an away win. Notice that for this match, the total amount wagered is £992, which is slightly lower than the nominal amount of £1000 that we allocated for each wager.

Now, we compare each of the possible match outcomes with the actual results for the identified arbitrage bets to compute the profit.

```
# Compute profits from arbitrage bets
profit <- matrix(0,nrow(dat),1)

for(i in 1:nrow(dat)){
  if(arb.bets$Arb[i] >0){
    if(arb.bets$FTR[i] == "H"){profit[i] <- (Hstake[i]*(arb.bets$Hodds[i]-1))-Dstake[i]-Astake[i]}
    if(arb.bets$FTR[i] == "D"){profit[i] <- (Dstake[i]*(arb.bets$Dodds[i]-1))-Hstake[i]-Astake[i]}
    if(arb.bets$FTR[i] == "A"){profit[i] <- (Astake[i]*(arb.bets$Aodds[i]-1))-Hstake[i]-Dstake[i]}
  }
}

# Display results
profit <- round(profit,2)
arb.res <- cbind(arb.bets[,c(1:10,12)],money.bet,profit)
head(arb.res,20)
```

```
##
         Date       HomeTeam        AwayTeam FTR Hodds Hbm  Dodds Dbm
1  10/08/2018     Man United       Leicester   H 1.580   4  4.000   2
2  11/08/2018    Bournemouth         Cardiff   H 1.908   5  3.636   4
3  11/08/2018         Fulham  Crystal Palace   A 2.500   1  3.460   4
4  11/08/2018   Huddersfield         Chelsea   A 6.494   1  4.016   4
5  11/08/2018      Newcastle       Tottenham   A 3.906   1  3.571   4
6  11/08/2018        Watford        Brighton   H 2.427   4  3.300   3
7  11/08/2018         Wolves         Everton   D 2.381   6  3.401   4
8  12/08/2018        Arsenal        Man City   A 4.000   1  4.000   6
9  12/08/2018      Liverpool        West Ham   H 1.271   4  6.757   2
10 12/08/2018    Southampton         Burnley   D 1.859   4  3.597   3
11 18/08/2018        Cardiff       Newcastle   D 3.247   1  3.205   3
12 18/08/2018        Chelsea         Arsenal   H 1.832   4  4.000   1
13 18/08/2018        Everton     Southampton   H 2.000   3  3.559   4
14 18/08/2018      Leicester          Wolves   H 2.101   3  3.559   4
15 18/08/2018      Tottenham          Fulham   H 1.290   6  6.289   4
16 18/08/2018       West Ham     Bournemouth   A 2.101   1  3.650   4
17 19/08/2018       Brighton      Man United   H 5.495   1  3.597   1
18 19/08/2018        Burnley         Watford   A 2.469   4  3.096   1
19 19/08/2018       Man City    Huddersfield   H 1.100   1 12.987   1
20 20/08/2018 Crystal Palace       Liverpool   A 7.353   4  5.102   4
    Aodds Abm Arb Hstake Dstake Astake wager profit
1   7.519   1   0    633    250    133  1016      0
2   4.739   6   0    524    275    211  1010      0
3   3.003   1   0    400    289    333  1022      0
4   1.621   4   0    154    249    617  1020      0
5   2.101   6   0    256    280    476  1012      0
6   3.401   1   0    412    303    294  1009      0
7   3.300   1   0    420    294    303  1017      0
8   2.000   3   0    250    250    500  1000      0
9  14.085   1   0    787    148     71  1006      0
10  5.208   6   0    538    278    192  1008      0
11  2.551   6   0    308    312    392  1012      0
12  4.505   1   0    546    250    222  1018      0
13  4.739   1   1    500    281    211   992      8
14  4.000   6   0    476    281    250  1007      0
15 12.500   4   0    775    159     80  1014      0
16  3.745   2   0    476    274    267  1017      0
17  1.812   4   0    182    278    552  1012      0
18  3.497   6   0    405    323    286  1014      0
19 34.483   6   0    909     77     29  1015      0
20  1.449   3   0    136    196    690  1022      0
```

From this, we see that the Everton versus Southampton arbitrage opportunity match yielded a profit of £8 for a wager of £992.

Finally, we compute some performance metrics for the arbitrage strategy over the entire season.

```
# Compute number of arbitrage bets made and profit for the season
nbets <-sum(arb.res$Arb) # This is number of arbitrage bets made.
print(nbets)
```

```
## [1] 33
```

```
tot.profit <- sum(profit) # This is the profit.
print(tot.profit)
```

```
## [1] 303.71
```

```
approx.yield <- round((100*tot.profit/(nom.wager*nbets)),2) # Approx. yield per bet.
print(approx.yield) # Expressed as a percentage.
```

```
## [1] 0.92
```

So over season 2018–2019, 33 arbitrage opportunities were identified, and these yielded a total profit of £303.71 for a nominal wager of £1000 on each match. So, in effect, £33,000 was wagered to recoup £303.71, which is a yield of about 0.92%. Having said this, this is somewhat misleading because in theory the bets involved no risk at all, with the $1000 stake simply being recycled over and over again – in which case it could be argued that the yield was 30.37%. However, this of course assumes that all the arbitrage bets were executed perfectly and that nothing went wrong when the bets were placed – something that cannot always be guaranteed.

From Example 7.4, it can be seen that even in a league as popular and familiar as the EPL, arbitrage opportunities exist, although they occur somewhat infrequently. If we apply the code in Example 7.4 to the eight EPL seasons from 2013 to 2021, we find that a total of 369 arbitrage opportunities occurred, yielding potentially a total profit of £3322.31 for a nominal wager of £1000 per match, as shown in Table 7.3. However, it is noticeable that the number of arbitrage opportunities greatly decreased in seasons 2019–2020 and 2020–2021, possibly due to the disruption caused by the COVID-19 pandemic.

TABLE 7.3

Profits Achieved by an Arbitrage Betting Strategy Applied to the English Premier League from Season 2013–2014 to Season 2020–2021

Season	Nominal Wager per Match (£)	Number of Arbitrage Bets	Arbitrage Profit (£)	Arbitrage Yield (%)	Pandemic
2013–2014	1000	58	396.05	0.68	No
2014–2015	1000	73	492.09	0.67	No
2015–2016	1000	17	139.14	0.82	No
2016–2017	1000	71	775.16	1.09	No
2017–2018	1000	107	1144.86	1.07	No
2018–2019	1000	33	303.71	0.92	No
2019–2020	1000	5	52.13	1.04	Yes
2020–2021	1000	5	19.17	0.38	Yes
Mean	**1000.00**	**46.1**	**415.29**	**0.83**	**n.a.**
Total	**n.a.**	**369**	**3322.31**	**n.a.**	**n.a.**

7.7 Money Management and Discipline

Although this chapter takes a statistical and data science approach to the subject of sports betting, it is perhaps worth adding a few words of caution about money management and the general discipline that should be employed when gambling in the soccer market. Sports betting can be extremely addictive [4,6,15] and can lead to disastrous losses being incurred by bettors. One reason for this is that sports betting differs from gambling products such as the lottery or roulette in that it is marketed as a combination of skills and chance, similar to poker. This combination encourages bettors to overestimate the role that their own sports knowledge plays in securing winnings while tending to make individuals underestimate the extent to which chance plays a role in the outcome of sporting events [15], something reinforced by extensive advertising [6]. Therefore, it is vital when undertaking sports betting to be highly disciplined and to manage your money carefully – that is of course if you are interested in making a profit.

From the analysis presented in Examples 7.3 and 7.4, it can be seen that it is possible to make a consistent profit from sports betting using R – provided, of course, that the bookmakers don't close your accounts because you are too successful [7]. However, betting using algorithms in R can be a rather boring process that requires lots of discipline, hard work, and time – often for a relatively modest return. Also, it must be undertaken in a rigorous and methodical manner, which is very far from the adrenalin-fuelled, 'get rich quick' picture portrayed in the adverts produced by the sports betting industry.

Key to being successful is the use of a robust betting strategy that has clearly defined rules, criteria, and thresholds that must be met before any bets are placed. Such a strategy should be tested and validated using historical data to ensure that it will produce a profit when used in real life. Once developed and refined, the betting strategy should be applied rigorously, with any temptation to 'gamble' outside the specified rules resisted.

Another thing that is essential when undertaking sports betting is money management. This is because it is all too easy to bet and lose lots of money, especially if you are trying to recover your losses, which of course is the gambler's fallacy! Therefore, if you want to take sports betting seriously, it is advisable to set aside a 'pot' of money, say £1000, specifically for the purposes of gambling, which you should try to grow in a similar way to someone growing an investment account. In addition, you should look to wager only about 2%–2.5% of your total pot on any one occasion. This means that if you want to bet on Saturday's EPL games, you will be restricted to a total of £20–25, assuming that your pot is £1000. By doing this, you will not only reduce the chances of losing all your money very quickly but also be forced to think carefully about the matches on which you plan to place bets. As such, good money management reinforces discipline and helps you resist the temptation to make reckless gambles.

Lastly, it is perhaps worth stating once again that being a successful sports bettor is much more akin to running an investment account rather than the fun-loving, adrenalin-fuelled lifestyle portrayed in the sports betting adverts. If you wish to pursue sports betting seriously, then the R code presented in this chapter should help you get started. However, it is important to remember that this chapter is only an introduction to the subject and that in order to be truly successful, you will have to develop, refine, and validate your own betting models using R.

References

1. Lock S: Global amount wagered on European soccer 2019–2020, by league. In: *Statistica*; https://www.statista.com/statistics/1263462/value-betting-on-european-soccer/; 2021.
2. Houghton S, Moss M, Casey E: Affiliate marketing of sports betting: A cause for concern? *International Gambling Studies* 2020, 20(2):240–245.
3. Levitt SD: How do markets function? An empirical analysis of gambling on the National Football League. National Bureau of Economic Research, Cambridge, MA; 2003.
4. Budimir I, Jelaska PM: Statistical analysis of long term betting phenomenon: Journey game to addiction. *Acta Kinesiologica* 2017, 11(Suppl. 2):28–32.
5. Buchdahl J: *Squares & Sharps, Suckers & Sharks: The Science, Psychology & Philosophy of Gambling*: Oldcastle Books, Harpenden; 2021.

6. Lopez-Gonzalez H, Estevez A, Griffiths MD: Marketing and advertising online sports betting: A problem gambling perspective. *Journal of Sport and Social Issues* 2017, 41(3):256–272.
7. Davies R: Revealed: How bookies clamp down on successful gamblers and exploit the rest. In: *The Guardian*; https://www.theguardian.com/society/2022/feb/19/stake-factoring-how-bookies-clamp-down-on-successful-gamblers; 2022.
8. Ramesh S, Mostofa R, Bornstein M, Dobelman J: Beating the House: Identifying Inefficiencies in Sports Betting Markets. *arXiv preprint arXiv:191008858*; 2019.
9. Sauer RD, Brajer V, Ferris SP, Marr MW: Hold your bets: Another look at the efficiency of the gambling market for National Football League games. *Journal of Political Economy* 1988, 96(1):206–213.
10. Snyder WW: Horse racing: Testing the efficient markets model. *The Journal of Finance* 1978, 33(4):1109–1118.
11. Constantinou AC, Fenton NE: Profiting from arbitrage and odds biases of the European football gambling market. *The Journal of Gambling Business and Economics* 2013, 7(2):41–70.
12. Vlastakis N, Dotsis G, Markellos RN: How efficient is the European football betting market? Evidence from arbitrage and trading strategies. *Journal of Forecasting* 2009, 28(5):426–444.
13. Buckle M, Huang C-S: The efficiency of sport betting markets: An analysis using arbitrage trading within Super Rugby. *International Journal of Sport Finance* 2018, 13(3):279–294.
14. Marshall BR: How quickly is temporary market inefficiency removed? *The Quarterly Review of Economics and Finance* 2009, 49(3):917–930.
15. Lopez-Gonzalez H, Griffiths MD, Estevez A: Why some sports bettors think gambling addiction prevented them from becoming winners? A qualitative approach to understanding the role of knowledge in sports betting products. *Journal of Gambling Studies* 2020, 36(3):903–920.

8

Who Are the Key Players? Using Passing Networks to Analyse Match Play

Network analysis has been used successfully for many years to model the complex social interactions that exist within organisations and societies. In sport, network analysis is becoming increasingly popular because it offers insights into the structure and performance of teams and organisations that might otherwise be hidden. One area in soccer where network analysis looks to have great potential is in analysing styles of match play using passing data. By producing a network of who passed to whom during a match, it is possible not only to assess the style of play of a team but also to identify the key players who influenced match play. As such, network analysis has considerable potential as a tool for evaluating on-the-pitch team performance.

The branch of mathematics that underpins network analysis is called 'graph theory'. Although many people have never heard of graph theory, they actually use it every day because it is the basis of the famous PageRank algorithm [1,2] that underpins the Google search engine. In this chapter, we explore how graph networks can be used in soccer to model interactions between clubs and also to evaluate the performance of teams and players during matches. As with the other chapters in this book, example scripts will be provided to illustrate how the various techniques employed can be executed using R.

8.1 Graph Theory and Networks

In soccer, we are often interested in relationships. For example, coaches and managers may want to know who their most influential players are or which partnerships work best on the pitch. They may also want to understand why the squad is performing in a particular way during matches so that remedial action can be taken to redress problems – something that can be particularly helpful when making important investment decisions such as purchasing new players. Rather surprisingly, graph theory can help in this respect, as it forms the basis of both network analysis and many of the techniques that can be used to rank teams and players (see Chapter 9).

DOI: 10.1201/9781003328568-8

In mathematics, a graph is a network that is made up of vertices (also called nodes), which are connected by edges (or lines) denoting the relationships between the various vertices. In graph theory, a distinction is made between undirected graphs (e.g., Figure 8.1), where the edges linking the vertices have no direction, and directed graphs (e.g., Figure 8.2), where the edge links are asymmetrical, indicating a net 'flow' in one direction or another. Using a graph, it is possible to convey a huge amount of information about the relationships that exist between the various parties in a network. For example, consider the undirected graph in Figure 8.1, which shows the towns and cities in a small country that are connected by air flights. From this, we can see that flight connections exist between City 1 and City 2, and also that City 1 is connected with towns 1, 2, and 3, whereas City 2 only has flights to towns 1 and 3. In addition, it can be seen that it is possible to fly between Town 2 and Town 3, whereas no flights at all go to Town 4, which is completely isolated.

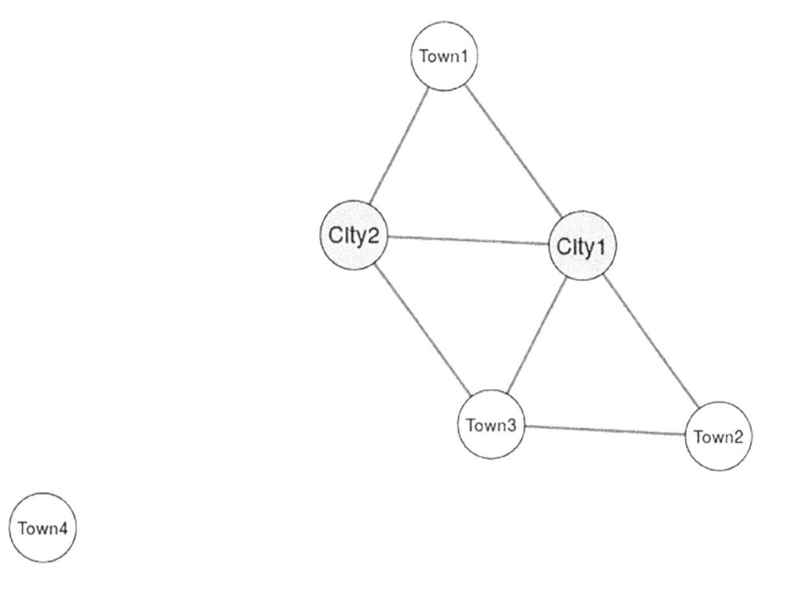

FIGURE 8.1
Example of a simple undirected graph showing the flight connections between four towns and two cities.

One important thing to notice about the network graph in Figure 8.1 is that the lines (edges) connecting the nodes (vertices) have no direction. This is because we are only concerned here as to whether or not the various towns and cities are connected by a flight and are thus making the simplistic assumption that every aeroplane that flies from one town to another always flies back again. Therefore, all we need to denote in the graph is that the two towns have a flight connection. Notwithstanding this, the graph as it stands does not give any indication of how many flights occur between the various towns and cities. If we were to add this information, then the graph would become 'weighted', with each edge link showing the number of flights occurring daily between the various destinations.

Figure 8.2 is an example of a weighted directed graph. It shows the total value of all the player transfers that occurred in a single season between six fictitious soccer clubs. Notice that the edge links are both directional and weighted. So, for example, we can see that Club 2 paid Club 4 a total of £120 million in player transfer fees during the season. We don't know how many players were involved in transfers between these two clubs – all we know is that the players went from Club 4 to Club 2 and that the money went in the opposite direction. However, we can also see that Club 2 paid Club 1 £80 million in transfer fees, but in the same season, Club 2 received £50 million from Club 1. Presumably, two separate transfer deals were undertaken between the two clubs in the same season, with players travelling in opposite directions.

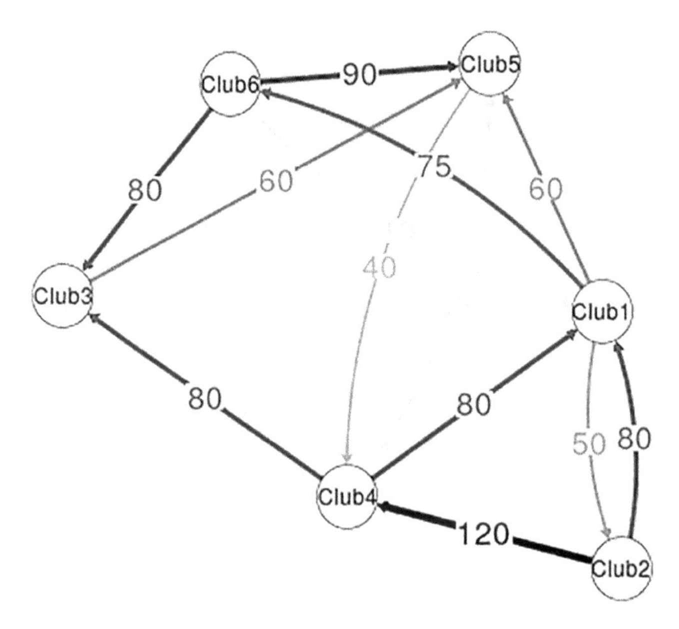

FIGURE 8.2
Example of a weighted directed graph showing the value of player transfers (in millions of pounds) between six fictitious soccer clubs.

The directed graph in Figure 8.2 was constructed in R using the adjacency matrix in Table 8.1, in which the rows represent the value of the transfer fees paid out by the clubs and the columns are the moneys received. This asymmetry is typical of the adjacency matrices used to produce directed graphs in which interactions occur in both directions, as is often the case in soccer. Once the adjacency matrix has been constructed, it is very easy to produce a network graph using R, as illustrated in Examples 8.1 and 8.2, which show

how to construct graphs for a hypothetical mini soccer tournament using various data types. In Example 8.1, only 'who-played-who' data is utilised whereas in Example 8.2, the match score lines are utilised.

TABLE 8.1

Adjacency Matrix for Directed Graph Shown in Figure 8.2

	Club 1	Club 2	Club 3	Club 4	Club 5	Club 6
Club 1	0	50	0	0	60	75
Club 2	80	0	0	120	0	0
Club 3	0	0	0	0	60	0
Club 4	80	0	80	0	10	0
Club 5	0	0	0	40	0	0
Club 6	10	0	80	0	90	0

Values represent millions of pounds.

EXAMPLE 8.1: SIMPLE TOURNAMENT NETWORK

In this example, we will look at how to construct a simple network graph in R for a hypothetical soccer competition between several fictitious clubs. To make things interesting, the mini-tournament has deliberately been poorly designed so that not all the teams play each other, with some teams playing more matches than others. In this first example, we restrict ourselves to constructing a simple graph of who-played-who.

The first thing that we need to do is enter all the match outcome results, which in this case we do by creating a separate vector for each match and then using the 'rbind' function to create a data frame, which we shall call 'mini'.

```
rm(list = ls())   # Clears all variables from workspace

# First we input the results for the matches in the mini-soccer competition.
match1 <- c("Midtown","Halton",3,2,"H") # Midtown v Halton (score: 3-2)
match2 <- c("Oldbury","Newtown",3,1,"H") # Oldbury v Newtown (score: 3-1)
match3 <- c("Longbury","Scotsway",4,2,"H") # Longbury v Scotsway (score: 1-3)
match4 <- c("Tilcome","Oldbury",0,1,"A") # Tilcome v Oldbury (score: 1-1)
match5 <- c("Scotsway","Tilcome",3,3,"D") # Scotsway v Tilcome (score: 3-3)
match6 <- c("Midtown","Longbury",2,2,"D") # Midtown v Longbury (score: 2-2)
match7 <- c("Halton","Midtown",1,2,"A") # Halton v Midtown (score: 1-2)
match8 <- c("Newtown","Longbury",1,3,"A") # Newtown v Longbury (score: 1-3)
match9 <- c("Halton","Scotsway",4,2,"H") # Halton v Scotsway (score: 4-2)
match10 <- c("Oldbury","Midtown",4,1,"H") # Oldbury v Midtown (score: 4-1)

mini <- rbind.data.frame(match1,match2,match3,match4,match5,match6,
                  match7,match8,match9,match10)
colnames(mini) <- c("HomeTeam","AwayTeam","HG","AG","Results")
mini$HG <- as.numeric(mini$HG) # Convert to integers.
mini$AG <- as.numeric(mini$AG) # Convert to integers.
print(mini)
```

This displays the following:

```
##
   HomeTeam AwayTeam HG AG Results
1   Midtown   Halton  3  2       H
2   Oldbury  Newtown  3  1       H
3  Longbury Scotsway  4  2       H
4   Tilcome  Oldbury  0  1       A
5  Scotsway  Tilcome  3  3       D
6   Midtown Longbury  2  2       D
7    Halton  Midtown  1  2       A
8   Newtown Longbury  1  3       A
9    Halton Scotsway  4  2       H
10  Oldbury  Midtown  4  1       H
```

Now we harvest team names using the 'unique' function and collate them into a vector.

```
teams <- unique(mini$HomeTeam)
teams <- sort(teams)
n <- length(teams) # This identifies the length of the 'teams' vector.
```

Next, we produce a simple adjacency table of who-played-who, in which the rows represent home matches and the columns away matches.

```
HomeTeam <- mini$HomeTeam
AwayTeam <- mini$AwayTeam
adj.tab1 <- table(HomeTeam,AwayTeam)
print(adj.tab1)
```

```
##
          AwayTeam
HomeTeam   Halton Longbury Midtown Newtown Oldbury Scotsway Tilcome
  Halton        0        0       1       0       0        1       0
  Longbury      0        0       0       0       0        1       0
  Midtown       1        1       0       0       0        0       0
  Newtown       0        1       0       0       0        0       0
  Oldbury       0        0       1       1       0        0       0
  Scotsway      0        0       0       0       0        0       1
  Tilcome       0        0       0       0       1        0       0
```

Now we can convert the table to an adjacency matrix of who-played-who using the following:

```
temp1 <- as.numeric(as.matrix(adj.tab1))
adj.mat1 <- matrix(temp1, nrow=n, ncol=n)
print(adj.mat1) # NB. This adjacency matrix is asymmetrical, indicating a directional graph.
```

```
##
     [,1] [,2] [,3] [,4] [,5] [,6] [,7]
[1,]    0    0    1    0    0    1    0
[2,]    0    0    0    0    0    1    0
[3,]    1    1    0    0    0    0    0
[4,]    0    1    0    0    0    0    0
[5,]    0    0    1    1    0    0    0
[6,]    0    0    0    0    0    0    1
[7,]    0    0    0    0    1    0    0
```

Having built the adjacency matrix, we can use the 'qgraph' package [3] to create the network graph. Here we draw the graph in black and white and specify the vertex and edge sizes along with the label size ('label.cex=1.3') using the commands prescribed in the 'qgraph' package [3].

```
# Build graph using 'qgraph' package.
install.packages("qgraph") # This installs the 'qgraph' package.
# NB. This command only needs to be executed once to install the package.
# Thereafter, the 'qgraph' library can be called using the 'library' command.

library(qgraph)
mini.graph1 <- qgraph(adj.mat1, labels=teams, label.cex=1.3, esize=5,
                  vsize=10, edge.color="black") # Who-played-who network
title("Matches played", adj=1, line=3) # This puts title in top right hand corner
```

The output of this code is presented in Figure 8.3, which shows the graph of the who-played-who network. Note that the graph is directed, with the arrows indicating who played at home and who played away – the arrows point from the home team towards the away team. So from the graph, we can see that Halton played Midtown both at home and away, whereas they only played Scotsway at home.

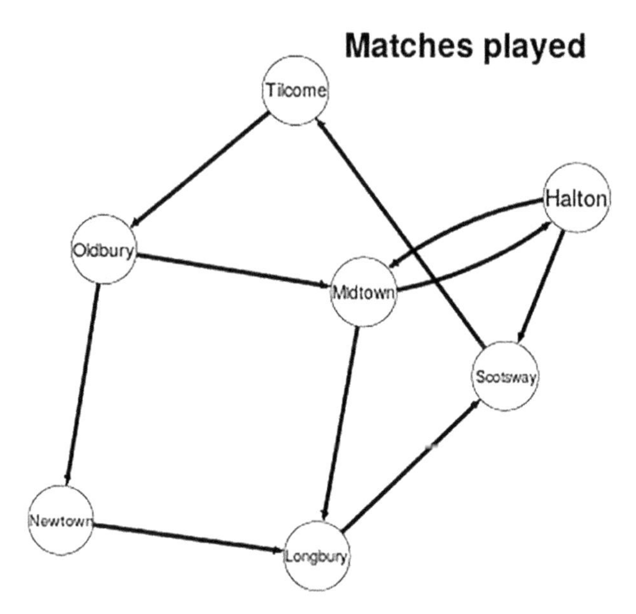

FIGURE 8.3
Directed graph showing who-played-who in the mini soccer tournament.

EXAMPLE 8.2: MINI-TOURNAMENT NETWORK BASED ON GOALS SCORED

In Example 8.1, we saw how a simple who-played-who graph can be constructed. In this example, we build on Example 8.1 and show how a weighted graph can be constructed using the goals scored data. In this graph, the edge lines are directional and represent goals scored by the various teams against other teams. To construct the who-scored-against-who graph, we shall use the 'mini' data frame created in Example 8.1.

One of the major challenges we have to overcome when constructing the goals scored adjacency matrix is correctly assigning the goals to the various teams. This is because the same teams appear multiple times in both the HomeTeam and AwayTeam columns on the 'mini' data frame. However, this problem can be overcome by assigning numerical levels to the teams in the HomeTeam and AwayTeam columns, which in effect gives each team listed its own unique identification number. We can do this using the following code:

```
# Assign numerical values to individual teams
HT <- as.factor(mini$HomeTeam)
levels(HT) <- 1:length(levels(HT))
HT <- as.numeric(HT)
print(HT)
```

```
##
[1] 3 5 2 7 6 3 1 4 1 5
```

```
AT <- as.factor(mini$AwayTeam)
levels(AT) <- 1:length(levels(AT))
AT <- as.numeric(AT)
print(AT)
```

```
##
[1] 1 4 6 5 7 2 3 2 6 3
```

Now we can create the adjacency matrix.

```
# Create new matrix
X <- cbind(HT,AT,mini[,3:4])
# Populate adjacency matrix with weights
adj1 <- matrix(0,n,n)
adj2 <- matrix(0,n,n)
p <- nrow(mini)

for (k in 1:p){
  i = X[k,1]
  j = X[k,2]
  if (adj1[i,j] == 0){adj1[i,j] <- X[k,3]}
  if (adj2[j,i] == 0){adj2[j,i] <- X[k,4]}
}

adj.mat2 <- adj1+adj2
rownames(adj.mat2) <- teams
colnames(adj.mat2) <- teams
print(adj.mat2) # NB. This adjacency matrix is asymmetrical, indicating a directional graph.
```

```
##
         Halton Longbury Midtown Newtown Oldbury Scotsway Tilcome
Halton        0        0       3       0       0        4       0
Longbury      0        0       2       3       0        4       0
Midtown       5        2       0       0       1        0       0
Newtown       0        1       0       0       1        0       0
Oldbury       0        0       4       3       0        0       1
Scotsway      2        2       0       0       0        0       3
Tilcome       0        0       0       0       0        3       0
```

In this adjacency matrix, the rows represent the goals scored by the various teams, and the columns are the goals conceded. As the goals scored and conceded are different from each other, this means the adjacency matrix is asymmetric.

Now we can create the network graph using the 'qgraph' package, as follows:

```
# Plot directional graph
library(qgraph)
mini.graph1 <- qgraph(adj.mat2, labels = teams, label.cex = 1.7, edge.labels = TRUE,
               edge.color="black", edge.label.cex = 2) # Who-scored-against-who network
title("Goals scored", adj=0.5, line=3)
```

This produces the weighted graph shown in Figure 8.4, in which the edge line arrows indicate the goals scored by the various teams against their opponents, and the edge weights are the numbers of goals scored. Note that the thicker and darker the edge lines, the greater the number of goals scored.

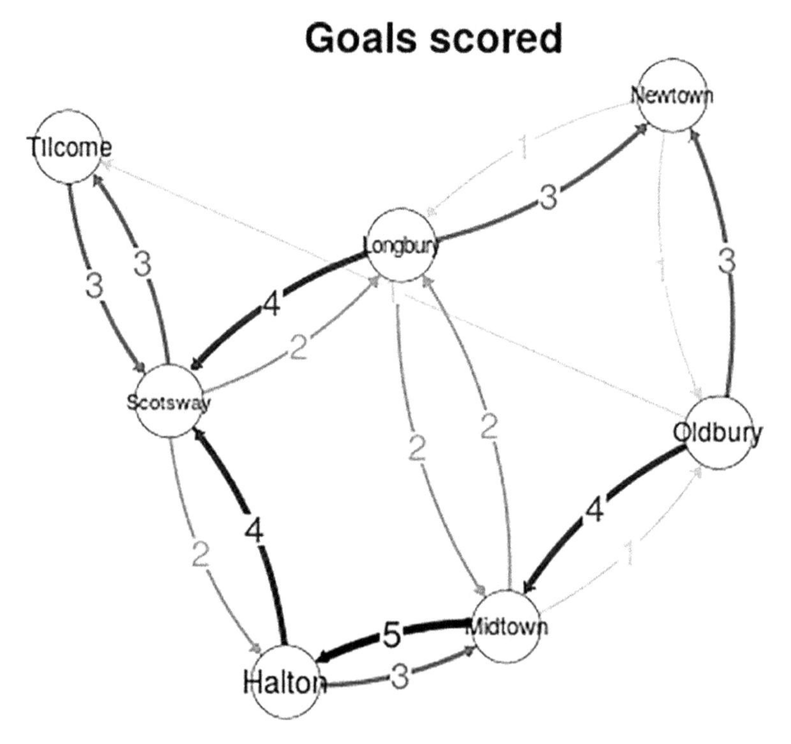

FIGURE 8.4
Weighted graph showing who-scored-against-who in the mini soccer tournament.

8.2 Passing Networks

Now that we have learned how to construct network graphs, let us extend the knowledge that we have acquired to analyse the passes made between players in an individual soccer match. Analysis of the passing networks of teams in matches has received much attention in recent years, with a number of scientific papers written on the subject [4–7]. This is because passing networks can yield insights into the patterns of play adopted by the respective teams during matches. Importantly, such networks can provide coaches with information regarding the performance of individual players and their influence on the match.

It is relatively easy to produce a passing network graph, provided, of course, that the appropriate data is available. To illustrate how such graphs can be constructed, in Example 8.3, we shall produce passing network graphs for the two teams, Spain and the Netherlands, who contested the 2010 FIFA World Cup final – a match that Spain won 1-0 in extra time.

EXAMPLE 8.3: 2010 WORLD CUP FINAL PASSING NETWORK

In this example, we show how to construct a network graph based on passing data for the various players who were involved in the 2010 World Cup final, in which Spain beat the Netherlands 1-0 after extra time. In order to do this, we will construct two adjacency matrices (i.e., one for Spain and the other for the Netherlands) using data produced by FIFA [8], which records the number of successful passes between the various players (including substitutes) on each team. To illustrate that alternative methods can be utilised, we will use both the 'qgraph' [3] and 'igraph' [9] packages in R to construct the respective network graphs. The source files for this example (Spain_2010_WC_final.csv and Netherlands_2010_WC_final.csv), which contain the data for the passing network, can be found in the GitHub repository at: https://github.com/cbbeggs/SoccerAnalytics.

First, we need to load the CSV files containing the respective adjacency matrix data for the two teams, which we have stored in a directory called 'Datasets'.

```
# Load the data
Spain <- read.csv("C:/Datasets/Spain_2010_WC_final.csv", sep=",")
Netherlands <- read.csv("C:/Datasets/Netherlands_2010_WC_final.csv", sep=",")

# Display the data
print(Spain)
```

```
##
    Player Casillas Pique Puyol Iniesta Villa Xavi Capdevila Alonso Ramos
1 Casillas        0     6     5       0     0    0         2      0     4
2    Pique        3     0     7       4     4    4         3      1     6
3    Puyol        6     8     0       1     0   12         7      5     1
```

```
4     Iniesta        0    0    1     0   1   7        8    7   3
5       Villa        0    0    0     4   0   2        1    0   2
6        Xavi        0   11    7     8  10   0        9    8   7
7  Capdevila         1    0    6    11   0  12        0    3   1
8      Alonso        0    4    3    11   1   8        1    0   3
9       Ramos        0    1    5     4   4   8        2    6   0
10  Busquets         0   10    9     7   2  11        6    8   6
11      Pedro        0    2    0     0   2   0        1    3   4
12     Torres        0    0    0     0   0   0        0    0   0
13  Fabregas         0    4    2     9   1   7        2    0   2
14      Navas        1    1    0     2   1   3        1    1   5
   Busquets Pedro Torres Fabregas Navas
1         2     1      1        1     0
2         7     4      0        4     4
3         5     2      0        6     1
4         6     2      0        5     1
5         0     2      0        0     2
6        10     4      0        4     2
7        12     2      0        5     0
8         4     4      0        0     2
9        10     4      1        2     7
10        0     2      0        6     3
11        1     0      0        0     0
12        1     0      0        0     0
13        3     0      0        0     1
14        0     0      1        1     0
```

Here, Casillas is the Spanish goalkeeper, and Navas, Fabregas, and Torres are the substitutes who replaced Pedro, Alonso, and Villa, respectively. The rows represent successful passes made, whereas the columns are the number of passes received.

```
print(Netherlands)
```

```
##
         Player Stekelenburg van_der_Wiel Heitinga Mathijsen Bronckhorst
1  Stekelenburg            0            4        8        13           5
2  van_der_Wiel            8            0        4         0           3
3      Heitinga            5            2        0         4           0
4     Mathijsen            8            5        2         0           4
5   Bronckhorst            3            2        2         4           0
6    van_Bommel            2            3        3         3           1
7          Kuyt            3            1        0         0           2
8       de_Jong            2            4        0         1           1
9    van_Persie            0            2        0         0           2
10     Sneijder            0            0        4         2           0
11       Robben            2            0        0         0           0
12    Braafheid            1            3        0         0           0
13         Elia            1            0        0         0           0
14 van_der_Vaart            0            0        1         1           0
   van_Bommel Kuyt de_Jong van_Persie Sneijder Robben Braafheid Elia
1           0    0       0          3        1      1         0    0
2           7    0       1          1        2      5         0    0
3          10    0       1          3        0      2         0    0
4           1    3       5          1        2      1         0    1
5           1    9       3          2        0      1         0    2
6           0    1       3          4        3      1         0    1
7           0    0       1          1        2      1         0    0
8           2    4       0          1        6      3         0    1
```

```
9           1   2     0         0       6     4         1    1
10          2   1     3         3       0     8         1    3
11          0   0     3         2       3     0         0    0
12          0   0     0         0       0     0         0    1
13          0   0     1         0       0     0         2    0
14          1   0     0         0       4     0         0    0
   van_der_Vaart
1              0
2              0
3              0
4              0
5              0
6              2
7              0
8              0
9              0
10             1
11             1
12             2
13             0
14             0
```

Here, Stekelenburg is the Dutch goalkeeper, and Ella, van der Vaart, and Braafheid are the substitutes who replaced Kuyt, de Jong, and van Bronckhorst, respectively. As with the data for the Spanish team, the rows represent successful passes made, and the columns are the number of passes received.

Having loaded the data frames into R, we now need to convert them into adjacency matrices, which we can do as follows:

```
# Create the adjacency matrices
# Make Spain team 1 (T1) and Netherlands team 2 (T2)
T1 <- Spain
T2 <- Netherlands

adj.T1 <- as.matrix(T1[1:14,2:15])
adj.T2 <- as.matrix(T2[1:14,2:15])
```

Now we can plot the passing network graphs for both teams using 'qgraph'.

```
# Spain
library(qgraph)
q.T1 <- qgraph(adj.T1, edge.color="black") # Spanish passing network
title("Spain", adj=0.1, line=-1)
```

This produces Figure 8.5, in which the edge weights have been omitted for clarity.

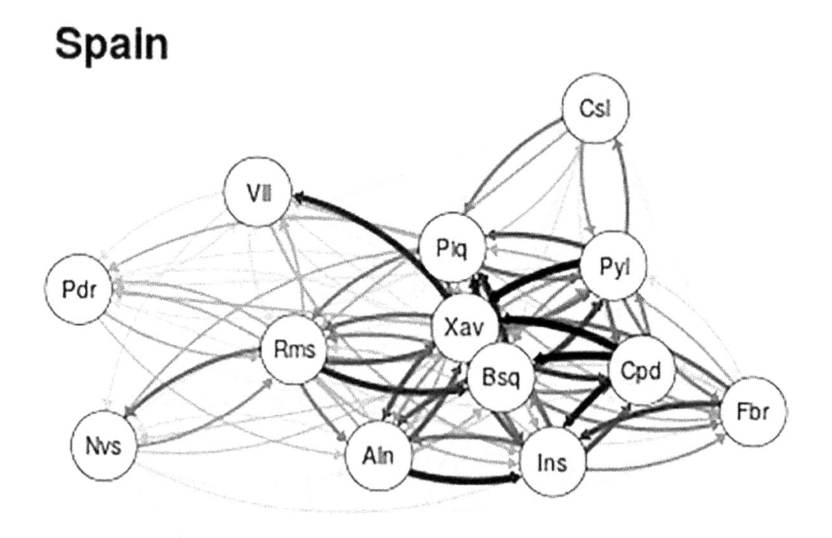

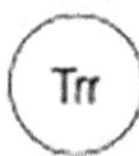

FIGURE 8.5
Weighted graph showing the passing network for Spain, produced using the 'qgraph' package. (NB. Thicker and darker lines are indicative of a greater number of passes.)

```
# Netherlands
q.T2 <- qgraph(adj.T2, edge.color="black") # Dutch passing network
title("Netherlands", adj=0.8, line=-1)
```

This produces Figure 8.6.

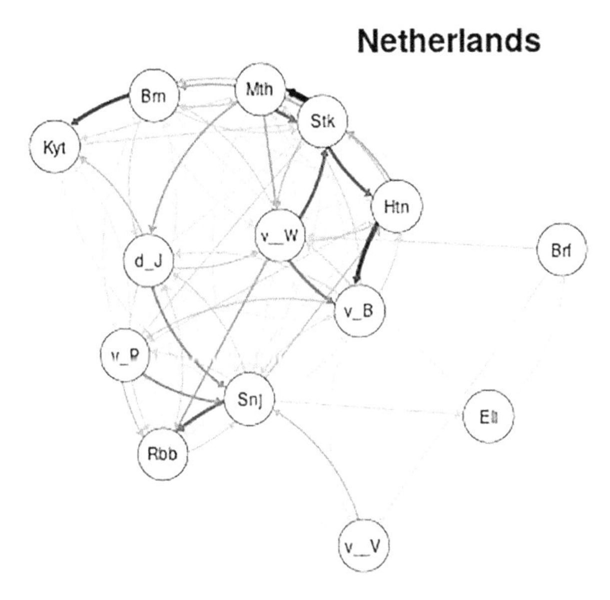

FIGURE 8.6
Weighted graph showing the passing network for the Netherlands, produced using the 'qgraph' package. (NB. Thicker and darker lines are indicative of a greater number of passes.)

Although in Figures 8.5 and 8.6, the edge weight values have been omitted for clarity, the thickness of the lines gives an indication of the strength of the passing relationships between the various players. In each graph, the thickest lines represent the strongest relationships, with the arrows representing the direction of the passes. So for example, it can be seen that Stekelenburg, the Dutch goalkeeper, made many passes to the centre-back, Mathijsen (i.e., 13 passes), but received many fewer passes in return from that player (i.e., 8 passes). Note also that 'qgraph' automatically abbreviates the player's names so that they fit nicely into the vertices.

Comparison between Figures 8.5 and 8.6 reveals that the two teams exhibited very different patterns of passing during the match. The graph for the Spanish team reveals that a large number of successful passes were completed during the match, with most of the players in the team involved – particularly the midfield players and the centre-backs. As a result, the graph for the Spanish team is very dense and cohesive. By comparison, the graph for the Netherlands is much sparser, indicating that far fewer successful passes were completed. Also, it is noticeable that the strongest passing relationships in the Netherlands team tended to involve the goalkeeper and the defenders, suggesting that for much of the match the Dutch team did not have possession, and when they did get it, possession was often restricted to their own half of the pitch. Indeed, the match statistics appear to bear this out, with Spain having 57% of the possession and the Netherlands only 43%.

Figures 8.5 and 8.6 were produced using 'qgraph'. However, an alternative method in R for producing network graphs is to use the 'igraph' package. The following code produces the equivalent passing network graphs for the two teams using 'igraph'.

```
# Alternative method using the 'igraph' package
install.packages("igraph") # This installs the 'igraph' package.
# NB. This command only needs to be executed once to install the package.
# Thereafter, the 'igraph' library can be called using the 'library' command.

# Spain
library(igraph)
g.T1 <- graph_from_adjacency_matrix(adj.T1, weighted=TRUE)
plot(g.T1, vertex.color = "white", vertex.label.color = "black", layout=layout.kamada.kawai)
title("Spain", adj=0.2, line=-1)
```

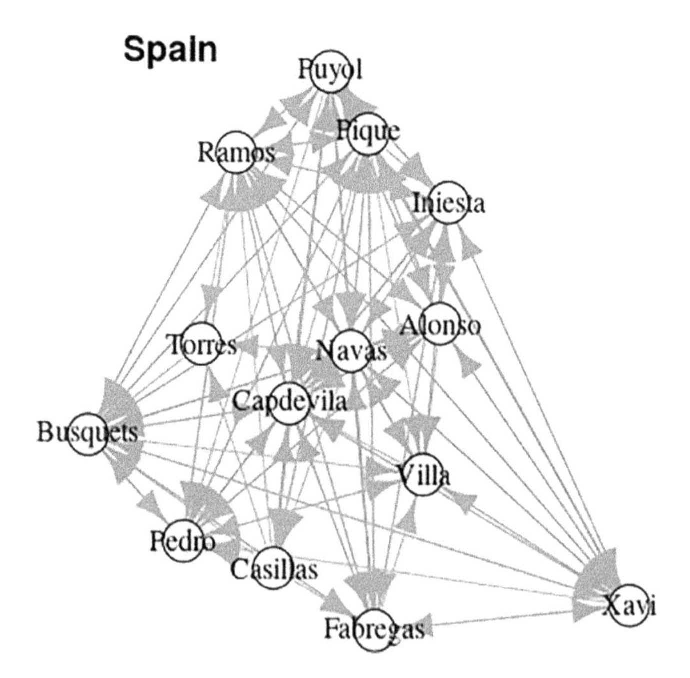

FIGURE 8.7
Weighted graph showing the passing network for Spain, produced using the 'igraph' package.

```
# Netherlands
library(igraph)
g.T2 <- graph_from_adjacency_matrix(adj.T2, weighted=TRUE)
plot(g.T2, vertex.color = "white", vertex.label.color = "black", layout=layout.kamada.kawai)
title("Netherlands", adj=0.8, line=-1)
```

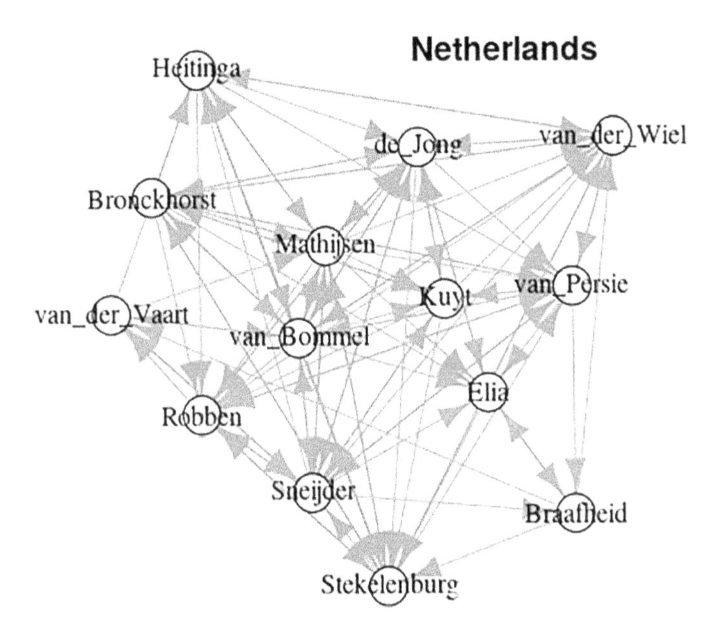

FIGURE 8.8
Weighted graph showing the passing network for the Netherlands, produced using the 'igraph' package.

8.3 Characterising Graphs and Producing Descriptive Statistics

While the network graphs in the examples above look impressive, they are so complex that it is difficult to describe them succinctly. This makes it difficult to compare graphs because numerical values are needed in order to make meaningful comparisons. So rather than relying on long-winded verbal explanations of the apparent characteristics of graphs, what is needed is a set of agreed-upon metrics that can be used to describe and succinctly characterise networks. Fortunately for us, many descriptive statistics have been developed that can be used to analyse and characterise network graphs. Some of these are briefly outlined below.

- **Size:** The most basic characteristic of a graph is its size, which is simply the number of vertices (nodes).
- **Density:** Density is a measure of graph interconnectivity and is the ratio (expressed as a fraction) of the number of observed edges in the

graph to the maximum number of possible edges between the vertices. Thus, the graph density always lies between zero and one, with graphs that are more interconnected having values that are closer to one. For a directed network, the maximum number of possible edges, E_{max}, between k nodes can be determined using Equation 8.1, while the equivalent value for an undirected graph can be determined using Equation 8.2, where k is the number of vertices (nodes) and L is the number of observed edges in the network.

$$\text{Directed network}: E_{max} = \frac{L}{k \times (k-1)} \quad (8.1)$$

$$\text{Undirected network}: E_{max} = \frac{2L}{k \times (k-1)} \quad (8.2)$$

- **Diameter:** The diameter of a graph is a useful measure of its compactness. In a graph, a path is the series of steps between one node and another. Paths can be short, involving few intermediate nodes, or long, involving many intermediate steps, depending on the compactness of the graph. The diameter of a graph is therefore the length of the shortest path between the most distant nodes. As such, the diameter is a measure of the compactness of the graph and reflects the worst-case scenario for sending information or goods across the network.
- **Average path length (APL):** The APL is related to the graph diameter and is defined as the average number of steps along the shortest paths for all the possible pairs of network nodes. APL is a measure of the efficiency with which information (or goods) can be transported around the network and, as such, distinguishes an easily negotiable network from one that is complicated and inefficient. A shorter APL is generally more desirable.

In addition, a number of graph centrality metrics have been developed, which are particularly useful. In graph theory and network analysis, indicators of centrality assign ranks to the individual nodes (vertices) within graphs that correspond to their importance within the network. So for example, centrality metrics can be used to identify influential persons or actors within, say, social networks [10], outbreaks of infectious disease [11], gene-associated diseases [12], or indeed, the performance of sports teams [5,13,14].

While it is true that graph centrality indices generally try to identify and characterise important vertices, there is some debate as to what actually constitutes an important node. Is an important node one that, for example, facilitates the efficient flow of information across a network, or is it one that provides cohesiveness to the network? The two can be very different. Consequently, there are many different measures of centrality that can be used, each with its own specific definition – something that can be very confusing. So in an attempt to keep things simple, we will restrict ourselves here to explaining just a few of the more important indices of centrality that can be used.

- **Degree centrality:** Degree centrality is probably the simplest measure of centrality and by far the easiest to understand. It is based on the simple notion that important nodes are those that have more direct ties (edges) with other nodes. For example, if during a match most of the players in a soccer team keep passing the ball to the same midfield player, then it is reasonable to assume that this player must be influential; otherwise, why would so many passes be directed towards him or her?

  In an undirected network, degree centrality is simply the number of edges linking each node to other adjacent nodes. So if a node is directly connected to five other nodes, then its degree centrality score is simply five. However, with directed graphs, things get a little more complicated because we have to separate incoming links from outgoing links. So, for example, in a soccer match's passing network, we need to distinguish successful passes made by players from those received by the same players. So, we use the term **in-degree centrality** to represent the links (edges) directed towards the node and **out-degree centrality** to represent the edges directed away from the node.
- **Closeness centrality:** Instead of concentrating on the direct connections between the nodes, we could instead focus on how close each node is to all the other nodes in the network. This leads to the concept of closeness centrality, where the important nodes are those that are closest to all the other nodes in the network. In this scenario, the normalised closeness of a node is the average length of the shortest paths between that node and all the other nodes in the graph.
- **Betweenness centrality:** Betweenness centrality is a measure of the extent to which a node sits 'between' pairs of other nodes in a network. As such, it is a measure of the number of times a node acts as a bridge along the shortest path between two other nodes.

EXAMPLE 8.4: NETWORK DESCRIPTIVE STATISTICS EXAMPLE

In this example, we shall produce a set of descriptive statistics to evaluate the passing network graphs for Spain and the Netherlands shown in Figures 8.5 and 8.6. It should be noted that the statistics produced in this example are by no means exhaustive. We have simply restricted ourselves here to some of the more important and useful metrics – there are plenty of others that we could have produced using R. Note also that for simplicity, we have restricted ourselves to using the packages 'igraph' [9] and 'qgraph' [9], which are two of the more popular network analysis packages in R. Both are substantial packages, and readers who are interested in pursuing the subject further are encouraged to explore for themselves the other functions in these packages, as this example is only intended to be a brief introduction.

First of all, we shall use the 'igraph' package to produce some simple descriptive statistics for the two graphs. This can be done using the following code, which produces a comparative table for the two teams.

```
library(igraph)
# Spain
nv.T1 <- vcount(g.T1) # Number of vertices
ne.T1 <- ecount(g.T1) # Number of edges
den.T1 <- graph.density(g.T1) # Graph density
dia.T1 <- diameter(g.T1) # Diameter of graph
apl.T1 <- average.path.length(g.T1) # Average path length
res.T1 <- c(nv.T1, ne.T1, den.T1, dia.T1, apl.T1) # Combine into vector

# Netherlands
nv.T2 <- vcount(g.T2) # Number of vertices
ne.T2 <- ecount(g.T2) # Number of edges
den.T2 <- graph.density(g.T2) # Graph density
dia.T2 <- diameter(g.T2) # Diameter of graph
apl.T2 <- average.path.length(g.T2) # Average path length
res.T2 <- c(nv.T2, ne.T2, den.T2, dia.T2, apl.T2) # Combine into vector

# Compile descriptive statistics for the graphs
results <- round(cbind.data.frame(res.T1,res.T2),2)
colnames(results) <- c("Spain","Netherlands")
rownames(results) <- c("Number of vertices","Number of edges","Graph density",
                "Graph diameter","Average path length")
print(results)
```

```
##
                      Spain Netherlands
Number of vertices    14.00       14.00
Number of edges      126.00      106.00
Graph density          0.69        0.58
Graph diameter         7.00        4.00
Average path length    2.77        2.20
```

From this, we see that the network graph for Spain is denser than that for the Netherlands. This indicates that: (i) in the Spanish team, more players were involved in the passing network compared with the Dutch team; and (ii) the Dutch team used fewer passing options compared with the Spanish team, which was more flexible and utilised more pathways. This fact is reflected in the graph diameter and APL metrics for the two teams, which reveal that many more players were involved in the passing moves made by Spain than was the case for the Netherlands team, and also that the passes between the players were on average much shorter for the Spanish team compared with the Dutch team.

Having compiled descriptive statistics for the two graphs, we can now turn our attention to evaluating the performance of the individual players. Here we produce two results data frames, one for Spain and one for the Netherlands, with both showing the out-degrees (i.e., the number of passes made), the in-degrees (i.e., the number of passes received), and the betweenness and closeness centralities (i.e., both measures of centrality in the graph) for each player. We can do this very easily using the 'qgraph' package and the following code. (NB. 'qgraph' automatically abbreviates the names of the players.)

```
# Compile descriptive statistics for the individual players
library(qgraph)

# Spain
outdeg.qT1 <- centrality(q.T1)$OutDegree
indeg.qT1 <- centrality(q.T1)$InDegree
between.qT1 <- centrality(q.T1)$Betweenness
close.qT1 <- centrality(q.T1)$Closeness
players.qT1 <- cbind.data.frame(outdeg.qT1,indeg.qT1,between.qT1,round(close.qT1,3))
colnames(players.qT1) <- c("OutDegrees","InDegrees","Betweenness","Closeness")
print(players.qT1)
```

```
##
    OutDegrees InDegrees Betweenness Closeness
Csl         22        11         1.0     0.203
Piq         51        47         2.0     0.253
Pyl         54        45        20.0     0.281
Ins         41        61        12.0     0.229
Vll         13        26         0.0     0.143
Xav         80        74        48.0     0.315
Cpd         53        43         0.0     0.272
Aln         41        42         1.5     0.233
Rms         54        44        37.5     0.284
Bsq         70        61        25.5     0.298
Pdr         13        27         0.0     0.157
Trr          1         3         0.0     0.066
Fbr         31        34         0.0     0.218
Nvs         17        23         0.5     0.173
```

```
# Netherlands
outdeg.qT2 <- centrality(q.T2)$OutDegree
indeg.qT2 <- centrality(q.T2)$InDegree
between.qT2 <- centrality(q.T2)$Betweenness
close.qT2 <- centrality(q.T2)$Closeness
players.qT2 <- cbind.data.frame(outdeg.qT2,indeg.qT2,between.qT2,round(close.qT2,3))
colnames(players.qT2) <- c("OutDegrees","InDegrees","Betweenness","Closeness")
print(players.qT2)
```

```
##
       OutDegrees InDegrees Betweenness Closeness
Stk            35        35        31.5     0.188
V__W           31        26        25.5     0.175
Htn            27        24        18.5     0.161
Mth            33        28         7.5     0.176
Brn            29        18         6.0     0.150
V_B            27        25        20.0     0.158
Kyt            11        20         0.0     0.118
d_J            25        21        15.0     0.165
V_P            19        21         0.0     0.140
Snj            28        29        41.5     0.162
Rbb            11        27         0.0     0.122
Brf             7         4         9.0     0.124
Eli             4        10        11.5     0.070
V__V            7         6         1.0     0.117
```

We can compare the overall performance of the two teams by summing the columns in the above tables.

```
# Comparison between the results for the two teams.
res.qT1 <- colSums(players.qT1)
res.qT2 <- colSums(players.qT2)
res <- rbind(res.qT1,res.qT2)
rownames(res) <- c("Spain","Netherlands")
print(res)
```

Which produces:

```
##
            OutDegrees InDegrees Betweenness Closeness
Spain              541       541         148     3.125
Netherlands        294       294         187     2.032
```

From this, we see that the Dutch team completed only about 54% of the passes made by the Spanish team. Furthermore, because closeness centrality is a measure of how close each node is to all other nodes in the network, this indicates that the players in the Spanish team were much more cohesive and better connected than those in the Dutch team. This opinion is shared by Cotta et al. [4], who concluded that the "tika-taka" passing style of the Spanish team exhibited small-world properties, implying that the team was highly connected. By comparison, however, the betweenness centrality of the Dutch players was generally higher than that of the Spanish team. While this might appear confusing at first sight, it is important to remember that betweenness centrality is a measure of the amount of influence that a node has over the flow of information in a graph and is often used to find nodes that serve as a bridge from one part of a graph to another. So in the context of the 2010 FIFA World Cup final, this suggests that the Netherlands were relying more on longer passes involving a few link players compared with Spain, who had many more passing options to choose from. Evidence for this comes from the fact that Stekelenburg (goalkeeper) and Sneijder (attacking midfielder) were the players with the highest betweenness scores in the Dutch team, whereas in the Spanish team, several players in the central midfield (Xavi, Ramos, and Busquets) were much more involved.

A measure of how influential players are to their team can be found using the functions 'page_rank' and 'authority.score' in the 'igraph' package. The former utilises the Google PageRank algorithm [1,15–17] to rate the performance of the players, while the latter uses the principal eigenvector of A^TA, where A is the adjacency matrix of the graph [18]. Both measures give an indication of the relative importance of individual players to the overall performance of their team. They can be determined as follows:

```
# igraph can be used to rate the performance of the players
library(igraph)
# Spain
nodes.T1 <- V(g.T1)$name # Names of Spanish players
pr.T1 <- page_rank(g.T1, directed = TRUE)$vector
auth.T1 <- authority.score(g.T1)$vector
players.gT1 <- round(cbind.data.frame(pr.T1,auth.T1),3)
rownames(players.gT1) <- nodes.T1
colnames(players.gT1) <- c("PageRank","Authority")
print(players.gT1)
```

```
##
           PageRank Authority
Casillas      0.025     0.143
Pique         0.078     0.663
Puyol         0.072     0.657
Iniesta       0.109     0.834
Villa         0.053     0.384
Xavi          0.126     1.000
Capdevila     0.080     0.596
Alonso        0.080     0.617
Ramos         0.090     0.504
Busquets      0.112     0.853
Pedro         0.053     0.366
Torres        0.015     0.024
Fabregas      0.060     0.502
Navas         0.048     0.310
```

```
# Netherlands
nodes.T2 <- V(g.T2)$name # Names of Spanish players
pr.T2 <- page_rank(g.T2, directed = TRUE)$vector
auth.T2 <- authority.score(g.T2)$vector
players.gT2 <- round(cbind.data.frame(pr.T2,auth.T2),3)
rownames(players.gT2) <- nodes.T2
colnames(players.gT2) <- c("PageRank","Authority")
print(players.gT2)
```

```
##
              PageRank Authority
Stekelenburg     0.112     1.000
van_der_Wiel     0.082     0.713
Heitinga         0.074     0.805
Mathijsen        0.083     0.960
Bronckhorst      0.060     0.571
van_Bommel       0.073     0.782
Kuyt             0.056     0.563
de_Jong          0.081     0.540
van_Persie       0.074     0.600
Sneijder         0.111     0.606
Robben           0.088     0.753
Braafheid        0.034     0.049
Elia             0.039     0.249
van_der_Vaart    0.034     0.100
```

From this, we see that both of these algorithms do a similar job and rate many of the players in approximately the same order. For the Spanish teams, Xavi, Busquets, and Iniesta appear to have been particularly influential, whereas for the Netherlands, Stekelenburg and Mathijsen were influential. Having said this, there are some noticeable discrepancies between the PageRank and authority rankings, particularly amongst the Dutch team. For example, Sneijder scored very highly using the PageRank algorithm but much lower using the authority algorithm.

8.4 Bipartite Network Graphs

Up until now we have only investigated network graphs in which all the vertices come from the same class (i.e., teams or players). However, it is often the case when investigating networks that they involve nodes from two or more classes. For example, in soccer, players and managers frequently move between different clubs. So any network seeking to model this scenario must be bipartite and contain nodes that represent clubs and nodes that represent managers or players. Therefore, we need to be able to incorporate two classes of nodes into our network models using R. In Example 8.5, we illustrate how this can be done using the 'igraph' package, which has lots of useful functions for creating and manipulating bipartite graphs.

EXAMPLE 8.5: BIPARTITE GRAPH EXAMPLE

In this example, we will create a small bipartite network graph showing the movement of seven fictitious managers (m1–m7) between six soccer clubs (A–F). To make things interesting, we will move the managers between several clubs so that the graph is more interconnected. Although the network is fictitious, it is by no means unlikely, given that the average tenure of a manager in the English Premier League (EPL) is about two years [19], with many managers moving between several clubs. For example, in recent years, David Moyes has managed four clubs in the EPL, while Sam Allardyce has managed eight EPL clubs during his career.

To produce the bipartite graph, we first create two vectors of the same length, one for the clubs and another matched vector corresponding to the managers associated with the respective clubs. We then combine them into a data frame, as follows:

```
# Create two vectors showing the relationships between managers and clubs.
V1 <- c("A","A","A", "B","B","B","C","D","D","E","F") # Clubs
V2 <- c("m1","m2","m7","m3","m4","m5","m5","m6","m7","m2","m6") # Managers

# Combine vectors into a data frame
d <- cbind.data.frame(V1,V2)
```

Having created the data frame, we use 'igraph' to convert it into a graph object.

```
library(igraph)
g <- graph_from_data_frame(d, directed=FALSE) # Create graph
V(g)$label <- V(g)$name # Set labels
print(V(g)$label) # Show unique nodes
```

This produces a vector containing the 13 unique names of the respective nodes.

```
##
[1] "A"  "B"  "C"  "D"  "E"  "F"  "m1" "m2" "m7" "m3" "m4" "m5" "m6"
```

Note that the first six nodes listed in this vector represent the clubs, and the last seven elements are the managers. Using the following code, we can assign numerical values to the respective nodes in such a way that the club nodes are classified as 1 and the manager nodes are 2. Here, the %in% operator is used to check whether each element in the vector belongs to either the set of clubs or the set of managers. It returns a logical vector indicating whether each element is a 1 or a 2.

```
# Set type
V(g)[name %in% V1]$type <- 1
V(g)[name %in% V2]$type <- 2
print(V(g)$type) # Display types
```

```
##
[1] 1 1 1 1 1 1 2 2 2 2 2 2 2
```

Finally, we specify the node colours and shapes and then plot the graph using the following code:

```
# Define colour and shape mappings.
col <- c("lightgray", "white")
shape <- c("square","circle")

# Plot bivariate graph
plot(g, vertex.label.color = "black",
    vertex.color = col[as.numeric(V(g)$type)],
    vertex.shape = shape[as.numeric(V(g)$type)]
)
```

This produces the bipartite graph shown in Figure 8.9, in which the club nodes are squares and the manager nodes are circles. From this, it can be seen that the graph divides into two distinct sub-groups, which are completely independent of each other. The first involves clubs A, D, E, and F, and managers m1, m2, m6, and m7, whereas the second involves clubs B and C, and managers m3, m4, and m5.

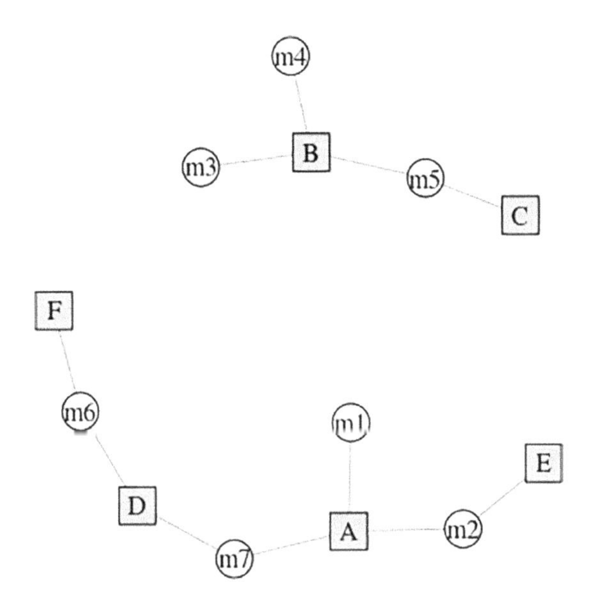

FIGURE 8.9
Bipartite network graph of managers and clubs.

8.5 Final Thoughts

Network analysis is a subject well worth pursuing because it enables complex social and sporting systems to be better understood. As such, there is much that can be learned in addition to the basics outlined in this chapter. The examples presented here are simply designed to provide an introduction to network analysis using R. However, if you have successfully mastered the code presented in this chapter and understand the basic concepts outlined, then you are well equipped to tackle more challenging network analysis problems. All you need to do now is practice and adapt the code presented here to suit your own particular challenges. In this way, you can, over time, hone your network analysis skills.

References

1. Page L, Brin S, Motwani R, Winograd T: The PageRank citation ranking: Bringing order to the web. Technical Report SIDL-WP-1999-0120, Stanford Digital Library Technologies Project; 1998.
2. Brin S, Page L: Reprint of: The anatomy of a large-scale hypertextual web search engine. *Computer Networks* 2012, 56(18):3825–3833.
3. Epskamp S, Cramer AOJ, Waldorp LJ, Schmittmann VD, Borsboom D: qgraph: Network visualizations of relationships in psychometric data. *Journal of Statistical Software* 2012, 48:1–18.
4. Cotta C, Mora AM, Merelo JJ, Merelo-Molina C: A network analysis of the 2010 FIFA world cup champion team play. *Journal of Systems Science and Complexity* 2013, 26(1):21–42.
5. Clemente FM, Sarmento H, Aquino R: Player position relationships with centrality in the passing network of world cup soccer teams: Win/loss match comparisons. *Chaos, Solitons & Fractals* 2020, 133:109625.
6. Rahnamai Barghi A: Analyzing dynamic football passing network. (Master's thesis) University of Ottawa; 2015.
7. Buldu JM, Busquets J, Martinez JH, Herrera-Diestra JL, Echegoyen I, Galeano J, Luque J: Using network science to analyse football passing networks: Dynamics, space, time, and the multilayer nature of the game. *Frontiers in Psychology* 2018, 9:1900.
8. 2010 FIFA World Cup South Africa, Passing Distribution, Final, Netherlands - Spain. https://www.scribd.com/doc/34196135/Netherlands-Spain-Final-FIFA-2010-Passing-Distribution; 2010.
9. Csardi G, Nepusz T: The igraph software package for complex network research. *InterJournal, Complex Systems* 2006, 1695(5):1–9.
10. Maji G, Dutta A, Malta MC, Sen S: Identifying and ranking super spreaders in real world complex networks without influence overlap. *Expert Systems with Applications* 2021, 179:115061
11. Dekker AH: Network centrality and super-spreaders in infectious disease epidemiology. In: *20th International Congress on Modelling and Simulation (MODSIM2013)*: Adelaide, Australia; 1–6 December 2013. www.mssanz.org.au/modsim2013
12. Ozgur A, Vu T, Erkan G, Radev DR: Identifying gene-disease associations using centrality on a literature mined gene-interaction network. *Bioinformatics* 2008, 24(13):i277–i285.
13. Pena JL, Touchette H: A network theory analysis of football strategies. In: *Sports Physics: Proceedings 2012 Euromech Physics of Sports Conference*; Editions de l'Ecole Polytechnique, Palaiseau; 2012: 517–528.
14. Clemente FM, Martins FML: Who are the prominent players in the UEFA champions league? An approach based on network analysis. *Walailak Journal of Science and Technology (WJST)* 2017, 14(8):627–636.
15. Brin S, Page L: The anatomy of a large-scale hypertextual web search engine. *Computer Networks and ISDN Systems* 1998, 33:107–117.
16. Lazova V, Basnarkov L: PageRank Approach to Ranking National Football Teams. *arXiv preprint arXiv:150301331* 2015.

17. Zack L, Lamb R, Ball S: An application of Google's PageRank to NFL rankings. *Involve, A Journal of Mathematics* 2012, 5(4):463–471.
18. Wilson KA, Green ND, Agrawal L, Gao X, Madhusoodanan D, Riley B, Sigmon JP: Graph-based proximity measures. *Practical Graph Mining with R* 2013:135.
19. Price S: Premier league managers are lasting longer in the job than European counterparts. In: *Forbes*; 2022. https://www.forbes.com/sites/steveprice/2022/03/22/premier-league-managers-are-lasting-longer-in-the-job-than-european-counterparts/

9

Which Is the Best Team? Ranking Systems in Soccer

In this chapter, we investigate ranking systems, which can be used to give an indication of the relative strengths of soccer teams and players. Such systems can be extremely helpful in situations where traditional league structures do not apply, such as in knockout and international tournaments where the competition is often fragmented. Ranking algorithms can also be helpful when making match predictions or trying to identify value in betting odds. This is because they can yield insights regarding the relative strengths of the competing teams – something that can be difficult to determine in fragmented competitions such as the FIFA World Cup. Ranking systems employ mathematical algorithms that utilise simple metrics to compute the relative strengths of the competing teams. In this chapter, we will investigate several of the more popular ranking systems and explore how they can be implemented using R. Before reading this chapter, it is advisable to familiarise yourself with Chapter 8, as many of the ranking systems discussed here are extensions of graph theory, which is introduced in that chapter.

9.1 Ranking Systems

One of the major challenges when making soccer predictions or identifying value in betting odds is knowing the true strength of the teams involved. This can be a major problem in competitions that do not employ a traditional league structure, or indeed, early in the season when most of the teams in a league have not yet had the chance to play each other, making it difficult to assess their relative strengths. Under such circumstances, league standings can be very misleading, with the positions of some teams overinflated due to having played easy opponents, while those of others may be suppressed for the opposite reason. Likewise, a similar problem exists in knockout and international tournaments where the competition is fragmented, with most teams only playing a few selected opponents – making it difficult to make comparisons. Therefore, any system that can accurately rate (or rank) teams according to their relative strength is going to be helpful when trying to predict the outcome of individual soccer matches.

DOI: 10.1201/9781003328568-9

In response to this challenge, a number of ranking (also called rating) systems have been developed that enable the relative strengths of the competing teams or athletes to be assessed in fragmented competitions [1,2]. Many of these ranking systems were developed in the USA, where sporting competitions (e.g., American football, basket ball), particularly at the college level, are fragmented, with not all teams playing each other [1–5]. However, rating systems also have considerable potential in soccer, not only to rank teams but also as tools with which to help predict the likely outcome of soccer matches, particularly in competitions where a traditional league structure does not exist.

In this chapter, we shall investigate how ranking systems can be implemented in R and applied to soccer competitions. Specifically, we will focus on three of the more popular methods: the Colley [1,2,6], Massey [1,2,7] and Elo [1,8] algorithms, each of which we will apply to a part-completed mini-soccer league involving eight teams (see Example 9.1). We will also explore how the Elo system can be applied in practice to predict the outcome of individual soccer matches.

While the ranking of teams might appear to have little in common with network analysis (see Chapter 8), in reality many ranking systems are an extension of graph theory [5], insomuch as they often utilise either an adjacency matrix or a modified adjacency matrix [2]. As such, network analysis and ranking should be considered linked subjects, with systems such as the PageRank and Keener algorithms in particular being closely related to graph theory [2]. In fact, as we have seen in Chapter 8, it is possible to implement the PageRank ranking algorithm [9,10] using the 'page_rank' function in the 'igraph' package [11]. However, while there is considerable interest in the PageRank algorithm in soccer [12–14], it is a mathematically complex subject and beyond the scope of this introductory text. Therefore, in order to keep things simple, the PageRank algorithm is not covered in this chapter; rather, we shall focus instead on the more established Colley, Massey, and Elo systems that are widely used in team sports [1,3,4]. Those readers interested in exploring the PageRank algorithm further are directed to the scientific literature on the subject (e.g., [1,2,4,9,10]).

EXAMPLE 9.1: NETWORK FOR A PART-COMPLETED MINI-SOCCER LEAGUE

In order to illustrate how the various ranking systems can be applied in R, we first need to set up a scenario to be analysed. Therefore, in this example, we will create a mini-soccer league in which eight fictional clubs (teams A–H) compete. However, because we deliberately want to simulate here an early-in-the-season scenario in which only a few matches have been played, we will assume that only 10 out of the possible 56 matches have been completed.

First, we clear any existing data from the workspace, and then we enter the match results, which in this case we do by creating a separate vector for each match and then using the 'rbind.data.frame' function to create a data frame, which we shall call 'mini'.

```
rm(list = ls())    # Clears all variables from workspace

# First we create the results for a mini-soccer league competition.
match1 <- c("Team A","Team E",3,2,"H") # Team A v Team E (score: 3-2)
match2 <- c("Team B","Team F",1,1,"D") # Team B v Team F (score: 1-1)
match3 <- c("Team C","Team G",5,2,"H") # Team C vs Team G (score: 5-2)
match4 <- c("Team D","Team H",0,1,"A") # Team D v Team H (score: 0-1)
match5 <- c("Team E","Team D",2,3,"A") # Team E v Team D (score: 2-3)
match6 <- c("Team F","Team C",2,1,"H") # Team F v Team C (score: 2-1)
match7 <- c("Team G","Team B",0,0,"D") # Team G v Team B (score: 0-0)
match8 <- c("Team H","Team A",1,3,"A") # Team H v Team A (score: 1-3)
match9 <- c("Team A","Team F",4,2,"H") # Team A v Team F (score: 4-2)
match10 <- c("Team G","Team D",2,2,"D") # Team G v Team D (score: 2-2)

mini <- rbind.data.frame(match1,match2,match3,match4,match5,
                  match6,match7,match8,match9,match10)
colnames(mini) <- c("HomeTeam","AwayTeam","HG","AG","Results")
mini$HG <- as.numeric(mini$HG) # Convert to integers.
mini$AG <- as.numeric(mini$AG) # Convert to integers.
print(mini)
```

This produces:

```
##
   HomeTeam AwayTeam HG AG Results
1    Team A   Team E  3  2       H
2    Team B   Team F  1  1       D
3    Team C   Team G  5  2       H
4    Team D   Team H  0  1       A
5    Team E   Team D  2  3       A
6    Team F   Team C  2  1       H
7    Team G   Team B  0  0       D
8    Team H   Team A  1  3       A
9    Team A   Team F  4  2       H
10   Team G   Team D  2  2       D
```

Now that we have the 'mini' data frame, we can go ahead and create the adjacency matrix ('adj.mat') using the following code. (NB. For more details on adjacency matrices, see Chapter 8.)

```
# Assign numerical values to individual teams
HT <- as.factor(mini$HomeTeam)
levels(HT) <- 1:length(levels(HT))
```

```
HT <- as.numeric(HT)
AT <- as.factor(mini$AwayTeam)
levels(AT) <- 1:length(levels(AT))
AT <- as.numeric(AT)

# Create new matrix
X <- cbind(HT,AT,mini[,3:4])

# Now we harvest team names and collate them into a vector
teams <- unique(mini$HomeTeam)
teams <- sort(teams)
n <- length(teams)

# Populate adjacency matrix with weights
adj1 <- matrix(0,n,n)
adj2 <- matrix(0,n,n)
p <- nrow(mini)

for (k in 1:p){
  i = X[k,1]
  j = X[k,2]
  if (adj1[i,j] == 0){adj1[i,j] <- X[k,3]}
  if (adj2[j,i] == 0){adj2[j,i] <- X[k,4]}
}

adj.mat <- adj1+adj2
rownames(adj.mat) <- teams
colnames(adj.mat) <- teams
print(adj.mat) # NB. This adjacency matrix is asymmetrical, indicating a directional graph.
```

```
##
        Team A Team B Team C Team D Team E Team F Team G Team H
Team A       0      0      0      0      3      4      0      3
Team B       0      0      0      0      0      1      0      0
Team C       0      0      0      0      0      1      5      0
Team D       0      0      0      0      3      0      2      0
Team E       2      0      0      2      0      0      0      0
Team F       2      1      2      0      0      0      0      0
Team G       0      0      2      2      0      0      0      0
Team H       1      0      0      1      0      0      0      0
```

In this adjacency matrix, the rows represent goals scored by the teams, and the columns represent goals conceded.

Finally, we can plot the directed network graph using the package 'qgraph' [15].

```
# Plot directional graph using 'qgraph' package.
install.packages("qgraph") # This installs the 'qgraph' package.
# NB. This command only needs to be executed once to install the package.
# Thereafter, the 'qgraph' library can be called using the 'library' command.

library(qgraph)
mini.graph1 <- qgraph(adj.mat, labels = teams, label.cex = 2, edge.labels = TRUE,
                edge.color="black", edge.label.cex = 2) # Who-scored-against-who network
title("Goals scored", adj=0.5, line=3)
```

This produces the graph shown in Figure 9.1.

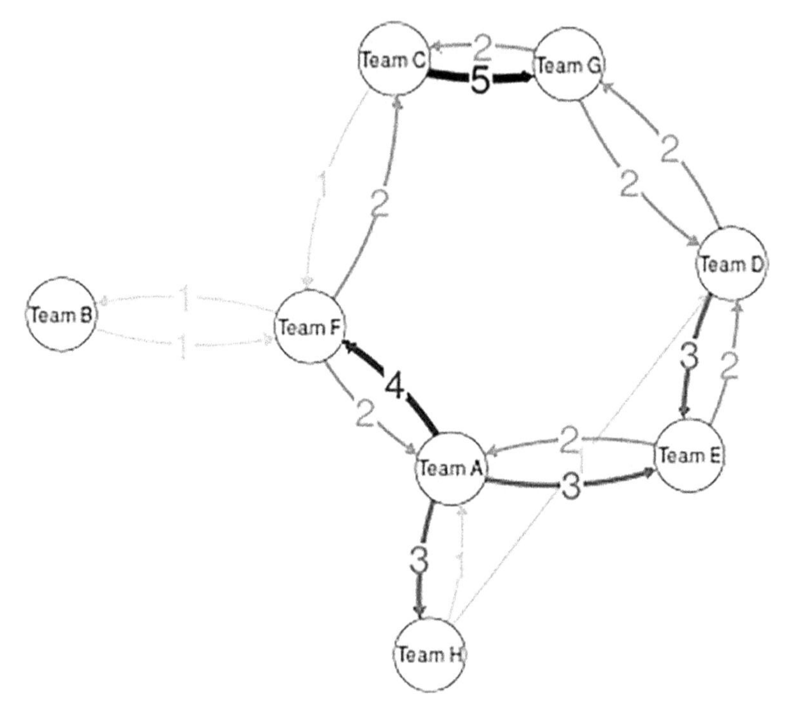

FIGURE 9.1
Directed and weighted network graph showing who-scored-against-who in the first ten matches of the mini-soccer league.

9.2 Colley Ranking Algorithm

The Colley ranking system was developed by Wesley Colley at Princeton University in the USA and aims to be a bias-free method of assessing the relative strengths of teams in match-oriented sports [1,6]. The system utilises only the total number of wins and losses as its input and ignores draws and the margin of victory in specific matches. As such, it is a methodology that is simple and easy to execute and which has been approved for use by the National Collegiate Athletic Association (NCAA) in North America [1].

The Colley method utilises Laplace's rule of succession [1] to produce a $[n \times 1]$ win-lose vector, v (where n is the number of competing teams) using Equation 9.1.

$$v_i = 1 + 0.5(w_i - l_i) \tag{9.1}$$

where v_i is the combined 'score' of the ith team; w_i is the number of wins of the ith team; and l_i is the number of losses of the ith team. So for example, if a team has won two matches and lost one, then the corresponding element in vector v will be 1.5, whereas if another team has lost two matches and won only one, the value of the element will be 0.5. Importantly, the Colley method ignores draws when constructing vector v. Notice also that the win-lose vector, v, takes no account of goals scored or conceded; rather, it simply equates to the win-lose ratio.

To determine the ranks of the various teams in the competition, Colley's method needs to solve the linear equation:

$$Cr_c = v \tag{9.2}$$

where r_c is the Colley rating vector, which defines the ranking of the teams, and C is the Colley coefficient matrix defined as [3]:

$$C_{ij} = \begin{cases} 2 + p_i & if\ i - j \\ -p_{ij} & if\ i \neq j \end{cases} \tag{9.3}$$

where p_i is the total number of times team i has played, and p_{ij} is the number of matches played between teams i and j. This produces a symmetrical matrix in which the diagonal is the total number of times each team has played plus two and all the other values are negative. Note that the Colley matrix makes no distinction between home and away matches.

Although all this might appear a bit daunting mathematically, it is actually quite easy to solve this system of equations in R, as we illustrated in Example 9.2, which applies the Colley method to rank the eight teams in the mini-soccer league after ten games.

EXAMPLE 9.2: COLLEY RANKING ALGORITHM APPLIED TO A PART-COMPLETED MINI-SOCCER LEAGUE

The following code can be used to rank the teams in the part-completed mini-league using the Colley method. Here, we apply it to the teams after ten games have been completed.

First, we need to create a symmetrical adjacency matrix representing who-has-played-who, as follows.

```
# Create who-has-played-who adjacency matrix
WPWadj.mat <- matrix(0,n,n)

for (k in 1:p){
  i = X[k,1]
```

```
  j = X[k,2]
  if (WPWadj.mat[j,i] == 0){WPWadj.mat[j,i] = 1}
  else {WPWadj.mat[j,i] = WPWadj.mat[j,i] + 1}
  if (WPWadj.mat[i,j] == 0){WPWadj.mat[i,j] = 1}
  else {WPWadj.mat[i,j] = WPWadj.mat[i,j] + 1}
}

rownames(WPWadj.mat) <- teams
colnames(WPWadj.mat) <- teams
print(WPWadj.mat) # NB. This adjacency matrix is symmetric and not directional.
```

```
##
       Team A Team B Team C Team D Team E Team F Team G Team H
Team A      0      0      0      0      1      1      0      1
Team B      0      0      0      0      0      1      1      0
Team C      0      0      0      0      0      1      1      0
Team D      0      0      0      0      1      0      1      1
Team E      1      0      0      1      0      0      0      0
Team F      1      1      1      0      0      0      0      0
Team G      0      1      1      1      0      0      0      0
Team H      1      0      0      1      0      0      0      0
```

Having produced the adjacency matrix, we can easily adapt it to create the Colley matrix as follows:

```
# Produce the Colley matrix
sumrowplus2 <- (rowSums(WPWadj.mat)) + 2   # Compute the sum of the rows
d <- diag(sumrowplus2)
C <- (-1*WPWadj.mat) + d  # This is the Colley matrix.
print(C)
```

```
##
       Team A Team B Team C Team D Team E Team F Team G Team H
Team A      5      0      0      0     -1     -1      0     -1
Team B      0      4      0      0      0     -1     -1      0
Team C      0      0      4      0      0     -1     -1      0
Team D      0      0      0      5     -1      0     -1     -1
Team E     -1      0      0     -1      4      0      0      0
Team F     -1     -1     -1      0      0      5      0      0
Team G      0     -1     -1     -1      0      0      5      0
Team H     -1      0      0     -1      0      0      0      4
```

From the diagonal of the Colley matrix, we can instantly see that four teams have played three matches and four have played only two matches.

Next, we can create the win-lose vector, v, using Equation 9.1 and the following code:

```
# Create win-lose vector
homeResults <- table(mini$HomeTeam,mini$Result)
colnames(homeResults) <- c("HL","HD","HW")
print(homeResults)
```

```
##
         HL HD HW
  Team A  0  0  2
  Team B  0  1  0
  Team C  0  0  1
  Team D  1  0  0
  Team E  1  0  0
  Team F  0  0  1
  Team G  0  2  0
  Team H  1  0  0
```

```
awayResults <- table(mini$AwayTeam,mini$Result)
colnames(awayResults) <- c("AW","AD","AL")
print(awayResults)
```

```
##
         AW AD AL
  Team A  1  0  0
  Team B  0  1  0
  Team C  0  0  1
  Team D  1  1  0
  Team E  0  0  1
  Team F  0  1  1
  Team G  0  0  1
  Team H  1  0  0
```

```
# Compute win-lose vector
w <- homeResults[,3] + awayResults[,1]
l <- homeResults[,1] + awayResults[,3]
e <- matrix(1,n,1)
v <- e + 0.5*(w-l)
print(v) # This is the win-lose vector.
```

```
##
     [,1]
[1,]  2.5
[2,]  1.0
[3,]  1.0
[4,]  1.0
[5,]  0.0
```

```
[6,]  1.0
[7,]  0.5
[8,]  1.0
```

Having created *C* and *v*, we are now in a position to solve Equation 9.2, which we can do using the 'ginv' function in the package 'MASS' as follows: The 'ginv' function is used here to calculate the Moore-Penrose generalised inverse of the Colley matrix *C*, which is then multiplied by the win-lose vector *v* using the matrix multiplication operator '%*%'. (NB. The mathematics behind this is complex and beyond the scope of this book. All that most readers need to know is that it works!)

```
# Solve equation using the 'ginv' function in the 'MASS' package
install.packages("MASS") # This installs the 'MASS' package.
# NB. This command only needs to be executed once to install the package.
# Thereafter, the 'MASS' library can be called using the 'library' command.

library(MASS)
Colley.r <- ginv(C) %*% v

# Compile team ratings
Colley.rat <- cbind.data.frame(teams, round(Colley.r,3))
colnames(Colley.rat) <- c("Team","Rating")
print(Colley.rat)
```

```
##
    Team Rating
1 Team A  0.783
2 Team B  0.483
3 Team C  0.483
4 Team D  0.450
5 Team E  0.308
6 Team F  0.550
7 Team G  0.383
8 Team H  0.558
```

Finally, we rank the teams according to their Colley rating score.

```
# Rank teams according to Colley rating score
Colley.temp <- Colley.rat[order(Colley.rat[,2]),]
Colley.rank <- Colley.temp[n:1,]
print(Colley.rank)
```

```
##
    Team Rating
1 Team A  0.783
8 Team H  0.558
6 Team F  0.550
```

```
3 Team C  0.483
2 Team B  0.483
4 Team D  0.450
7 Team G  0.383
5 Team E  0.308
```

Often, it is helpful to present such information graphically. This can easily be done in base R using the 'barplot' function together with the 'horiz = TRUE' argument, which produces a horizontal bar plot. Here the names of the teams are grouped into a vector called 'teams'.

```
# Plot bar chart
teams <- Colley.rank$Team
Rating <- Colley.rank$Rating
par(las=2) # Makes text labels perpendicular to axis
barplot(Rating, horiz=TRUE, names.arg = teams, xlim=c(0,0.8), xlab="Rating")
title("Colley ratings of teams")
```

This produces Figure 9.2, which shows the Colley ratings, ranked in order, for each team in the mini-league.

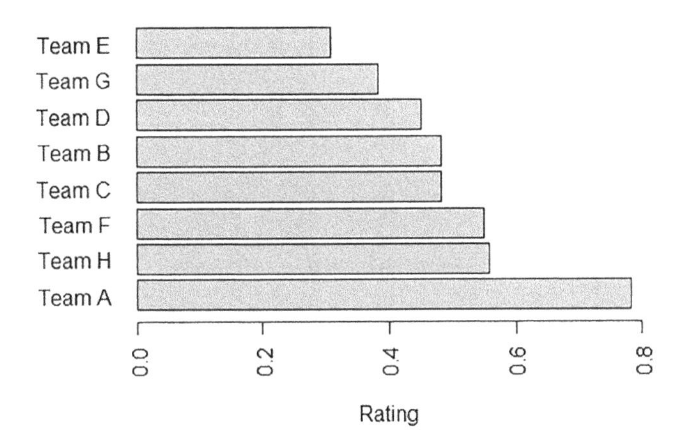

FIGURE 9.2
Horizontal bar plot of Colley ratings for the teams in the mini-soccer league.

From this, we see that the Colley algorithm ranks Team A, who won three matches, the highest, while Team E, who lost twice, gets the lowest rank, which is perhaps not surprising.

9.3 Massey Ranking Algorithm

One of the criticisms of the Colley method is that it takes no account of goal difference, which is an important indicator of how well teams are performing [16,17]. For example, it does not matter whether a team loses 1-0 or 8-0. In the Colley method, both of these results are treated as equal, despite the fact that the latter is a humiliating defeat compared with the former. One solution to this problem was developed by Kenneth Massey, who proposed a ranking method that used a least squares approach to solve a system of linear equations that expressed the relationship between team ratings and the margin of victory [1,7]. Massey's method involved constructing a $[m \times n]$ matrix, W, and recording the outcomes of m matches between n teams. The matrix, W, is populated according to the following rules [3], where w_{ki} is an indicator variable for the outcome of the kth game for team T_i.

$$W_{ki} = \begin{cases} 1 \text{ if team } T_i \text{ won the } k^{\text{th}} \text{ game} \\ -1 \text{ if team } T_i \text{ lost the } k^{\text{th}} \text{ game} \\ 0 \qquad \text{otherwise} \end{cases} \tag{9.4}$$

Massey used the matrix, W, and a vector, y, containing the margins of victory to solve the following equation, in which r_m is the vector of unknown ratings.

$$W^T W r_m = W^T y \tag{9.5}$$

Conveniently, Massey was able to simplify Equation 9.5 to:

$$M r_m = d \tag{9.6}$$

where the Massey matrix, M, is:

$$M_{ij} = \left(W^T W\right)_{ij} = \begin{cases} p_i \text{ if } i = j \\ -p_{ij} \text{ if } i \neq j \end{cases} \tag{9.7}$$

p_i is the total number of games played by team i; p_{ij} is the number of matches played between teams i and j, and d is the vector of cumulative goal differentials.

From Equations 9.5, 9.6, and 9.7, we can see that:

$$d = W^T y \tag{9.8}$$

Rather conveniently, the Massey matrix, M, is simply the Colley matrix, C, with two subtracted from each value in the diagonal (i.e., making the diagonal values the actual number of matches played by the respective teams) [1].

To illustrate how the algorithm can be executed in R, in Example 9.3, the Massey system is applied to the results for the eight teams in the mini-league.

EXAMPLE 9.3: APPLICATION OF THE MASSEY RANKING ALGORITHM TO A PART-COMPLETED MINI-SOCCER LEAGUE

As with the Colley example above, in this example we shall use the following code to rank the teams in the part-completed mini-league using the Massey method. To do this, we first need to create the Massey matrix, which we can do by altering the Colley matrix as follows:

```
# Create identify matrix x2
i2 <- diag(n)*2

# Create Massey matrix
M <- C - i2
print(M)
```

```
##
        Team A Team B Team C Team D Team E Team F Team G Team H
Team A       3      0      0      0     -1     -1      0     -1
Team B       0      2      0      0      0     -1     -1      0
Team C       0      0      2      0      0     -1     -1      0
Team D       0      0      0      3     -1      0     -1     -1
Team E      -1      0      0     -1      2      0      0      0
Team F      -1     -1     -1      0      0      3      0      0
Team G       0     -1     -1     -1      0      0      3      0
Team H      -1      0      0     -1      0      0      0      2
```

Now we need to compute overall goal differences for each club and produce a vector, y.

```
y <- matrix(0,n,1)

for (i in 1:n){
  tempH <- mini[mini$HomeTeam == teams[i],]
  tempA <- mini[mini$AwayTeam == teams[i],]
  GFH <- sum(tempH$HG)
  GFA <- sum(tempA$AG)
  GAH <- sum(tempH$AG)
  GAA <- sum(tempA$HG)
  GF <- GFH+GFA
```

```
  GA <- GAH+GAA
  y[i] <- GF-GA
}

print(y) # This is the vector of goal differences.
```

```
##
     [,1]
[1,]    5
[2,]    0
[3,]    2
[4,]    0
[5,]   -2
[6,]   -1
[7,]   -3
[8,]   -1
```

Having done this, we can then put everything together to compute the Massey rankings. As with the Colley method, we can solve Equation 9.6 using the 'ginv' function in the 'MASS' package.

```
# Compute Massey ratings
library(MASS)
Massey.r <- ginv(M) %*% y

# Compile team ratings
Massey.rat <- cbind.data.frame(teams, round(Massey.r,3))
colnames(Massey.rat) <- c("Team","Rating")
print(Massey.rat)
```

```
##
    Team Rating
1 Team A  1.625
2 Team B -0.438
3 Team C  0.562
4 Team D -0.500
5 Team E -0.437
6 Team F  0.250
7 Team G -1.125
8 Team H  0.063
```

```
# Rank teams according to Massey rating score
Massey.temp <- Massey.rat[order(Massey.rat[,2]),]
Massey.rank <- Massey.temp[n:1,]
print(Massey.rank)
```

```
##
    Team Rating
1 Team A  1.625
3 Team C  0.562
6 Team F  0.250
8 Team H  0.063
5 Team E -0.437
2 Team B -0.438
4 Team D -0.500
7 Team G -1.125
```

From this, we see that, rather surprisingly, Team E is not ranked bottom, despite having lost both matches played. The reason for this is that Team E only narrowly lost in the two matches in which it was involved (i.e., total goal difference of –2). By comparison, Team G (ranked bottom) had a worse total goal difference (i.e., –3 goals) after three matches, despite having drawn two of the matches. As with the Colley method, the Massey algorithm ranked Team A as the best performer.

9.4 Elo Ranking Algorithm

The Elo rating system was originally developed by the physicist Ardad Elo to assess the relative skill levels of chess players [1,8]. However, it has subsequently been developed and applied to a wide range of team sports, including soccer, American football, and basketball [1]. Indeed, FIFA now uses a modified version of the Elo algorithm to rank international soccer teams [18,19]. Unlike the Colley and Massey algorithms, which make no allowance for the order in which matches are played or who plays whom, the Elo system works on a who-beat-who basis and therefore updates the rating after every round of competition. As such, the Elo rating reflects the current performance of soccer teams and is generally considered to be a better predictor of match performance compared with the Colley and Massey algorithms [19].

The central premise of Elo's system, as originally developed, was that each chess player's performance is assumed to be normally distributed about a mean (the expected value), μ, which changes slowly over time as the player either improves or becomes worse. This means that in the short term (i.e., for each individual game), μ can be considered a constant, while in the long term it can gradually change. Consequently, once a player has become established, the only thing that can change their Elo rating is the extent to which his or her performance is above or below their mean or expected value. To this end, Elo came up with the formula shown in Equations 9.9 and 9.10, which can be applied to each player in the chess match (i.e., player i and player j) to compute new ratings, r_{new}, for both players after the match has been completed.

$$\text{Player } i: r_{i\text{new}} = r_{i\text{old}} + K\left(S_{ij} - \mu_{ij}\right) \tag{9.9}$$

$$\text{Player } j: r_{j\text{new}} = r_{j\text{old}} + K\left(S_{ji} - \mu_{ji}\right) \tag{9.10}$$

where $r_{i\text{old}}$ and $r_{j\text{old}}$ are the old Elo ratings for each player; $r_{i\text{new}}$ and $r_{j\text{new}}$ are the new Elo ratings for each player; μ_{ij} and μ_{ji} are the expected win probabilities (Elo probabilities) for the respective players; K is a constant determined by the nature of the competition and the sport [1]; and S_{ij} and S_{ji} are constants that reflect the outcome of the match (i.e., 1 for a win, 0.5 for a draw, and 0 for a loss) as formally expressed in Equation 9.11.

$$S_{ij} = \begin{cases} 1 \text{ if } i \text{ beats } j, \\ 0 \text{ if } i \text{ loses to } j, \\ 0.5 \text{ if } i \text{ and } j \text{ draw} \end{cases} \tag{9.11}$$

The clever thing about the Elo system is that it automatically adjusts to reflect the abilities of the teams/players competing in the various matches. It does this by using a logistic function (Equations 9.12 and 9.13) to compute the expected win (Elo) probabilities (μ_{ij} and μ_{ji}) for each team (player) based on their respective Elo ratings before the match (i.e., $r_{i\text{old}}$ and $r_{j\text{old}}$).

$$\text{Player } i: \mu_{ij} = \frac{1}{1+10^{-d_{ij}/\phi}} \tag{9.12}$$

$$\text{Player } j: \mu_{ji} = \frac{1}{1+10^{-d_{ji}/\phi}} \tag{9.13}$$

where: ϕ is the logistic parameter, which is generally set at $\phi = 400$ for chess [1] and soccer [20], although $\phi = 1000$ has been suggested for American football [1]; and;

$$d_{ij} = r_{i\text{old}} - r_{j\text{old}} \tag{9.14}$$

and;

$$d_{ji} = r_{j\text{old}} - r_{i\text{old}} \tag{9.15}$$

So for example, in a chess match between players A and B, if Player A has a rating of 1800 and Player B has a rating of 1500, then, assuming $\phi = 400$, the Elo probabilities for the respective players will be:

$$\text{Player A}: \mu_{AB} = \frac{1}{1+10^{-(1800-1500)/400}} = 0.849$$

$$\text{Player B: } \mu_{BA} = \frac{1}{1+10^{-(1500-1800)/400}} = 0.151$$

This tells us that the probability of Player A winning is 0.849 (84.9%), and the probability of Player B winning is just 0.151 (15.1%). So according to the Elo algorithm, Player A is much more likely to win the match compared with Player B.

So, assuming that $K=15$, if Player A wins, then their reward will be:

$$\text{Player A: } r_{Anew} = 1800 + 15(1 - 0.849) = 1802.3 \ \ (\text{NB. An additional 2.3 Elo points.})$$

Whereas, if Player B wins the match, their new Elo rating will be:

$$\text{Player B: } r_{Bnew} = 1500 + 15(1 - 0.151) = 1512.7 \ \ (\text{NB. An additional 12.7 Elo points.})$$

From this, we can see that the quality of the opponent greatly affects the Elo rating, with a win for a stronger player having much less impact than would be the case if a weaker player defeated a stronger opponent.

9.4.1 Elo K-Factor

In Equations 9.9 and 9.10, the *K*-factor is a constant that reflects the nature of the competition and the sport. As such, choosing an appropriate value for *K* is a matter of judgement. If *K* is too large, then too much weight will be given to the difference between the actual and expected scores, with the result that the ratings will tend to become volatile [1]. Conversely, if *K* is too small, then the Elo algorithm will lose its ability to detect finer changes in player or team performance. With respect to this, there is still a debate as to the appropriate value of *K* to use in soccer. The World Football Elo rating system [21] recommends:

- $K=60$ for the FIFA World Cup finals;
- $K=50$ for continental championships and major intercontinental tournaments;
- $K=40$ for the FIFA World Cup, continental qualifiers, and major tournaments;
- $K=30$ for all other tournaments;
- $K=20$ for friendly matches.

However, Sullivan and Cronin [20] found $K=25$ to be more appropriate for the English Premier League (EPL), whereas opisthokonta.net [22] found $K=18.5$ to be optimum for the EPL.

EXAMPLE 9.4: APPLICATION OF THE ELO RANKING ALGORITHM TO A PART-COMPLETED MINI-SOCCER LEAGUE

To illustrate how the Elo system can be utilised in soccer to rank teams, in this example we shall apply the algorithm to the mini-soccer league shown in Example 9.1. In particular, we will make $K=18.5$ in line with the findings of opisthokonta.net [22] and set the Elo probability for all the teams at an initial value of 0.5. We will also give the Elo ratings for the teams an initial arbitrary value of 1200, as specified by Sullivan and Cronin [20].

First we need to enhance the 'mini' data frame by adding two new columns, home team Elo probability (HTEP) and away team Elo probability (ATEP), as follows:

```
Elo.dat <- mini
nobs = nrow(Elo.dat) # Here nobs is the number of observations.

Elo.dat["HTEP"] <- NA # That creates the new column for the home team Elo points
Elo.dat["ATEP"] <- NA # That creates the new column for the away team Elo points

for(i in 1:nobs){
  if(Elo.dat$Result[i] == "H"){Elo.dat$HTEP[i] <- 1; Elo.dat$ATEP[i] <- 0}
  else if(Elo.dat$Result[i] == "D"){Elo.dat$HTEP[i] <- 0.5; Elo.dat$ATEP[i] <- 0.5}
  else {Elo.dat$HTEP[i] <- 0; Elo.dat$ATEP[i] <- 1}
}

print(Elo.dat)
```

```
##
   HomeTeam AwayTeam HG AG Results HTEP ATEP
1    Team A   Team E  3  2       H  1.0  0.0
2    Team B   Team F  1  1       D  0.5  0.5
3    Team C   Team G  5  2       H  1.0  0.0
4    Team D   Team H  0  1       A  0.0  1.0
5    Team E   Team D  2  3       A  0.0  1.0
6    Team F   Team C  2  1       H  1.0  0.0
7    Team G   Team B  0  0       D  0.5  0.5
8    Team H   Team A  1  3       A  0.0  1.0
9    Team A   Team F  4  2       H  1.0  0.0
10   Team G   Team D  2  2       D  0.5  0.5
```

Next, we create an empty matrix to store the updated Elo rating scores after each match played. Given that ten matches have been played, we specify eleven columns in this matrix and populate the matrices with zeros. Finally, we initialise the matrix by populating the first column with a score of 1200.

```
# Create a ranking matrix and set all teams' initial ranking to zero
team.A <- Elo.dat$HomeTeam
team.B <- Elo.dat$AwayTeam
```

```
# Create empty matrix to store Elo rank results
ranks <- matrix(0,n,(nrow(Elo.dat)+1))
row.names(ranks) <- teams
ranks[,1] <- 1200   # This primes the ranks matrix with an initial Elo score of 1200
m <- ncol(ranks)
```

Here we specify $K=18.5$, which is the optimum K value determined by opisthokonta.net [22].

```
# Specify K-factor.
K = 18.5
```

We also add some extra columns to the Elo.dat data frame, as follows:

```
# Create some new columns to store Elo results.
Elo.dat["HomeProb"] <- NA # To store home win expected probability
Elo.dat["AwayProb"] <- NA # To store away win expected probability
Elo.dat["HomeRating"] <- NA # To store updated home team Elo rating
Elo.dat["AwayRating"] <- NA # To store updated away team Elo rating
```

Having done this, we can now populate the respective result matrices using the following R code:

```
# Now we populate the ranks matrix using the Elo algorithm.
for(i in 1:(nrow(Elo.dat))){
  fA <- match(Elo.dat[i,1],teams) # Assign numerical identifier to home team.
  fB <- match(Elo.dat[i,2],teams) # Assign numerical identifier to away team.
  if(ranks[fA,i] == ranks[fB,i]){
    expA <- 0.5 # Expected probability of home team
    expB <- 0.5 # Expected probability of home team
  }
  else{
    expA <- 1/(1+10^(-((ranks[fA,i]-ranks[fB,i])/400))) # Expected probability of A
    expB <- 1/(1+10^(-((ranks[fB,i]-ranks[fA,i])/400))) # Expected probability of B
  }
  rA <- ranks[fA,i] + K*(Elo.dat[i,6]-expA) # Elo algorithm applied to home team.
  rB <- ranks[fB,i] + K*(Elo.dat[i,7]-expB) # Elo algorithm applied to away team.
  ranks[,(i+1)] <- ranks[,i]
  ranks[fA,(i+1)] <- rA
  ranks[fB,(i+1)] <- rB
  Elo.dat$HomeProb[i] <- round(expA,3)
  Elo.dat$AwayProb[i] <- round(expB,3)
  Elo.dat$HomeRating[i] <- rA
  Elo.dat$AwayRating[i] <- rB
}
```

Now we can display the populated result matrices.

```
# Display results
print(Elo.dat) # Overall results
```

```
##
   HomeTeam AwayTeam HG AG Results HTEP ATEP HomeProb AwayProb
1    Team A   Team E  3  2       H  1.0  0.0    0.500    0.500
2    Team B   Team F  1  1       D  0.5  0.5    0.500    0.500
3    Team C   Team G  5  2       H  1.0  0.0    0.500    0.500
4    Team D   Team H  0  1       A  0.0  1.0    0.500    0.500
5    Team E   Team D  2  3       A  0.0  1.0    0.500    0.500
6    Team F   Team C  2  1       H  1.0  0.0    0.487    0.513
7    Team G   Team B  0  0       D  0.5  0.5    0.487    0.513
8    Team H   Team A  1  3       A  0.0  1.0    0.500    0.500
9    Team A   Team F  4  2       H  1.0  0.0    0.513    0.487
10   Team G   Team D  2  2       D  0.5  0.5    0.487    0.513
   HomeRating AwayRating
1    1209.250   1190.750
2    1200.000   1200.000
3    1209.250   1190.750
4    1190.750   1209.250
5    1181.500   1200.000
6    1209.496   1199.754
7    1190.996   1199.754
8    1200.000   1218.500
9    1227.510   1200.486
10   1191.236   1199.760
```

This shows us the updated Elo ratings for the respective teams after every match, together with the expected win probabilities computed using Equations 9.12 and 9.13.

By viewing the 'ranks' matrix, we can see the updated Elo ratings for the respective teams after every round of competition.

```
print(round(ranks,3)) # Elo ranks
```

```
##
       [,1]    [,2]    [,3]    [,4]    [,5]    [,6]     [,7]     [,8]
Team A 1200 1209.25 1209.25 1209.25 1209.25 1209.25 1209.250 1209.250
Team H 1200 1200.00 1200.00 1200.00 1209.25 1209.25 1209.250 1209.250
Team F 1200 1200.00 1200.00 1200.00 1200.00 1200.00 1209.496 1209.496
Team C 1200 1200.00 1200.00 1209.25 1209.25 1209.25 1199.754 1199.754
Team B 1200 1200.00 1200.00 1200.00 1200.00 1200.00 1200.000 1199.754
Team D 1200 1200.00 1200.00 1200.00 1190.75 1200.00 1200.000 1200.000
Team G 1200 1200.00 1200.00 1190.75 1190.75 1190.75 1190.750 1190.996
Team E 1200 1190.75 1190.75 1190.75 1190.75 1181.50 1181.500 1181.500
           [,9]    [,10]    [,11]
Team A 1218.500 1227.510 1227.510
Team H 1200.000 1200.000 1200.000
Team F 1209.496 1200.486 1200.486
Team C 1199.754 1199.754 1199.754
Team B 1199.754 1199.754 1199.754
Team D 1200.000 1200.000 1199.760
Team G 1190.996 1190.996 1191.236
Team E 1181.500 1181.500 1181.500
```

The populated 'ranks' matrix displays the revised Elo score for each team after every match that has been played. From this, we see that all the teams have an initial value of 1200. However, as matches are played, this value either increases or decreases, allowing the teams to be easily ranked after every match in the competition.

Finally, we can compile the Elo rankings after ten matches.

```
# Compile results
er <- as.data.frame(ranks[,m]) # Select final ratings only
colnames(er) <- c("Elo_Rating")

# Now we use the 'dplyr' package to arrange the teams in descending order.
install.packages("dplyr") # This installs the 'dplyr' package.
# NB. This command only needs to be executed once to install the package.
# Thereafter, the 'dplyr' library can be called using the 'library' command.

library(dplyr)
Elo.rank <- er %>% arrange(desc(Elo_Rating))
print(Elo.rank)
```

```
##
        Elo_Rating
Team A    1227.510
Team F    1200.486
Team H    1200.000
Team D    1199.760
Team C    1199.754
Team B    1199.754
Team G    1191.236
Team E    1181.500
```

From this, we see that the Elo algorithm ranks Team A as the best team with 1227.51 Elo points and Team E as the worst team with just 1181.5 points after ten matches.

For educational purposes, the Elo rating code in Example 9.4 has been written 'long-hand' so that the reader can appreciate how the Elo algorithm works. However, there is a much faster way to perform the Elo calculation, and that is to use the dedicated 'elo' package in R [23]. To illustrate this, in Example 9.5, we repeat the analysis performed in Example 9.4, but this time with the 'elo' package.

EXAMPLE 9.5: ELO RANKING EXAMPLE USING THE 'ELO' PACKAGE

In this example, we repeat the analysis performed in Example 9.4, but this time using the 'elo' package. The code necessary to do this is presented below:

```
# Now we use the 'elo' package to arrange the teams in descending order.
install.packages("elo") # This installs the 'elo' package.
# NB. This command only needs to be executed once to install the package.
# Thereafter, the 'elo' library can be called using the 'library' command.

library(elo)
mini_EloMod <- elo.run(data = Elo.dat,
               formula = HTEP ~ HomeTeam + AwayTeam,
               k=18.5, initial.elos=1200)

mini_EloRes <- mini_EloMod %>% as.data.frame()
print(mini_EloRes)
```

```
##
   team.A team.B       p.A wins.A   update.A   update.B    elo.A    elo.B
1  Team A Team E 0.5000000    1.0  9.2500000 -9.2500000 1209.250 1190.750
2  Team B Team F 0.5000000    0.5  0.0000000  0.0000000 1200.000 1200.000
3  Team C Team G 0.5000000    1.0  9.2500000 -9.2500000 1209.250 1190.750
4  Team D Team H 0.5000000    0.0 -9.2500000  9.2500000 1190.750 1209.250
5  Team E Team D 0.5000000    0.0 -9.2500000  9.2500000 1181.500 1200.000
6  Team F Team C 0.4866913    1.0  9.4962105 -9.4962105 1209.496 1199.754
7  Team G Team B 0.4866913    0.5  0.2462105 -0.2462105 1190.996 1199.754
8  Team H Team A 0.5000000    0.0 -9.2500000  9.2500000 1200.000 1218.500
9  Team A Team F 0.5129546    1.0  9.0103400 -9.0103400 1227.510 1200.486
10 Team G Team D 0.4870454    0.5  0.2396600 -0.2396600 1191.236 1199.760
```

As we can see, using the 'elo' package, it is possible to execute the Elo algorithm with just a few lines of code. However, in order to establish the final ranking order, we need to execute the following code:

```
# Now we rank the teams.
rank.teams(mini_EloMod, ties.method = "min",)
```

```
##
Team A Team B Team C Team D Team E Team F Team G Team H
     1      5      5      4      8      2      7      3
```

9.5 Comparison between the Colley, Massey, and Elo Ranking Algorithms

Table 9.1 shows the results of applying the three ranking methods after the first ten matches of the mini-league competition. From this, we see that although there is some agreement between the ranking orders produced by the various systems, there are nonetheless some notable discrepancies between the ranking orders produced by the three algorithms. Of the three systems, the Elo algorithm is in closest agreement with the league standings (arising from three points for a win and one point for a draw), with four out of the eight teams ranked in the same positions and a normalised tau distance of 0.107 (see Key Concept Box 9.1 for more details on the tau distance metric). By comparison, the Colley algorithm performed less well, exhibiting a greater tau distance of 0.214, with the Massey algorithm being even less in agreement (tau=0.286), which is perhaps not surprising given that the latter utilises goal difference, which does not contribute to the points total in a typical soccer league.

In order to assess the capabilities of the three ranking systems with real league data, the Colley, Massey, and Elo algorithms were applied to historical match data from the EPL for season 2020–21 after 10, 19, 29, and 38 rounds of competition. The results of this analysis are presented in Table 9.2, which ranks the teams according to their league position and the ratings produced using the three ranking algorithms. From this, we see that the Colley and Elo algorithms produce ranking orders that are very close to the partial league standings achieved in real life. Indeed, given that the Colley algorithm does not accommodate draws, it is quite remarkable how closely the Colley ranking order resembles that of the actual league standings, with the normalised tau distance between the two being only 0.021 after 38 rounds of competition. By comparison, however, the Massey system exhibited much less agreement with the partial league standings. As such, this appears to demonstrate the validity of the Colley and Elo ranking algorithms in a soccer league context, suggesting that, with regard to ranking teams, both systems should perform reasonably well in competitions that do not involve a traditional league structure.

TABLE 9.1

Comparison of the Rankings Produced by the Colley, Massey, and Elo systems for the First Ten Matches of the Mini-Soccer League

Team	Played	League Points	League Standing	Colley Rating	Colley Rank	Massey Rating	Massey Rank	Elo Rating	Elo Rank
Team A	3	9	1	0.783	1	1.625	1	1227.510	1
Team B	2	2	6	0.483	5	−0.438	6	1199.754	5=
Team C	2	3	4	0.483	4	0.562	2	1199.754	5=
Team D	3	4	2	0.483	6	−0.500	7	1199.760	4
Team E	2	0	8	0.308	8	−0.437	5	1181.500	8
Team F	3	4	3	0.550	3	0.250	3	1200.486	2
Team G	3	2	7	0.383	7	−1.125	8	1191.236	7
Team H	2	3	5	0.558	2	0.063	4	1200.000	3
Tau distance	n.a.	n.a	n.a	n.a	0.214	n.a	0.286	n.a	0.107

TABLE 9.2

Team Rankings after 10, 19, 29, and 38 Rounds of Competition for the English Premier League Season 2020–2021

	League Position	Colley	Massey	Elo	League position	Colley	Massey	Elo
	10 Rounds	**10 Rounds**	**10 Rounds**	**10 Rounds**	**19 Rounds**	**19 Rounds**	**19 Rounds**	**19 Rounds**
Rank	**98 Matches**	**98 Matches**	**98 Matches**	**98 Matches**	**185 Matches**	**185 Matches**	**185 Matches**	**185 Matches**
1	Tottenham	Tottenham	Tottenham	Liverpool	Man United	Man City	Man City	Man United
2	Liverpool	Liverpool	Chelsea	Tottenham	Man City	Man United	Tottenham	Man City
3	Chelsea	Chelsea	Aston Villa	Chelsea	Leicester	Leicester	Liverpool	Leicester
4	Leicester	Man United	West Ham	Man United	Liverpool	Tottenham	Leicester	Tottenham
5	West Ham	Southampton	Liverpool	West Ham	Tottenham	Liverpool	Aston Villa	Liverpool
6	Southampton	Man City	Man City	Southampton	Everton	Everton	Man United	Everton
7	Wolves	West Ham	Leicester	Wolves	West Ham	West Ham	Chelsea	West Ham
8	Everton	Leicester	Southampton	Man City	Aston Villa	Southampton	Everton	Southampton
9	Man United	Wolves	Everton	Leicester	Chelsea	Aston Villa	Southampton	Aston Villa
10	Aston Villa	Everton	Man United	Aston Villa	Southampton	Chelsea	West Ham	Chelsea
11	Man City	Newcastle	Wolves	Everton	Arsenal	Arsenal	Arsenal	Arsenal
12	Leeds	Aston Villa	Brighton	Leeds	Leeds	Crystal Palace	Leeds	Leeds
13	Newcastle	Leeds	Arsenal	Newcastle	Crystal Palace	Leeds	Brighton	Crystal Palace
14	Arsenal	Arsenal	Newcastle	Arsenal	Wolves	Wolves	Wolves	Burnley
15	Crystal Palace	Crystal Palace	Leeds	Crystal Palace	Burnley	Burnley	Crystal Palace	Wolves
16	Brighton	Brighton	Crystal Palace	Brighton	Newcastle	Brighton	Fulham	Brighton
17	Fulham	Burnley	Fulham	Fulham	Brighton	Newcastle	Burnley	Newcastle
18	West Brom	West Brom	Sheffield United	Burnley	Fulham	Fulham	Newcastle	Fulham

(Continued)

TABLE 9.2 (*Continued*)

Team Rankings after 10, 19, 29, and 38 Rounds of Competition for the English Premier League Season 2020–2021

	League Position	**Colley**	**Massey**	**Elo**	**League position**	**Colley**	**Massey**	**Elo**
19	Burnley	Fulham	West Brom	West Brom	West Brom	West Brom	Sheffield United	West Brom
20	Sheffield United	Sheffield United	Burnley	Sheffield United	Sheffield United	Sheffield United	West Brom	Sheffield United
Tau distance	**n.a.**	**0.095**	**0.137**	**0.074**	**n.a.**	**0.032**	**0.111**	**0.026**

	League Position	**Colley**	**Massey**	**Elo**	**League position**	**Colley**	**Massey**	**Elo**
	29 Rounds	**29 Rounds**	**29 Rounds**	**29 Rounds**	**38 Rounds**	**38 Rounds**	**38 Rounds**	**38 Rounds**
Rank	**290 Matches**	**290 Matches**	**290 Matches**	**290 Matches**	**380 Matches**	**380 Matches**	**380 Matches**	**380 Matches**
1	Man City	Man City	Man City	Man City	Man City	Man City	Man City	Man City
2	Man United	Man United	Man United	Man United	Man United	Man United	Man United	Man United
3	Leicester	Leicester	Leicester	Leicester	Liverpool	Liverpool	Liverpool	Liverpool
4	Chelsea	Chelsea	Tottenham	Chelsea	Chelsea	Chelsea	Tottenham	Chelsea
5	West Ham	West Ham	Chelsea	West Ham	Leicester	West Ham	Chelsea	West Ham
6	Tottenham	Tottenham	Liverpool	Tottenham	West Ham	Leicester	Leicester	Arsenal
7	Liverpool	Liverpool	West Ham	Everton	Tottenham	Tottenham	Arsenal	Leicester
8	Everton	Everton	Arsenal	Arsenal	Arsenal	Arsenal	West Ham	Leeds
9	Arsenal	Arsenal	Aston Villa	Liverpool	Leeds	Everton	Aston Villa	Tottenham
10	Aston Villa	Aston Villa	Everton	Aston Villa	Everton	Leeds	Leeds	Everton
11	Leeds	Leeds	Leeds	Leeds	Aston Villa	Aston Villa	Everton	Aston Villa

(Continued)

TABLE 9.2 (*Continued*)

Team Rankings after 10, 19, 29, and 38 Rounds of Competition for the English Premier League Season 2020–2021

	League Position	Colley	Massey	Elo	League position	Colley	Massey	Elo
12	Crystal Palace	Wolves	Brighton	Crystal Palace	Newcastle	Wolves	Brighton	Newcastle
13	Wolves	Crystal Palace	Wolves	Burnley	Wolves	Newcastle	Newcastle	Brighton
14	Southampton	Burnley	Burnley	Wolves	Crystal Palace	Crystal Palace	Wolves	Wolves
15	Burnley	Southampton	Southampton	Brighton	Southampton	Brighton	Southampton	Crystal Palace
16	Brighton	Brighton	Fulham	Southampton	Brighton	Southampton	Burnley	Southampton
17	Newcastle	Newcastle	Crystal Palace	Fulham	Burnley	Burnley	Crystal Palace	Burnley
18	Fulham	Fulham	Newcastle	Newcastle	Fulham	Fulham	Fulham	Fulham
19	West Brom	West Brom	Sheffield United	West Brom	West Brom	West Brom	West Brom	West Brom
20	Sheffield United	Sheffield United	West Brom	Sheffield United	Sheffield United	Sheffield United	Sheffield United	Sheffield United
Tau distance	**n.a.**	**0.011**	**0.084**	**0.032**	**n.a.**	**0.021**	**0.063**	**0.037**

Note: Rankings determined using the Colley, Massey and Elo algorithms.

KEY CONCEPT BOX 9.1: TAU DISTANCE

One of the main challenges when comparing the results produced by different ranking systems is trying to quantify the observed differences. For example, consider the following rank orders produced by three ranking systems for five soccer teams:

Ranking algorithm 1 produces: 1. Team C; 2. Team D; 3. Team A; 4. Team B; and 5. Team E

Ranking algorithm 2 produces: 1. Team E; 2. Team C; 3. Team B; 4. Team D; and 5. Team A

Ranking algorithm 3 produces: 1. Team C; 2. Team A; 3. Team D; 4. Team B; and 5. Team E

From this, we see that algorithms 1 and 3 produce very similar ranking orders for the teams, with only the positions of two teams, A and D, swapped around. By comparison, algorithm 2 produces a very different ranking order from algorithms 1 and 3. So the mathematical challenge is how do we quantify these differences so that we can make a meaningful comparison? Well, one way in which this can be done is to compute Kendall's tau distance, which is a metric that counts the number of pair-order disagreements between two ranking lists. As such, the tau distance between any two rankings is simply the number of pairs that are in a different order in the two lists. So for example, the tau distance between two ranking lists, 1 3 7 6 2 5 4 and 7 1 3 2 6 5 4 is three because the pairs 7-1, 3-7, and 6-2 are in a different order, whereas all the other pairs are in the same order.

While the tau distance is a useful metric, it has the weakness that as the ranking lists get longer, there is a tendency for Kendall's tau distance to also increase because there is likely to be an increased number of discordant pairs. So to overcome this problem, it is good practice to normalise the tau distance in order for comparisons to be made easily. The normalised tau distance can be calculated as follows.

$$\tau = \frac{n_d}{n(n-1)/2} \tag{9.16}$$

where τ is the normalised Kendall's tau distance; n is the number of pairs of observations; and n_d is the number of discordant pairs.

If we compute the normalised tau distances between the ranking orders for the three algorithms above, we find that:

- For algorithms 1 and 2: tau distance = 6 and normalised tau distance = 0.6
- For algorithms 1 and 3: tau distance = 1 and normalised tau distance = 0.1
- For algorithms 2 and 3: tau distance = 7 and normalised tau distance = 0.7

9.6 Using the Elo Ratings to Predict the Outcome of Soccer Matches

Using Equations 9.12 and 9.13, it is possible to calculate the expected win (Elo) probabilities for the two teams competing in any given soccer match. This means that as well as ranking teams, the Elo algorithm should in theory be able to predict outcomes of individual soccer matches, provided of course, that the Elo ratings of the competing teams are known beforehand. Indeed, all the evidence suggests that the Elo algorithm is quite good at making match outcome predictions in team sports [1,19]. For example, using a modified version of the Elo algorithm to predict the outcomes of American Football (NFL) games, Langville and Meyer [1] reported a hindsight accuracy of 75.3% and a foresight accuracy of 62.2%. Likewise, using a modified Elo algorithm, Sullivan and Cronin [20] were able to correctly predict about 60% of EPL match outcomes. Having said this, the Elo algorithm is better suited to making predictions in sports such as tennis, where draws do not occur. In sports such as soccer, where draws are relatively frequent and home advantage has a strong effect, it is necessary to adapt and tune the Elo algorithm in order to make accurate predictions [20] – something that is beyond the scope of this book. So here we will content ourselves with explaining how to make soccer match predictions using a relatively simple Elo model – although we will adapt the model to accommodate home advantage.

One way to use the Elo algorithm to make match predictions is to use the computed Elo probabilities and then apply decision thresholds. For example, Sullivan and Cronin [20] suggest that if the Elo model gives a team a win probability of 0.60 (i.e., 60% probability that this team will win) or greater, then our prediction should be that the team will win. Likewise, if a team's Elo probability is less than or equal to 0.40 (i.e., 40% probability), then we should assume that it will lose. However, if the Elo probability is between

0.40 and 0.60, then Sullivan and Cronin suggest that we should assume that the match will be drawn. Unfortunately, while this logic has some validity, it results in far too many draws being predicted (as illustrated in Example 9.6). So perhaps a better strategy is to only make predictions on matches where the Elo probability is either less than 0.40 or greater than or equal to 0.60, as illustrated in Example 9.6. This way, we have a much higher chance of making correct predictions about the outcome of certain matches.

To illustrate how Sullivan and Cronin's strategy can be applied in practice and also to show how home advantage can be accommodated, the R code in Example 9.6 uses the 'elo' package to predict the outcome of 20 consecutive matches in the EPL during season 2020–21.

EXAMPLE 9.6: ELO MATCH PREDICTION EXAMPLE

In this example, we use the Elo prediction methodology to forecast the outcome of 20 consecutive EPL matches that occurred from 3 to 6 February 2021. To produce the Elo ratings, we used data from the 210 EPL matches that occurred immediately prior to these dates. Again, we use the 'elo' package in R, but unlike the previous examples, we will build in a factor for home advantage into the Elo model.

First, we shall clear all the existing data from the workspace.

```
rm(list = ls())  # Clears all variables from workspace
```

Now we load the historical match data for the EPL season 2020–2021.

```
# Load data
mydata <- head(read.csv('https://www.football-data.co.uk/mmz4281/2021/E0.csv'),380)[,1:11]
names(mydata)
```

```
##
 [1] "Div"      "Date"     "Time"     "HomeTeam" "AwayTeam" "FTHG"
 [7] "FTAG"     "FTR"      "HTHG"     "HTAG"     "HTR"
```

From the 'mydata' data set, we now select columns four to eight, which we rename for convenience. In addition, we add two new empty vectors called "HTEP' and "ATEP", in which we shall list the Elo points (i.e., 1 for a win, 0.5 for a draw, and 0 for a loss) for the home and away teams.

```
dat <- mydata[,c(4:8)]
colnames(dat) <- c("HomeTeam","AwayTeam","HG","AG","Results")
dat["HTEP"] <- NA # That creates the new column for the home team Elo points
dat["ATEP"] <- NA # That creates the new column for the away team Elo points
ng <- nrow(dat) # Determine number of matches
```

Now, we can populate the "HTEP" and "ATEP" columns using the following code and display the first ten rows.

```
# Populate empty columns
for(i in 1:ng){
  if(dat$Result[i] == "H"){dat$HTEP[i] <- 1; dat$ATEP[i] <- 0}
  else if(dat$Result[i] == "D"){dat$HTEP[i] <- 0.5; dat$ATEP[i] <- 0.5}
  else {dat$HTEP[i] <- 0; dat$ATEP[i] <- 1}
}

head(dat, 10)
```

```
##
           HomeTeam    AwayTeam HG AG Results HTEP ATEP
1            Fulham     Arsenal  0  3       A    0    1
2    Crystal Palace Southampton  1  0       H    1    0
3         Liverpool       Leeds  4  3       H    1    0
4          West Ham   Newcastle  0  2       A    0    1
5         West Brom   Leicester  0  3       A    0    1
6         Tottenham     Everton  0  1       A    0    1
7          Brighton     Chelsea  1  3       A    0    1
8  Sheffield United      Wolves  0  2       A    0    1
9           Everton   West Brom  5  2       H    1    0
10            Leeds      Fulham  4  3       H    1    0
```

Having produced a working data set, we now put the first 210 matches into a training data frame, which we will use to build and tune the Elo model, and matches 211–230 into a testing data frame, which we shall use to make the actual match outcome predictions.

```
# Divide data into two sub-groups: training data and testing data.
dat.train <- dat[1:210,]
dat.test <- dat[211:230,]
```

Now we can build the Elo model using the 'elo' package. Note that we use the 'adjust(HomeTeam, HGA)' statement to allocate a home ground advantage (HGA) to the home teams, which in this case is 30. We also make $K=18.5$ and set the initial Elo rating value for the clubs at 1200. (NB. All these parameters can be tuned to produce the optimum results using the training data set.)

```
# Build model using the training data set, allowing for home advantage.
library(elo)
HGA <- 30 # Home ground advantage
EPL_mod <- elo.run(formula = HTEP ~ adjust(HomeTeam, HGA) + AwayTeam,
            data = dat.train, k=18.5, initial.elos=1200)

EPL_EloRes <- EPL_mod %>% as.data.frame()
head(EPL_EloRes,15) # This displays the first 15 matches.
```

This displays the Elo results for the first 15 matches. (NB. For convenience, we only show the results for the first 15 matches.) Note also that the 'elo' package automatically calls the home team, team.A, and the away team, team.B.

```
##
             team.A          team.B       p.A wins.A   update.A  update.B
1            Fulham         Arsenal 0.5430665      0 -10.046730 10.046730
2    Crystal Palace     Southampton 0.5430665      1   8.453270 -8.453270
3         Liverpool           Leeds 0.5430665      1   8.453270 -8.453270
4          West Ham       Newcastle 0.5430665      0 -10.046730 10.046730
5         West Brom       Leicester 0.5430665      0 -10.046730 10.046730
6         Tottenham         Everton 0.5430665      0 -10.046730 10.046730
7          Brighton         Chelsea 0.5430665      0 -10.046730 10.046730
8  Sheffield United          Wolves 0.5430665      0 -10.046730 10.046730
9           Everton       West Brom 0.5715949      1   7.925495 -7.925495
10            Leeds          Fulham 0.5453417      1   8.411178 -8.411178
11       Man United  Crystal Palace 0.5309685      0  -9.822918  9.822918
12          Arsenal        West Ham 0.5715949      1   7.925495 -7.925495
13      Southampton       Tottenham 0.5453417      0 -10.088822 10.088822
14        Newcastle        Brighton 0.5715949      0 -10.574505 10.574505
15          Chelsea       Liverpool 0.5453417      0 -10.088822 10.088822
      elo.A    elo.B
1  1189.953 1210.047
2  1208.453 1191.547
3  1208.453 1191.547
4  1189.953 1210.047
5  1189.953 1210.047
6  1189.953 1210.047
7  1189.953 1210.047
8  1189.953 1210.047
9  1217.972 1182.028
10 1199.958 1181.542
11 1190.177 1218.276
12 1217.972 1182.028
13 1181.458 1200.042
14 1199.472 1200.528
15 1199.958 1218.542
```

Alternatively, we can use the 'predict' function in the 'elo' package to produce the Elo probabilities of the home team. Note that this produces exactly the same values as 'p.A' in the table above.

```
# Predictions
pred.train <- predict(EPL_mod) # Training data set
head(pred.train,15) # These are the predicted win probabilities for Team A.
```

```
##
 [1] 0.5430665 0.5430665 0.5430665 0.5430665 0.5430665 0.5430665 0.5430665
 [8] 0.5430665 0.5715949 0.5453417 0.5309685 0.5715949 0.5453417 0.5715949
[15] 0.5453417
```

Having produced the updated Elo ratings for the matches in the training data set, we next need to specify some decision thresholds to apply to the Elo probabilities. Here we apply Sullivan and Cronin's [20] strategy: greater than or equal to 0.60 equates to a win, and less than or equal to 0.40 equates to a loss, with in between 0.40 and 0.60 predicting a draw.

```
# Specify thresholds
p.win <- 0.6
p.lose <- 0.4

# Compile match prediction results for training data set
train.pred <- cbind.data.frame(dat.train[,1:5], pred.train)
train.pred["Prediction"] <- NA # That creates the new column for the home team Elo points
train.pred["Outcome"] <- NA # That creates the new column for the away team Elo points
n.train <- nrow(train.pred)

for(i in 1:n.train){
  if(train.pred$pred.train[i] >= 0.6){train.pred$Prediction[i] <- "H"}
  else if(train.pred$pred.train[i] <= 0.4){train.pred$Prediction[i] <- "A"}
  else {train.pred$Prediction[i] <- "D"}
  if(train.pred$Results[i] == train.pred$Prediction[i]){train.pred$Outcome[i] <- 1}
  else{train.pred$Outcome[i] <- 0}
}
```

For convenience, we will only display the predictions for the first ten and last ten matches in the training data set.

```
# Display the prediction results
head(train.pred,10)
```

```
##
          HomeTeam    AwayTeam HG AG Results pred.train Prediction Outcome
1          Fulham     Arsenal  0  3       A  0.5430665          D       0
2  Crystal Palace Southampton  1  0       H  0.5430665          D       0
3       Liverpool       Leeds  4  3       H  0.5430665          D       0
4        West Ham   Newcastle  0  2       A  0.5430665          D       0
5       West Brom   Leicester  0  3       A  0.5430665          D       0
6       Tottenham     Everton  0  1       A  0.5430665          D       0
7        Brighton     Chelsea  1  3       A  0.5430665          D       0
8 Sheffield United     Wolves  0  2       A  0.5430665          D       0
9         Everton   West Brom  5  2       H  0.5715949          D       0
10          Leeds      Fulham  4  3       H  0.5453417          D       0
```

```
tail(train.pred,10)
```

```
##
             HomeTeam        AwayTeam HG AG Results pred.train Prediction
201       Southampton     Aston Villa  0  1       A  0.5421614          D
202           Chelsea         Burnley  2  0       H  0.5742104          D
203         Leicester           Leeds  1  3       A  0.6284936          H
204          West Ham       Liverpool  1  3       A  0.5328307          D
205          Brighton       Tottenham  1  0       H  0.4461556          D
206 Sheffield United       West Brom  2  1       H  0.5151296          D
207            Wolves         Arsenal  2  1       H  0.4757141          D
208        Man United     Southampton  9  0       H  0.6238458          H
209         Newcastle Crystal Palace  1  2       A  0.5077615          D
210           Burnley        Man City  0  2       A  0.3974088          A
    Outcome
201       0
202       0
203       0
204       0
205       0
206       0
207       0
208       1
209       0
210       1
```

From this, we observe that although the code has executed Sullivan and Cronin's selection criteria correctly, it results in far too many drawn matches being predicted. However, when we remove the draw predictions and decline to predict the outcome of matches for which the Elo probability is between 0.4 and 0.6, the predictions become more accurate, as we can see using the following code:

```
# Check performance of the model.
temp <- train.pred[,c(7,8)]
train.check <- temp[!(temp$Prediction == "D"),] # This removes the 'draws'.
n.pred <- nrow(train.check)
n.correct <- sum(train.check$Outcome)
train.accuracy <- n.correct/n.pred
print(train.accuracy) # This is prediction accuracy with the training data set.
```

```
##
[1] 0.6511628
```

This reveals that with the training data set, the revised Elo algorithm predicted the outcome of 28 out of 43 matches correctly, giving an overall prediction accuracy of 65.1%.

Having built an Elo model that appears to give reasonable results, we can now apply it to the testing data set using the following code:

```
# Now we use the mode to predict the outcome of the matches in the testing data set.
pred.test <- predict(EPL_mod, newdata = dat.test) # Testing data set

# Compile match prediction results
```

```
test.pred <- cbind.data.frame(dat.test[,1:5], pred.test)
test.pred["Prediction"] <- NA # That creates the new column for the home team Elo points
test.pred["Outcome"] <- NA # That creates the new column for the away team Elo points
nm <- nrow(test.pred)

for(i in 1:nm){
  if(test.pred$pred.test[i] >= 0.6){test.pred$Prediction[i] <- "H"}
  else if(test.pred$pred.test[i] <= 0.4){test.pred$Prediction[i] <- "A"}
  else {test.pred$Prediction[i] <- "D"}
  if(test.pred$Results[i] == test.pred$Prediction[i]){test.pred$Outcome[i] <- 1}
  else{test.pred$Outcome[i] <- 0}
}

# Display test data results
print(test.pred)
```

```
##
           HomeTeam       AwayTeam HG AG Results pred.test Prediction
211          Fulham      Leicester  0  2       A 0.4055987          D
212           Leeds        Everton  1  2       A 0.5172463          D
213     Aston Villa       West Ham  1  3       A 0.5289752          D
214       Liverpool       Brighton  0  1       A 0.6593032          H
215       Tottenham        Chelsea  0  1       A 0.5482845          D
216     Aston Villa        Arsenal  1  0       H 0.5706588          D
217         Burnley       Brighton  1  1       D 0.5407161          D
218       Newcastle    Southampton  3  2       H 0.4786518          D
219          Fulham       West Ham  0  0       D 0.4212529          D
220      Man United        Everton  3  3       D 0.6037564          H
221       Tottenham      West Brom  2  0       H 0.6890600          H
222          Wolves      Leicester  0  0       D 0.4455618          D
223       Liverpool       Man City  1  4       A 0.4990119          D
224 Sheffield United       Chelsea  1  2       A 0.3921753          A
225           Leeds Crystal Palace  2  0       H 0.5602649          D
226       Leicester      Liverpool  3  1       H 0.5206340          D
227  Crystal Palace        Burnley  0  3       A 0.5721923          D
228        Man City      Tottenham  3  0       H 0.6378142          H
229        Brighton    Aston Villa  0  0       D 0.4741553          D
230     Southampton         Wolves  1  2       A 0.5704301          D
    Outcome
211       0
212       0
213       0
214       0
215       0
216       0
217       1
218       0
219       1
220       0
221       1
222       1
223       0
224       1
225       0
226       0
227       0
228       1
229       1
230       0
```

As with the training data set, we see from this that with the testing data set, the algorithm wrongly predicts that many of the matches will be drawn. However, if these 'draw' predictions are ignored, then we see that the Elo algorithm correctly predicts the outcome of 60% of the matches.

9.7 Concluding Remarks

The aim of this chapter has been to provide an introduction to ranking systems in soccer. Having worked through the examples, you should now have a reasonable understanding of how such algorithms work and be able to use them to rank soccer teams and predict match outcomes using R. Having said this, it is important to recognise that this chapter is only an introduction to the subject and that, because of time constraints, there is far more that we have not covered here. For example, the Keener [1,24] and PageRank [9,10] algorithms are gaining traction in sport [12–14]. These more advanced ranking systems have potential but are generally trickier to use and more difficult to apply than the Colley, Massey, and Elo systems discussed in this chapter. With specific regard to the use of rating systems to make match predictions, it is also important to realise that the Elo system presented here is a fairly basic version and that more complex adaptations have been developed, which reportedly produce better results [19,20].

As clearly illustrated in Example 9.6, dichotomous ranking systems such as the Elo algorithm struggle with draws, which occur fairly frequently in soccer leagues. However, this is generally not the case with knockout cup competitions, where tied matches are often decided using a penalty shootout. The Elo system might therefore be better suited to knockout soccer tournaments than leagues. Notwithstanding this, because the Elo algorithm calculates the expected win probabilities for the competing teams, it has the potential to identify value in the odds offered by bookmakers on league matches and may still be a very useful tool in this context. It is therefore worth exploring the whole subject of ranking systems in more depth, as this will yield new insights that should help anyone with an interest in predicting the outcome of soccer matches.

References

1. Langville AN, Meyer CD: *Who's # 1? The Science of Rating and Ranking.* Princeton: Princeton University Press; 2012.
2. Beggs CB, Shepherd SJ, Emmonds S, Jones B: A novel application of PageRank and user preference algorithms for assessing the relative performance of track athletes in competition. *PLoS One* 2017, 12(6):e0178458.
3. Govan AY: *Ranking Theory with Application to Popular Sports.* Raleigh: North Carolina State University; 2008.
4. Govan AY, Meyer CD: Ranking national football league teams using google's pagerank. In: *AA Markov Anniversary Meeting: 2006; Charleston*: Boson Books; 2006.
5. Devlin S, Treloar T: A network diffusion ranking family that includes the methods of Markov, Massey, and Colley. *Journal of Quantitative Analysis in Sports* 2018, 14(3):91–101.
6. Colley WN: *Colley's Bias Free College Football Ranking Method: The Colley Matrix Explained.* Princeton: Princeton University; 2002.
7. Massey K: *Statistical Models Applied to the Rating of Sports Teams.* Bluefield College, Virginia; 1997.
8. Elo AE: The rating of chessplayers, past and present. Ishi Press International, Bronx, New York; 2008.
9. Page L, Brin S, Motwani R, Winograd T: The PageRank citation ranking: Bringing order to the web. In: *Stanford InfoLab*; 1999. http://www.eecs.harvard.edu/~michaelm/CS222/pagerank.pdf
10. Brin S, Page L: The anatomy of a large-scale hypertextual web search engine. *Computer Networks and ISDN Systems* 1998, 33:107–117.
11. Csardi G, Nepusz T: The igraph software package for complex network research. *InterJournal, Complex Systems* 2006, 1695(5):1–9.
12. Lazova V, Basnarkov L: PageRank Approach to Ranking National Football Teams. *arXiv preprint arXiv:150301331*, 2015.
13. Zhou Y, Wang R, Zhang Y-C, Zeng A, Medo M: Improving PageRank using sports results modeling. *Knowledge-Based Systems* 2022, 241:108168.
14. Rojas-Mora J, Chavez-Bustamante F, Rio-Andrade Jd, Medina-Valdebenito N: A methodology for the analysis of soccer matches based on pagerank centrality. In: *Sports Management as an Emerging Economic Activity.* Springer New York; 2017: 257–272.
15. Epskamp S, Cramer AOJ, Waldorp LJ, Schmittmann VD, Borsboom D: qgraph: Network visualizations of relationships in psychometric data. *Journal of Statistical Software* 2012, 48:1–18.
16. Beggs C, Bond AJ: A CUSUM tool for retrospectively evaluating team performance: The case of the English Premier League. *Sport, Business and Management: An International Journal* 2020, 10(3):263–289.
17. Heuer A, Rubner O: Fitness, chance, and myths: An objective view on soccer results. *The European Physical Journal B* 2009, 67(3):445–458.

18. FIFA: 2026 FIFA World Cup(tm): FIFA Council designates bids for final voting by the FIFA Congress. https://www.fifa.com/tournaments/mens/worldcup/canadamexicousa2026/media-releases/2026-fifa-world-cuptm-fifa-council-designates-bids-for-final-voting-by-the-fifa-
19. Lasek J, Szlavik Z, Bhulai S: The predictive power of ranking systems in association football. *International Journal of Applied Pattern Recognition* 2013, 1(1):27–46.
20. Sullivan C, Cronin C: Improving Elo rankings for sports experimenting on the english premier league. In *Virginia Tech CSx824/ECEx424 Technical Report*, VA, USA; 2016. https://courses.cs.vt.edu/cs5824/Fall15/project_reports/sullivan_cronin.pdf
21. World football Elo ratings. 2023. https://www.eloratings.net
22. opisthokonta.net: Tuning the Elo ratings: The K-factor and home field advantage. In: *opisthokonta.net*. https://opisthokonta.net/?p=1387; 2016.
23. Heinzen E: Ranking teams by Elo rating and comparable methods. In: *Package elo*. https://cran.r-project.org/web/packages/elo/elo.pdf; 2022.
24. Keener JP: The Perron-Frobenius theorem and the ranking of football teams. *SIAM Review* 1993, 35(1):80–93.

10

Using Linear Regression to Analyse Match Performance Data

In this chapter, we will build on the introduction to the linear regression analysis outlined in Chapters 1 and 2 and show how ordinary least-squares (OLS) regression can be used to analyse match performance data. Linear regression is a widely used technique in statistics that can be applied to explain relationships in data and also to make predictions. This makes it a useful technique, which can be extremely helpful to analysts working in sport given that the relationships encountered are often linear. Here we will show how linear regression can be used to identify key performance metrics that influence the success of soccer teams on the pitch an issue that should be of great interest to many working in professional soccer.

With most top-flight football matches televised, and also with the advent of specialist sports data collection companies, masses of high-quality performance data are now readily available (either for free or for a fee) via the Internet. Therefore, for most people, the challenge is not acquiring relevant data but rather knowing how to utilise and interpret it for some useful purpose. For example, while it is interesting to know the number of tackles, dribbles, and crosses that teams make during a season, what do we do with this information? This is a major challenge for all those analysts working in professional soccer, who are confronted daily with masses of performance data collected from the training ground and during matches. They are charged with not only making sense of the data but also communicating their insights to management in such a way that effective interventions can be made, which will improve team performance. Given this, soccer analysts need an armoury of tools and techniques at their disposal with which to make their load lighter. One such technique is linear (OLS) regression, which, although not as advanced as some other nonlinear and machine learning techniques, can still be surprisingly effective.

Generally speaking, in order for information to be useful, it must either help explain something that is not obvious or assist in predicting an outcome of some nature. This is where regression analysis comes in, because it can be used both to make predictions and to help explain observed results. As such, regression analysis can be an extremely useful tool for sports analysts. While many different types of regression analysis exist, because this book is an introductory text, we will content ourselves here with only looking at linear regression, which, although relatively simple, is still one of the most useful

DOI: 10.1201/9781003328568-10

and widely used techniques in statistics and data science. Hopefully, by the end of this chapter, you will feel confident enough with linear regression to be able to explore some of the more advanced techniques, such as polynomial and lasso regression, which are beyond the scope of this text.

10.1 What Is Regression?

Often in finance and economics, we want to understand the factors that influence important metrics like the value of a currency, the price of oil and gas, or the share price of a company. If, for example, we understand the relationship between the price of natural gas and the various factors that influence it, then we can confidently make predictions about how the gas price might behave and act accordingly. So, for example, if we observe that political instability is greatly reducing global gas production, we can be pretty confident that the market price of natural gas is likely to rise, especially during the winter months when demand for gas is high.

But by how much will the price of gas rise? Well, that is a difficult question to answer, and so economists build statistical models to try to predict what might happen. These models can be highly complex (way beyond the scope of this chapter), but at their heart, many have a very simple idea known as regression, which involves describing mathematically the relationship between a single response variable (sometimes called the dependent variable) and one or more predictor variables, known as independent variables. So in the case of natural gas, the gas price would be the response variable, and air temperature, number of daylight hours, global gas production, economic activity, etc. might be the predictors or independent variables.

The general idea with regression models is that the independent variables (e.g., air temperature, global gas production, economic activity, etc.) can be used to predict the response variable (e.g., the price of natural gas), which is why such models are so popular with economists. They can also be used to explain the variance in a response variable by telling us how much each predictor variable contributes to the behaviour of the whole system. This can be extremely useful because it enables us to identify those predictor variables that are important and those that are not. Consequently, regression models can be very helpful and are thus widely used in economics, statistics, and data science.

Although there are many different types of regression model, they can be broadly classified as being either linear or nonlinear. While both have their place, linear regression models are much more widely used because they are simpler and easier to implement compared with their nonlinear counterparts. For this reason, in this chapter we will focus only on linear regression. Because this is a book about soccer analytics, in this chapter we will apply regression models to real-life performance data collected from the teams in the English Premier League (EPL) for seasons 2020–2021 and 2021–2022. In each example presented, the response variable will be points earned (awarded) during the season, while the predictor variables will be metrics like shots-on-target (SoT), pass completion (PassComp), number of tackles, etc. Because all the examples in this chapter use the same data set (contained in the EPL_regression_data_2020_2021.csv file, which can be downloaded from GitHub at https://github.com/cbbeggs/SoccerAnalytics), before performing any regression analysis, we will first (in Example 10.1) spend a little time looking at the data itself and comparing any differences between the two seasons.

EXAMPLE 10.1: LOAD THE DATA SET AND PRODUCE DESCRIPTIVE STATISTICS

The data used in this example is stored in a CSV file (EPL_regression_data_2020_2021.csv, which can be downloaded from GitHub at https://github.com/cbbeggs/SoccerAnalytics), which was compiled using selected data from https://fbref.com. This CSV file contains on-the-pitch performance data (i.e., ten performance metrics) for the clubs in the EPL during seasons 2020–2021 and 2021–2022. The variables in this data set are as follows:

- "League" – the league from which the data is collected (i.e., the EPL)
- "Season" – the year in which the season starts (i.e., 2020 or 2021)
- "Team" – the name of the team
- "Points" – the total number of league points accrued in the season by each team
- "Shots" – the total number of shots made by each team during the season
- "SoT" – the total number of shots on target made by each team during the season
- "ShotDist" – the average distance, in yards, from goal of all shots taken
- "PassComp" – the total number of passes completed by each team during the season
- "Dribbles" – the total number of dribbles made by each team during the season
- "Tackles" – the total number of tackles made by each team during the season
- "Crosses" – the total number of crosses made by each team during the season
- "Intercepts" – the total number of intercepts made by each team during the season
- "AerialWon" – the total number of aerial battles won by each team during the season
- "AerialLost" – the total number of aerial battles lost by each team during the season

The following code splits the data set into two data frames, one for each season, and produces descriptive statistics. Finally, an independent *t*-test is performed on all the variables to identify any statistically significant differences (i.e., $p<0.05$) between the data collected for the two seasons.

First, we clear any existing data from the RStudio workspace.

```
rm(list = ls())   # This clears all variables from the work space
```

Then we input the data in the form of a CSV file, which we have stored here in a directory called 'Datasets'.

```
regdata <- read.csv("C:/Datasets/EPL_regression_data_2020_2021.csv", sep=",")
```

Now we can inspect the data.

```
head(regdata) # Displays the first six rows
```

```
##
  League Season        Team Points Shots SoT ShotDist PassComp Dribbles
1    EPL   2021     Arsenal     69   580 185     17.3    16110      295
2    EPL   2021 Aston Villa     45   461 159     17.1    12521      275
3    EPL   2021   Brentford     46   441 139     15.6    11551      233
4    EPL   2021    Brighton     51   483 140     16.8    16342      334
5    EPL   2021     Burnley     35   406 117     16.4     9508      255
6    EPL   2021     Chelsea     74   585 199     16.9    20878      360
  Tackles Crosses Intercepts AerialWon AerialLost
1     573     433        476       553        662
2     686     452        587       599        639
3     693     433        589       830        824
4     722     485        558       650        629
5     629     449        617       984        912
6     660     496        557       623        547
```

Next, we split the data set into two data frames, one for each season, and remove the first three columns containing the league, season, and team identifiers.

```
# Split data into 2020-21 and 2021-22 seasons
season1 <- regdata[regdata$Season == 2020,]   # 2020-21
season2 <- regdata[regdata$Season == 2021,]   # 2021-22


# Remove the first three columns
s1 <- season1[,c(4:14)]
s2 <- season2[,c(4:14)]
```

We can check that we have selected the right variables by simply typing:

```
names(s1)
```

```
##
[1] "Points"   "Shots"   "SoT"   "ShotDist"   "PassComp"   "Dribbles"   "Tackles"
[8] "Crosses"   "Intercepts" "AerialWon" "AerialLost"
```

Having done this, we can then produce the descriptive statistics and perform the *t*-tests using the following code, the results of which produce a single table.

```
# Season 2020-21 descriptive statistics
library(psych)
s1.temp <- describeBy(s1) # These are the descriptive statistics
s1.stats <- cbind.data.frame(s1.temp$n, s1.temp$mean, s1.temp$sd)
colnames(s1.stats) <- c("S1.n", "S1.mean", "S1.SD") # Rename variables

# Season 2021-22 descriptive statistics
s2.temp <- describeBy(s2) # These are the descriptive statistics
s2.stats <- cbind.data.frame(s2.temp$n, s2.temp$mean, s2.temp$sd)
colnames(s2.stats) <- c("S2.n", "S2.mean", "S2.SD") # Rename variables

# Perform independent t-test
ttresults <- sapply(c(1:11), function(i) {t.test(s1[,i],s2[,i], paired=FALSE)}) # NB. The default is
na.rm=TRUE
ttres <- as.data.frame(t(ttresults[3,])) # This selects just the p-values.
pval <- round(as.numeric(t(ttres)),3)

# Compile descriptive statistics results
Variables <- rownames(s1.temp)
stats <- cbind.data.frame(s1.stats,s2.stats)
stats <- round(stats,1)
stats.res <- cbind.data.frame(Variables,stats,pval)
print(stats.res)
```

```
##
    Variables S1.n S1.mean  S1.SD S2.n S2.mean  S2.SD  pval
1      Points   20    52.9   16.9   20    52.6   19.3 0.966
2       Shots   20   454.7   80.1   20   484.0   95.7 0.300
3         SoT   20   155.1   34.0   20   157.9   36.6 0.800
4    ShotDist   20    16.9    0.6   20    16.8    0.8 0.777
5    PassComp   20 15355.9 3854.8   20 14717.9 3906.7 0.606
6    Dribbles   20   370.7   68.6   20   320.5   49.2 0.012
7     Tackles   20   645.4   81.2   20   674.7   80.1 0.257
8     Crosses   20   448.6   70.7   20   444.5   63.1 0.849
9  Intercepts   20   416.2   43.7   20   574.6   58.3 0.000
10  AerialWon   20   655.8  111.0   20   675.2  116.1 0.592
11 AerialLost   20   655.8  122.6   20   675.2  137.0 0.640
```

From this, see that the only statistically detectable difference between the performance metrics for seasons 2020–2021 and 2021–2022 is associated with the number of dribbles and intercepts. For dribbles, the mean was 370.7 in season 2020–2021, whereas it fell to 320.5 in season 2021–2022 ($p=0.012$). By comparison, for intercepts, the situation was reversed, with a mean of 416.2 in season 2020–2021 and 574.2 in 2021–2022 ($p<0.001$).

10.2 Ordinary Least-Squares Regression

In Chapter 5, we introduced the concept of the correlation coefficient, r, which quantifies the relationship between two continuous variables. The r-value ranges from –1 to 1, with strong positive correlations having an r-value approaching 1 and strong inverse (negative) correlations having a value approaching –1. While correlation analysis can be very useful, it does have some major limitations, chief of which is the fact that it does not fully describe the relationship between two variables. For example, if we consider two vectors of the same length,

$$A = [2, 3, 2, 4, 1, 3, 5, 4]$$

$$B = [5, 7, 5, 9, 3, 7, 11, 9]$$

We can instantly see that the elements of B all conform to the function ($2A+1$). However, if we perform Pearson correlation analysis on these two vectors, we get the answer $r=1$, which, although true, is not the whole story. While the correlation analysis tells us how closely the two variables 'move' in relation to each other, it does not quantify the nature of that relationship. It cannot show that the elements in B are twice those in A plus one. In short, correlation analysis tells us nothing about the multiplier or the offset in the relationship. To find out about these, we have to turn instead to linear regression, which we introduced in Chapter 1. Linear regression, as typically used, takes the following generic form:

$$\text{Linear regression model}: y = b_0 + b_1x_1 + b_2x_2 + \ldots + e \tag{10.1}$$

where y is the response variable; x_1, x_2, ... are the predictor variables; b_0, b_1, b_2, ... are coefficients applied to the intercept and the predictor variables; and e is the residual error.

So for example, y might be the amount of energy consumed per day by an office building, and x_1 and x_2 could be the average daily outside air temperature and hours of daylight, respectively. Given this, it's not difficult to see that the value of y will be influenced by both x_1 and x_2, because buildings consume more energy on heating and lighting when it is cold and dark outside. However, the energy consumption of the building cannot be fully explained simply by how cold or dark it is outside because other factors, such as occupancy patterns and behaviour, also play a role. Therefore, Equation 10.1 incorporates a residual error term, e, which is a sort of 'fiddle factor' used to make up for the shortcomings of the predictor variables, x_1 and x_2.

OLS regression (the most widely used linear regression variant) is so-called because it uses a least-squares approach to calculate the values of the respective variable coefficients, b_0, b_1, b_2, etc. A full discussion of the linear algebra underpinning this is beyond the scope of this chapter, but suffice to say that the OLS approach minimises the sum of the squared errors of the residuals to produce a best-fit line through the data, which can be described using Equation 10.2.

While Equation 10.1 perfectly describes (i.e., 100%) the observed response variable, y, the way it is used in practice to perform linear regression is in a truncated form without the residual error term, e (Equation 10.2). In this form, the y term is replaced with a $\hat{y}$ term, which is the predicted value of the response variable. Importantly, all the $\hat{y}$ values predicted using Equation 10.2 lie along the best-fit line through the observed data.

$$\text{Multiple linear regression}: \hat{y} = b_0 + b_1x_1 + b_2x_2 + \cdots \tag{10.2}$$

When there are several predictor variables, as in Equation 10.2, we call it multiple linear regression analysis. However, when there is only one predictor variable (as shown in Equation 10.3), we generally use the term simple linear regression.

$$\text{Simple linear regression}: \hat{y} = b_0 + b_1x_1 \tag{10.3}$$

When Equation 10.3 is applied and plotted on a scatter plot, the predicted (fitted) values lie along a least-squares best-fit line that runs straight through the middle of the observed data, as illustrated in Figure 10.1, where the intercept at $x=0$ is b_0 and b_1 is the gradient of the best-fit line. In Figure 10.1, the value of b_0 is 2.14, and the value of b_1 is 1.296. So, if we consider, for example, the data point where $x=1.7$, then the OLS regression model predicts that $\hat{y}$ is 4.343, whereas in reality the observed value of y is 6.2. This means that for this particular data point, the residual error is 1.857.

If the model is a good predictor of the observed data, then the residual errors will be small, whereas if it is not that good at prediction, the errors will be large. The goodness of fit of a linear regression model can be assessed using the coefficient of determination, R^2, which is a measure of the amount of variance in the observed response variable that is explained by the model. The R^2 value is in fact identical to the square of the correlation coefficient, r.

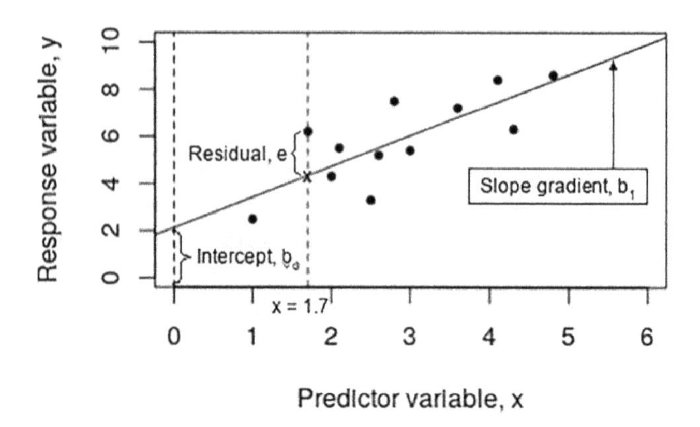

FIGURE 10.1
Scatter plot of x and y for a simple ordinary least-squares regression model.

In the example shown in Figure 10.1, R^2=0.602 and the r-value is 0.776, which is the square root of R^2. This implies that the model is able to explain 60.2% of the variance in the value of the response variable.

Another useful measure of goodness of fit that is often used is the mean absolute error (MAE), which is the average of the absolute values of the residual errors produced by the model. It is necessary to use the absolute value because some of the residual errors are positive and some are negative, and they would otherwise just cancel themselves out mathematically.

Simple OLS regression is explained in Example 10.2, which uses the predictor variables 'Shots' and 'SoT' to illustrate the process.

EXAMPLE 10.2: CORRELATION ANALYSIS AND SIMPLE LINEAR REGRESSION

In this example, we will compare correlation analysis with simple linear regression. With regard to the latter, we will produce two models, one with 'Shots' as the predictor variable and the other with 'SoT'. In each case, 'Points' will be the response variable.

To perform the correlation analysis, we will utilise the 'rcorr' function in the 'Hmisc' package, which we will use to analyse the data for the seasons 2020–2021 and 2021–2022.

```
# Install 'Hmisc' package
install.packages("Hmisc") # This installs the 'Hmisc' package.
# NB. This command only needs to be executed once to install the package.
# Thereafter, the 'Hmisc' library can be called using the 'library' command.

library("Hmisc")
```

```
# Season 2020-21
s1.cor <- rcorr(as.matrix(s1))
s1.corres <- round(as.data.frame(s1.cor[1]),3)
colnames(s1.corres) <- c("Points","Shots","SoT","ShotDist","PassComp","Dribbles",
                "Tackles","Crosses","Intercepts","AerialWon","AerialLost")
print(s1.corres)
```

```
##
            Points  Shots    SoT ShotDist PassComp Dribbles Tackles
Points       1.000  0.781  0.892   -0.327    0.750    0.289  -0.254
Shots        0.781  1.000  0.905   -0.333    0.847    0.492  -0.207
SoT          0.892  0.905  1.000   -0.272    0.778    0.424  -0.039
ShotDist    -0.327 -0.333 -0.272    1.000   -0.163    0.234   0.254
PassComp     0.750  0.847  0.778   -0.163    1.000    0.511  -0.139
Dribbles     0.289  0.492  0.424    0.234    0.511    1.000  -0.087
Tackles     -0.254 -0.207 -0.039    0.254   -0.139    0.087   1.000
Crosses      0.203  0.582  0.302   -0.511    0.348    0.122  -0.435
Intercepts  -0.422 -0.286 -0.215    0.143   -0.305   -0.051   0.595
AerialWon   -0.491 -0.573 -0.512   -0.176   -0.739   -0.456  -0.091
AerialLost  -0.762 -0.695 -0.704    0.073   -0.799   -0.554   0.081
            Crosses Intercepts AerialWon AerialLost
Points        0.203     -0.422    -0.491     -0.762
Shots         0.582     -0.286    -0.573     -0.695
SoT           0.302     -0.215    -0.512     -0.704
ShotDist     -0.511      0.143    -0.176      0.073
PassComp      0.348     -0.305    -0.739     -0.799
Dribbles      0.122     -0.051    -0.456     -0.554
Tackles      -0.435      0.595    -0.091      0.081
Crosses       1.000     -0.125    -0.121     -0.206
Intercepts   -0.125      1.000     0.177      0.206
AerialWon    -0.121      0.177     1.000      0.770
AerialLost   -0.206      0.206     0.770      1.000
```

```
# Season 2021-22
s2.cor <- rcorr(as.matrix(s2))
s2.corres <- round(as.data.frame(s2.cor[1]),3)
colnames(s2.corres) <- c("Points","Shots","SoT","ShotDist","PassComp","Dribbles",
              "Tackles","Crosses","Intercepts","AerialWon","AerialLost")
print(s2.corres)
```

```
##
            Points  Shots    SoT ShotDist PassComp Dribbles Tackles
Points       1.000  0.888  0.935   -0.511    0.887    0.338  -0.573
Shots        0.888  1.000  0.943   -0.370    0.835    0.317  -0.552
SoT          0.935  0.943  1.000   -0.414    0.881    0.385  -0.546
ShotDist    -0.511 -0.370 -0.414    1.000   -0.344    0.109   0.486
PassComp     0.887  0.835  0.881   -0.344    1.000    0.478  -0.467
Dribbles     0.338  0.317  0.385    0.109    0.478    1.000  -0.044
Tackles     -0.573 -0.552 -0.546    0.486   -0.467   -0.044   1.000
Crosses      0.691  0.781  0.665   -0.585    0.648    0.121  -0.646
Intercepts  -0.802 -0.603 -0.655    0.413   -0.738   -0.133   0.535
AerialWon   -0.448 -0.444 -0.464   -0.298   -0.632   -0.320   0.031
AerialLost  -0.710 -0.575 -0.675    0.070   -0.842   -0.468   0.265
```

```
            Crosses Intercepts AerialWon AerialLost
Points        0.691     -0.802    -0.448     -0.710
Shots         0.781     -0.603    -0.444     -0.575
SoT           0.665     -0.655    -0.464     -0.675
ShotDist     -0.585      0.413    -0.298      0.070
PassComp      0.648     -0.738    -0.632     -0.842
Dribbles      0.121     -0.133    -0.320     -0.468
Tackles      -0.646      0.535     0.031      0.265
Crosses       1.000     -0.489    -0.189     -0.349
Intercepts   -0.489      1.000     0.517      0.694
AerialWon    -0.189      0.517     1.000      0.779
AerialLost   -0.349      0.694     0.779      1.000
```

From this, we see that several predictor variables are strongly correlated with each other. For example, for both seasons, the correlation between 'Shots' and 'SoT' is $r>0.9$, which is of course to be expected given that one is a sub-set of the other. It can also be seen that 'PassComp' is strongly correlated ($r>0.7$) with several other variables, including 'Shots', 'SoT' and 'AerialLost'. As such, the data appears to exhibit a degree of multicollinearity, which is a term that data scientists use when variables in the data are strongly correlated with each other. Multicollinearity can cause considerable problems when performing multiple linear regression, which may cause models to become unstable and make interpretation difficult [1].

Now that we have a feel for the data, we shall proceed to build two simple linear models (i.e., one using 'Shots' and the other using 'SoT') utilising the data for season 2020–2021. Here we use the 'lm' function to build the models and the 'summary' command to display the results for each model. We will also compute the respective *Akaike information criterion* (*AIC*) values for the two models. The AIC value is a measure of the relative quality of the respective predictive models. In effect, it is an estimate of the relative amount of information lost by a given model – the less information lost, the higher the quality of the model. Therefore, the lower the AIC value, the better the model is at predicting the response variable.

First, we build the model with 'Shots' as the predictor.

```
# Build OLS regression models for season 2020-21
# Using Shots
shots.lm <- lm(Points ~ Shots, data = s1)
summary(shots.lm)
```

```
##
Call:
lm(formula = Points ~ Shots, data = s1)

Residuals:
    Min      1Q  Median      3Q     Max
-22.430  -7.576  -2.039  10.446  15.977
```

```
Coefficients:
            Estimate Std. Error t value Pr(>|t|)    
(Intercept) -21.9996    14.3042  -1.538    0.141    
Shots         0.1646     0.0310   5.309 4.78e-05 ***
---
Signif. codes:  0 '***' 0.001 '**' 0.01 '*' 0.05 '.' 0.1 ' ' 1

Residual standard error: 10.83 on 18 degrees of freedom
Multiple R-squared:  0.6103,	Adjusted R-squared:  0.5886 
F-statistic: 28.19 on 1 and 18 DF,  p-value: 4.776e-05
```

The AIC value can be displayed simply by typing:

```
AIC(shots.lm)
```

```
##
[1] 155.9471
```

Next, we build the model with 'SoT' as the predictor.

```
# Using SoT
SoT.lm <- lm(Points ~ SoT, data = s1)
summary(SoT.lm)
```

```
##
Call:
lm(formula = Points ~ SoT, data = s1)

Residuals:
     Min       1Q   Median       3Q      Max 
-11.9627  -5.2486  -0.4058   4.6586  14.3942 

Coefficients:
             Estimate Std. Error t value Pr(>|t|)    
(Intercept) -15.83666    8.40027  -1.885   0.0756 .  
SoT           0.44285    0.05296   8.361   1.3e-07 ***
---
Signif. codes:  0 '***' 0.001 '**' 0.01 '*' 0.05 '.' 0.1 ' ' 1

Residual standard error: 7.851 on 18 degrees of freedom
Multiple R-squared:  0.7953,	Adjusted R-squared:  0.7839 
F-statistic: 69.91 on 1 and 18 DF,  p-value: 1.299e-07
```

```
AIC(SoT.lm)
```

```
##
[1] 143.0752
```

From this, we see that using 'SoT' as the predictor variable produces the better model, with an R^2 value of 0.795 compared with 0.610 for the 'Shots' regression model. The computed AIC values also tell a similar story, with the AIC value (i.e., AIC=143.1) for the SoT.lm model much lower than that for the shots.lm model (i.e., AIC=155.9). The superiority of the SoT.lm model is graphically illustrated in Figure 10.2, which shows that the data points are clustered more tightly around the best-fit line for this model compared with that constructed using the 'Shots' predictor variable. Figure 10.2 can be produced using the following code.

```
# Scatter plot with best-fit lines
plot(s1$Shots, s1$Points, pch=4, col="black", xlim=c(0,800),
    ylim=c(-40,100),       ylab="Points", xlab="Shots & SoT")
points(s1$SoT, s1$Points, pch=20)
abline(lm(s1$Points ~ s1$Shots), lty=1)
abline(lm(s1$Points ~ s1$SoT), lty=2)
legend(450,15, c("Shots","Shots bestfit line","SoT","SoT bestfit line"),
      cex=0.8, col=c("black","black","black","black"),
      lty=c(0,1,0,2), pch=c(4,NA,20,NA), bty = "n")
```

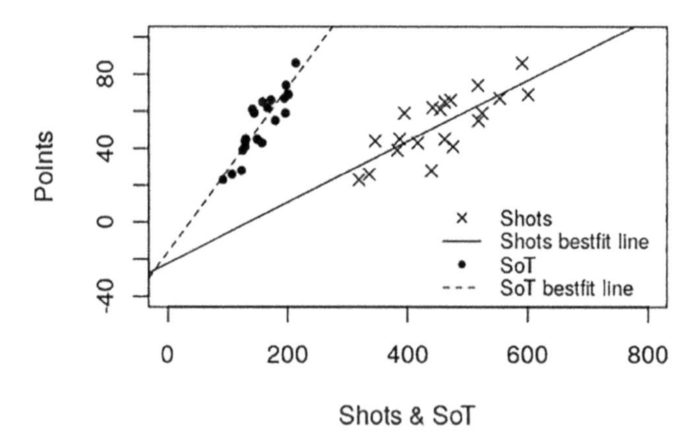

FIGURE 10.2
Scatter plot of ordinary least-squares regression models using 'Shots' and 'SoT' to predict 'Points' for season 2020–2021 of the English Premier League.

As expected, the results of the simple linear regression are completely unsurprising and mirror the correlation analysis results. Clearly, teams that score more goals earn more league points, and teams that put more shots on target are much more likely to score goals compared with teams whose shots are mostly off-target. What is perhaps more surprising is that although the points awarded per game are inherently nonlinear (i.e., 3 points for a win; 1 point for a draw; and zero points for a loss), the relationship between total points and 'Shots' and 'SoT' is clearly strongly linear, indicating that linear regression is an appropriate tool to use.

10.3 Multiple Linear Regression

Now that we have seen how R handles simple linear regression, we can advance things by including more predictor variables and utilising Equation 10.2 in a process known as multiple linear regression analysis. The clever thing about multiple linear regression is that it utilises the OLS methodology to compute a set of coefficients that, when applied to the predictor variables, minimise the sum of the squares of the residual errors. In other words, no matter how many variables are included as predictors, the method automatically produces the optimum solution. Having said this, just because regression analysis produces a solution that is mathematically sound does not necessarily mean that it should be trusted! In fact, statisticians encourage users not to trust the results that linear regression produces because they might not be representative of reality and therefore may not be generally applicable – even if the mathematics is technically correct. Accordingly, statistical methodologies (beyond the scope of this chapter) have been developed to test whether or not the variable coefficients in linear regression models are statistically significant. If the variable coefficients do not reach significance (i.e., generally set at $p<0.1$ for most regression models), then these variables are frequently removed from the regression model (although this is not always the case when the AIC is used for elimination purposes), leaving only those variables that are deemed to be significant. In this way, a linear regression model is produced that contains only those predictor variables that are statistically significant. As such, the refined model will be more parsimonious, which is something that statisticians generally like because it reduces the chance of Type 1 errors occurring (see Key Concept Box 11.1). While this helps to prevent over-fitting, the downside is that as we exclude more variables, the prediction accuracy of the model tends to reduce.

There are several well-known methods that can be employed to eliminate (or add) predictor variables from (or to) linear regression models (i.e., backward, forward, and bi-directional elimination). Furthermore, the variable elimination process can either utilise the coefficient p-value or, alternatively, the AIC value. In Example 10.3, we investigate both of these approaches and utilise a backward exclusion methodology to produce optimal models for predicting the points earned over a season using the performance metrics in the data set downloaded in Example 10.1.

Unlike correlation analysis and simple linear regression, both of which investigate relationships in data on a bipartite basis, multiple linear regression looks at the combined effect of all the predictor variables on the response variable. This makes the technique a particularly powerful tool when analysing the large multivariate data sets that are typically found in sport. However, multiple linear regression does have one major weakness, and that is multicollinearity, which can cause instability and make interpretation of the data difficult, with multiple competing models potentially being produced [1,2]. For this reason, when employing multiple linear regression, it is often advisable beforehand to remove any variables from the data that are strongly correlated (e.g., $p > 0.8$) with other predictor variables so that only one of the correlated variables is retained in the model. From a data science point of view, the information in these correlated variables is largely redundant, and retaining them will only cause multicollinearity problems when performing multiple linear regression analysis. Accordingly, in Example 10.3, we have removed the 'Shots' variable from the predictor variables because it is so strongly correlated with 'SoT'.

EXAMPLE 10.3: MULTIPLE LINEAR REGRESSION

Building on Example 10.2, in this example, we will apply multiple linear regression analysis to the EPL performance data set. However, because of potential multicollinearity issues, we will not include 'Shots' in the predictor variables for the reasons explained above. As in Example 10.2, we will use the data for season 2020–21 to build the model, from which we select the following variables.

```
# Remove Shots variables from data sets
build.dat <- s1[,c(1,3:11)] # Here we use season 2020-21 to build the linear model.
names(build.dat)
```

```
##
[1] "Points" "SoT"  "ShotDist"  "PassComp"  "Dribbles"  "Tackles"
[7] "Crosses" "Intercepts"  "AerialWon"  "AerialLost"
```

Now we can build the base model by making 'Points' the response variable and all the other variables the predictors. As with the simple regression model in Example 10.2, we use the 'lm' function in R to build the regression model.

```
# Create multiple linear model for season 2020-21 using build.dat
s1.lm1 <- lm(Points ~ SoT + ShotDist + PassComp + Dribbles +
        Tackles + Crosses + Intercepts + AerialWon + AerialLost, data=build.dat)
```

Here we have specified in longhand all the predictor variables to be included in the model. Alternatively, however, we can write the same thing in shorthand using a '.' term, which indicates to R that all the variables in the 'build.dat' data frame should be included in the model as predictors.

```
s1.lm1 <- lm(Points ~ ., data=build.dat)
```

Having built the model, we can evaluate its structure and performance using the 'summary' function.

```
summary(s1.lm1)
```

```
##
Call:
lm(formula = Points ~ SoT + ShotDist + PassComp + Dribbles +
    Tackles + Crosses + Intercepts + AerialWon + AerialLost,
    data = build.dat)

Residuals:
    Min      1Q  Median      3Q     Max
-4.4452 -2.6771  0.0826  1.6821  8.6780

Coefficients:
              Estimate Std. Error t value Pr(>|t|)
(Intercept)  1.467e+02  5.578e+01   2.629 0.025203 *
SoT          3.656e-01  5.912e-02   6.184 0.000104 ***
ShotDist    -2.194e+00  2.525e+00  -0.869 0.405294
PassComp     1.220e-04  6.708e-04   0.182 0.859272
Dribbles    -4.909e-02  2.239e-02  -2.193 0.053098 .
Tackles     -4.966e-02  2.200e-02  -2.257 0.047577 *
Crosses     -5.553e-02  2.273e-02  -2.443 0.034675 *
Intercepts  -2.754e-02  3.672e-02  -0.750 0.470564
AerialWon    9.806e-03  2.080e-02   0.471 0.647478
AerialLost  -5.363e-02  1.950e-02  -2.751 0.020457 *
---
Signif. codes:  0 '***' 0.001 '**' 0.01 '*' 0.05 '.' 0.1 ' ' 1

Residual standard error: 4.896 on 10 degrees of freedom
Multiple R-squared:  0.9558,	Adjusted R-squared:  0.916
F-statistic: 24.01 on 9 and 10 DF,  p-value: 1.269e-05
```

From this summary we see that the base model with all the predictors included achieves $R^2=0.956$, which is very impressive. However, we can also see that many of the predictor variables did not reach significance, implying that they probably should be excluded from the model. If we retain them, then we might produce an 'over-fitted' model that may perform very well with 2020–2021 data but poorly with data collected from other seasons, which would of course be a highly undesirable situation.

So how do we deal with this problem? Well, one solution is to exclude variables from the model one at a time, using the largest *p*-value as the selection criteria, until all the variables become statistically significant (i.e., $p<0.1$). This method is called backward exclusion and is one of the more widely used variable removal techniques.

From the summary above, we can see that 'PassComp' has the largest *p*-value (i.e., $p=0.859$). So we will exclude that variable first.

```
s1.lm2 <- lm(Points ~ SoT + ShotDist + Dribbles +
            Tackles + Crosses + Intercepts + AerialWon + AerialLost, data=build.dat)
summary(s1.lm2)
```

```
##
Call:
lm(formula = Points ~ SoT + ShotDist + Dribbles + Tackles + Crosses +
    Intercepts + AerialWon + AerialLost, data = build.dat)

Residuals:
    Min      1Q  Median      3Q     Max
-4.5134 -2.8551 -0.1129  1.6374  8.8217

Coefficients:
             Estimate Std. Error t value Pr(>|t|)
(Intercept) 150.15333   50.01940   3.002   0.0120 *
SoT           0.37077    0.04951   7.488 1.22e-05 ***
ShotDist     -2.25255    2.39159  -0.942   0.3665
Dribbles     -0.04841    0.02108  -2.297   0.0423 *
Tackles      -0.04987    0.02098  -2.377   0.0367 *
Crosses      -0.05500    0.02153  -2.555   0.0268 *
Intercepts   -0.02839    0.03479  -0.816   0.4318
AerialWon     0.00816    0.01789   0.456   0.6572
AerialLost   -0.05417    0.01841  -2.943   0.0134 *
---
Signif. codes:  0 '***' 0.001 '**' 0.01 '*' 0.05 '.' 0.1 ' ' 1

Residual standard error: 4.676 on 11 degrees of freedom
Multiple R-squared:  0.9556,     Adjusted R-squared:  0.9233
F-statistic: 29.61 on 8 and 11 DF,  p-value: 2.234e-06
```

From this, we see that despite the fact that the variable 'PassComp' is strongly correlated with 'Points' ($r=0.750$) (see Example 10.2), removing this variable from the model has very little impact on the overall R^2 value. This is because 'PassComp' and 'SoT' are highly correlated with each other ($r=0.778$), indicating that considerable redundancy exists between these two variables. Therefore, one of these variables can safely be eliminated without greatly impairing the overall performance of the model. On this occasion, it is 'PassComp' that is selected for elimination.

Next, we eliminate 'AerialWon' because it has the highest *p*-value ($p=0.657$). As before, eliminating 'AerialWon' from the model has minimal impact on its overall performance.

```
s1.lm3 <- lm(Points ~ SoT + ShotDist + Dribbles +
          Tackles + Crosses + Intercepts + AerialLost, data=build.dat)
summary(s1.lm3)
```

```
##
Call:
lm(formula = Points ~ SoT + ShotDist + Dribbles + Tackles + Crosses +
    Intercepts + AerialLost, data = build.dat)

Residuals:
    Min      1Q  Median      3Q     Max
-5.3925 -2.2807  0.1714  1.5871  9.1881

Coefficients:
             Estimate Std. Error t value Pr(>|t|)
(Intercept) 160.41553   43.17467   3.716  0.00295 **
SoT           0.37180    0.04780   7.778 5.01e-06 ***
ShotDist     -2.68725    2.11988  -1.268  0.22897
Dribbles     -0.04785    0.02034  -2.353  0.03653 *
Tackles      -0.05334    0.01890  -2.822  0.01541 *
Crosses      -0.05813    0.01972  -2.947  0.01221 *
Intercepts   -0.02369    0.03211  -0.738  0.47493
AerialLost   -0.04846    0.01305  -3.713  0.00296 **
---
Signif. codes:  0 '***' 0.001 '**' 0.01 '*' 0.05 '.' 0.1 ' ' 1

Residual standard error: 4.519 on 12 degrees of freedom
Multiple R-squared:  0.9548,	Adjusted R-squared:  0.9284
F-statistic:  36.2 on 7 and 12 DF,  p-value: 3.862e-07
```

Hopefully, you get the idea by now. With each refinement cycle, we exclude the variable with the highest *p*-value and repeat the process until all the remaining variables in the model become significant at $p<0.1$.

To save time here, we will not bother presenting the output (i.e., the model summaries) for the following lines of code. Suffice to say that we next exclude the variable 'Intercepts', which in this case should not be confused with the intercept of the model.

```
# Next, we eliminate 'Intercepts' (p = 0.475).
s1.lm4 <- lm(Points ~ SoT + ShotDist + Dribbles +
          Tackles + Crosses + AerialLost, data=build.dat)
summary(s1.lm4)
```

Finally, we eliminate 'ShotDist', which gives us the final refined model in which all the predictor variables are significant.

```
s1.lm5 <- lm(Points ~ SoT + Dribbles + Tackles + Crosses + AerialLost, data=build.dat)
summary(s1.lm5)
```

```
##
Call:
lm(formula = Points ~ SoT + Dribbles + Tackles + Crosses + AerialLost,
    data = build.dat)

Residuals:
   Min     1Q Median     3Q    Max
-6.999 -1.863  0.012  2.278  7.435

Coefficients:
             Estimate Std. Error t value Pr(>|t|)
(Intercept) 110.60942   21.01531   5.263 0.000120 ***
SoT           0.39670    0.04474   8.868 4.05e-07 ***
Dribbles     -0.06076    0.01829  -3.321 0.005046 **
Tackles      -0.06434    0.01441  -4.466 0.000533 ***
Crosses      -0.05150    0.01726  -2.983 0.009882 **
AerialLost   -0.04902    0.01305  -3.757 0.002123 **
---
Signif. codes:  0 '***' 0.001 '**' 0.01 '*' 0.05 '.' 0.1 ' ' 1

Residual standard error: 4.537 on 14 degrees of freedom
Multiple R-squared:  0.9468,     Adjusted R-squared:  0.9278
F-statistic: 49.86 on 5 and 14 DF,  p-value: 1.993e-08
```

From this, it can be seen that we now have a model in which all the predictor variables are significant and that exhibits an R^2 value of 0.947, which is still very impressive. This looks to be a pretty good model, and so we will stick with it, and refer to it now as the refined model. The refined model can be summarised using the mathematical expression:

$$\text{Points} = 110.609 + (0.3967 \times \text{SoT}) + (-0.0608 \times \text{Dribbles}) + (-0.0643 \times \text{Tackles}) +$$
$$(-0.0515 \times \text{Crosses}) + (-0.0490 \times \text{AerialLost})$$

We can display the AIC value for the refined model and the 95% confidence intervals for the variable coefficients by simply typing:

```
AIC(s1.lm5)
```

```
##
[1] 124.1112
```

```
# 95% confidence intervals of coefficients
confint(s1.lm5)
```

```
##
                   2.5 %       97.5 %
(Intercept) 65.53606175 155.68277616
SoT          0.30074666   0.49264432
Dribbles    -0.09999116  -0.02151890
Tackles     -0.09523574  -0.03343680
Crosses     -0.08852767  -0.01446911
AerialLost  -0.07700056  -0.02103538
```

We can also compare this refined model with the original base model to see if removing the variables has had a serious effect on its performance. This can be done using the 'anova' function.

```
anova(s1.lm1, s1.lm5)  # This test uses the F-statistic
```

```
##
Analysis of Variance Table

Model 1: Points ~ SoT + ShotDist + PassComp + Dribbles + Tackles + Crosses +
    Intercepts + AerialWon + AerialLost
Model 2: Points ~ SoT + Dribbles + Tackles + Crosses + AerialLost
  Res.Df    RSS Df Sum of Sq      F Pr(>F)
1     10 239.69
2     14 288.13 -4   -48.445 0.5053 0.7332
```

From this, we see that $p=0.733$, implying that there is no significant difference between the performance of the base and refined models.

As we can see, the whole backward exclusion process is a rather long-winded affair. However, R has an automatic 'step' function, which can perform the whole process much more quickly. Having said this, rather than using the coefficient *p*-values, the 'step' function seeks to minimise the AIC value, as follows: (NB. For convenience, the full output from the 'step' function is not presented here; only the summary of the resulting model is.)

```
s1.lm6 <- step(s1.lm1, direction="backward") # This uses a backwards exclusion methodology.
summary(s1.lm6)
```

```
##
Call:
lm(formula = Points ~ SoT + ShotDist + Dribbles + Tackles + Crosses +
    AerialLost, data = build.dat)

Residuals:
    Min      1Q  Median      3Q     Max
-5.7481 -2.0065  0.2219  1.2862  9.1009

Coefficients:
             Estimate Std. Error t value Pr(>|t|)
(Intercept) 157.64095   42.24954   3.731 0.002516 **
SoT           0.37974    0.04575   8.301 1.49e-06 ***
ShotDist     -2.65290    2.08188  -1.274 0.224872
```

```
Dribbles     -0.05015   0.01974  -2.541 0.024630 *
Tackles      -0.06233   0.01418  -4.395 0.000725 ***
Crosses      -0.06183   0.01874  -3.300 0.005750 **
AerialLost   -0.04934   0.01277  -3.864 0.001953 **
---
Signif. codes:  0 '***' 0.001 '**' 0.01 '*' 0.05 '.' 0.1 ' ' 1

Residual standard error: 4.439 on 13 degrees of freedom
Multiple R-squared:  0.9527,  Adjusted R-squared:  0.9309
F-statistic: 43.67 on 6 and 13 DF,  p-value: 7.112e-08
```

And the AIC value can be displayed using:

```
AIC(s1.lm6)
```

```
##
[1] 123.7572
```

From this, we can see that the 'step' function does indeed produce a model that has a lower AIC value than the refined model. The model also exhibits a slightly greater R^2 value compared with the refined model. However, it is noticeable that the model produced using the 'step' function (i.e., s1.lm6) also includes the variable 'ShotDist', which failed to reach significance. Notwithstanding this, we can see that the variables 'SoT', 'Dribbles', 'Tackles', 'Crosses', and 'AerialLost' are all significant predictors of the total number of points earned by a club over a season. Of these, it is noticeable that 'SoT' has a positive coefficient, implying that the more shots there are on target, the greater the number of points that will be accumulated – which is not an unexpected finding! Similarly, 'AerialLost' has a negative coefficient, implying that the fewer aerial battles lost, the more points that will be accumulated. However, rather unexpectedly, 'Dribbles', 'Tackles', and 'Crosses' all have negative coefficients, which is something that is difficult to interpret. Remember, the model is a mathematical construct that is designed to minimise the square of the residual errors. Therefore, it is advisable to take care when interpreting the signs ascribed to the various variable coefficients.

One of the take-home messages from Example 10.3 is that the variables included in the final regression model are very much dictated by the method employed to refine the base model. Different variable exclusion methods can result in several competing models (all slightly different from each other), which can make interpretation difficult. Refinement and tuning of regression models should therefore be considered something of an art rather than a pure science. Discretion is therefore required when refining and interpreting regression models. It is important to remember that such models are mathematical constructs and that they have their limitations. Having said this, the p-value variable exclusion method is generally considered superior when the model is primarily used for explanatory purposes, while AIC is thought to be better with models that are aimed at prediction [3].

10.4 Variable Importance

When multiple regression is used for explanatory purposes, investigators are often interested in the extent to which each independent variable contributes to predicting the response variable. Therefore, in order to draw meaningful inferences, investigators frequently want to quantify the relative importance of the predictor variables in terms of the proportionate contribution that each makes to the overall R^2 value of the model. Traditionally, this can be done by standardising the regression coefficients in the model using Equation 10.4.

$$\beta = b \times \frac{\sigma_r}{\sigma_p} \tag{10.4}$$

where β is the standardised variable coefficient; b is the variable coefficient; σ_r is the standard deviation of the response variable; and σ_p is the standard deviation of the predictor variable.

Variable standardisation can be helpful when interpreting linear regression models because it eliminates comparison problems associated with predictor variables that have different units (e.g., mm and tonnes, etc.). However, in practice, β standardisation is difficult to interpret because it refers to standard deviations, which are a tricky concept to grasp. Also, the method can be problematic when predictor variables are strongly correlated. However, this problem can be overcome by using the relative weights methodology proposed by Johnson [4], which involves mean centering and standardising to unit variance the predictor and response variables, and then performing singular value decomposition (SVD) of the standardised predictor variables [5]. This relative weight methodology has the great advantage that it is immune to any multicollinearity problems.

Example 10.4 illustrates how Johnson's relative weights method [4] can be used to evaluate the relative importance of the predictor variables in the linear regression models outlined in Example 10.3. The relative weights function used in Example 10.4 is that presented in Kabacoff [6]. The mathematics behind this function is complex and well beyond the scope of this book. However, for those who are interested, a detailed mathematical explanation of the relative weights code can be found in Beggs et al. [7].

EXAMPLE 10.4: VARIABLE IMPORTANCE

In this example, we will use the relative weights function presented in Kabacoff [6] to assess the relative importance of the predictor variables in the base and refined models in Example 10.3.

```
# relweights function
relweights <- function(fit,...){
  R <- cor(fit$model)
  nvar <- ncol(R)
  rxx <- R[2:nvar, 2:nvar]
  rxy <- R[2:nvar, 1]
  svd <- eigen(rxx)
  evec <- svd$vectors
  ev <- svd$values
  delta <- diag(sqrt(ev))
  lambda <- evec %*% delta %*% t(evec)
  lambdasq <- lambda ^ 2
  beta <- solve(lambda) %*% rxy
  rsquare <- colSums(beta ^ 2)
  rawwgt <- lambdasq %*% beta ^ 2
  import <- (rawwgt / rsquare) * 100
  lbls <- names(fit$model[2:nvar])
  rownames(import) <- lbls
  colnames(import) <- "Weights"
  barplot(t(import),names.arg=lbls,
        ylab="% of R-Square",
        xlab="Predictor Variables",
        main="Relative Importance of Predictor Variables",
        sub=paste("R-Square=", round(rsquare, digits=3)),
        ...)
  return(import)
}
```

NB. The code for the 'relweights' function listed above is as presented in Kabacoff [6] (pages 209–210) and is based on the work of Johnson [4].

First, we will use the 'relweights' function to compute the relative weights of the predictor variables in the refined linear regression model, as follows:

```
# Relative weights for the refined model
relweights(s1.lm5)
```

This produces a list of the weights below for the refined regression model (s1.lm5) and Figure 10.3.

```
##
            Weights
SoT       55.838016
Dribbles   4.051378
Tackles    6.278236
Crosses    1.825402
AerialLost 32.006968
```

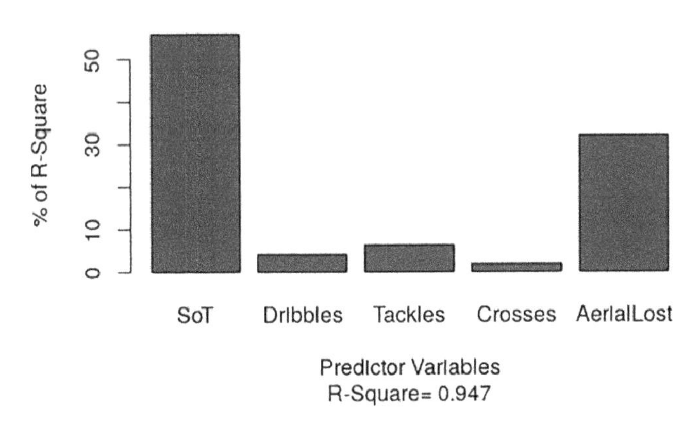

FIGURE 10.3
Relative weights of the predictor variables in the refined linear model (s1.lm5)

From this, we can see that for the refined regression model s1.lm5;

- 'SoT' accounts for approximately 52.9% (i.e., 0.947 × 55.83) of the variance in 'Points'.
- 'AerialLost' accounts for approximately 30.3% (i.e., 0.947 × 32.0) of the variance in 'Points'.
- 'Tackles' accounts for approximately 5.9% (i.e., 0.947 × 6.28) of the variance in 'Points'.
- 'Dribbles' accounts for approximately 3.8% (i.e., 0.947 × 4.05) of the variance in 'Points'.
- 'Crosses' accounts for approximately 1.7% (i.e., 0.947 × 1.83) of the variance in 'Points'.
- 5.3% of the variance in 'Points' is not explained by the model.

Now, we apply the 'relweights' function to the base model (s1.lm1).

```
# Relative weights for the base model
relweights(s1.lm1)
```

This produces:

```
##
              Weights
SoT         37.837914
ShotDist     4.950284
PassComp    15.849489
Dribbles     3.110640
Tackles      3.733533
Crosses      1.565777
Intercepts   5.769490
AerialWon    7.303419
AerialLost  19.879454
```

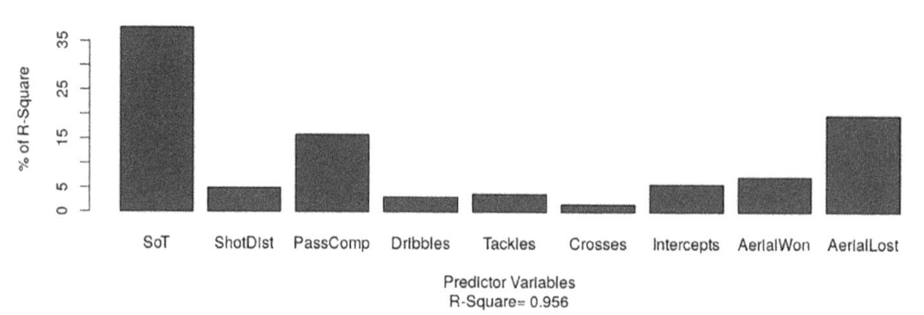

FIGURE 10.4
Relative weights of the predictor variables in the base linear model (s1.lm1)

Comparison between Figures 10.3 and 10.4 reveals something surprising. The variable 'PassComp', which we thought was unimportant because it was excluded early on in Example 10.3, appears to be more important after all, being the third most important predictor. Why?

Well, the short answer is that linear regression has trouble coping with predictor variables that are strongly correlated with each other. Thus, variables like 'PassComp' may get removed when the *p*-value exclusion method is employed – even though they are potentially important.

If we compute the Pearson correlation *r*-values for 'Points', 'SoT', 'PassComp', and 'AerialLost', we see that they are all strongly correlated, with 'PassComp' positively correlated with 'SoT' (r=0.778) and inversely correlated with 'AerialLost' (r=−0.799).

```
# Pearson correlation analysis of strongly correlated variables
cor.temp <-
round(cor(cbind(build.dat$Points,build.dat$SoT,build.dat$PassComp,build.dat$AerialLost)),3)
colnames(cor.temp) <- c("Points","SoT","PassComp","AerialLost")
rownames(cor.temp) <- c("Points","SoT","PassComp","AerialLost")
print(cor.temp)
```

```
##
           Points    SoT PassComp AerialLost
Points      1.000  0.892    0.750     -0.762
SoT         0.892  1.000    0.778     -0.704
PassComp    0.750  0.778    1.000     -0.799
AerialLost -0.762 -0.704   -0.799      1.000
```

This tells us that 'PassComp' is obviously an important predictor (explanatory) variable, something confirmed when we use it in a simple regression model, as follows:

```
# Regression model using PassComp as the predictor
s1.lm7 <- lm(Points ~ PassComp, data=build.dat)
summary(s1.lm7)
```

```
##
Call:
lm(formula = Points ~ PassComp, data = build.dat)

Residuals:
    Min      1Q  Median      3Q     Max
-24.282  -7.559   2.840   7.892  21.426

Coefficients:
             Estimate Std. Error t value Pr(>|t|)
(Intercept) 2.391e+00  1.080e+01   0.221  0.82721
PassComp    3.286e-03  6.829e-04   4.812  0.00014 ***
---
Signif. codes:  0 '***' 0.001 '**' 0.01 '*' 0.05 '.' 0.1 ' ' 1

Residual standard error: 11.47 on 18 degrees of freedom
Multiple R-squared:  0.5626,    Adjusted R-squared:  0.5383
F-statistic: 23.15 on 1 and 18 DF,  p-value: 0.0001398
```

From this, we can see that 'PassComp' alone can explain 56.3% of the variance in the variable 'Points'. Therefore, 'PassComp' is an important performance metric. Unfortunately, many students and inexperienced analysts are unaware that linear regression can result in important variables being excluded from a model simply because they are strongly correlated with other variables, something that can easily result in important key performance metrics being overlooked.

One useful diagnostic metric relating to regression analysis is the variance inflation factor (VIF), which is an indicator of multicollinearity in the data. If multicollinearity is present in the data, then the modelling process will tend to become unstable and may produce unreliable results [1]. The general rule is that if VIF>4, then multicollinearity may be an issue, and if VIF>10, it is a serious problem. Indeed, when VIF>10, it is probably best not to use multiple linear regression analysis at all and instead turn to other preferable and more advanced techniques such as lasso regression or principal component regression.

The VIF values should ideally be assessed using the base regression model. These can be produced using the 'car' package in R, as follows:

```
# Compute VIF values using the 'car' package
install.packages("car") # This installs the 'car' package.
# NB. This command only needs to be executed once to install the package.
# Thereafter, the 'car' library can be called using the 'library' command.

library(car)
vif(s1.lm1) # Calculate the variance inflation factors.
```

This produces the following, from which it can be seen that 'PassComp' exhibits the largest VIF with 5.3, while 'AerialWon' and 'AerialLost' both are >4.

```
##
        SoT    ShotDist    PassComp    Dribbles     Tackles     Crosses
   3.203703    2.120400    5.300942    1.869381    2.527237    2.047230
 Intercepts   AerialWon  AerialLost
   2.043736    4.227496    4.529653
```

By comparison, for the refined model (s1.lm5) the 'vif' function produces results that show that the multicollinearity issue no longer exists.

```
vif(s1.lm5) # Calculate the VIF values for refined regression model.
```

```
##
       SoT   Dribbles    Tackles    Crosses AerialLost
  2.136560   1.453460   1.262354   1.375181   2.362125
```

While Example 10.4 illustrates very well the explanatory power of the 'relweights' function, it also illustrates the pitfalls that can arise due to multicollinearity in the data. While backward exclusion might produce an explanatory regression model that looks great, with all the predictor variables being significant and a high R^2 value exhibited, it does not necessarily mean that it tells the whole story – especially if multicollinearity exists in the data, as is often the case in sport. Other excluded variables, such as 'PassComp', might still be important for explanatory purposes. So, it is important to be aware of this when analysing soccer data. Sadly, many researchers are not aware of this fact, and this can result, on occasions, in erroneous conclusions being reached.

So the take-home message is that great care should be taken when using linear regression to interpret sport performance data because this data can often be highly correlated. With regard to the current analysis, it appears that three on-the-pitch performance variables ('SoT, 'AerialLost', and 'PassComp') stand out above all the others as being influential in determining the number of points that a team will achieve in a season. Presumably, the greater the number of successful passes made by teams during matches, the greater the number of shooting opportunities (particularly on target) that will arise and the more goals that will be scored. Conversely, if teams lose fewer aerial battles, then the opposition is less likely to score goals against them.

10.5 Model Prediction

Until now, we have concentrated on using linear regression models for explanatory purposes. Now we turn our attention to prediction, which is an area of great interest to many people. However, before we look at this, it is perhaps worth clarifying what we mean by 'prediction'. In this context, we are not talking about trying to predict the future, as we did in Chapter 6; rather, the term 'prediction' refers here to predicting the value of the response variable using a set of independent variables in a regression model. This is often referred to as the model fit. If the predicted (fitted) values are close to the observed values, then the model is accurate. However, if this is not the case, then it's back to the drawing board to produce another, hopefully superior model that better predicts the response variable.

In this chapter, we have deliberately used a data set that is relevant to soccer to illustrate how linear regression works. While this data set can be used to illustrate the prediction process in R, it is worth pointing out that, in reality, few people would be interested in predicting the number of points earned by soccer clubs because everybody knows how many points each team has been awarded. However, for educational purposes, the data set is useful because everyone understands the response variable 'Points'.

Normally, when regression models are used for prediction purposes, the response variable is something that is difficult to measure, whereas the predictor variables are often relatively easy to measure. Thus, linear regression models use predictors that are easy to measure in order to estimate the value of something that is hard to measure. This can be extremely useful. For example, in medicine, regression models have been developed to predict the central venous pressure in the heart, something that normally requires an invasive surgical procedure, using blood flow information from jugular veins that can be easily collected non-invasively [8]. In this way, the model potentially estimates the central venous pressure of patients without the need for them to endure a difficult and risky medical procedure.

When building a regression model for prediction purposes, it is important that it be generally applicable. It is no point constructing an over-fitted prediction model that is great for the 2020–2021 EPL season if it does not work well with data from other seasons. So in Example 10.5 we will test the refined regression model (s1.lm5) that we built using the data for season 2020–2021 against the data for season 2021–2022 to see how well it performs. By testing the model against data from season 2021–2022, we are validating it to see if it has general applicability. In Example 10.5, we will use two metrics, the coefficient of determination, R^2, and the MAE, to assess the goodness of fit of the predictions.

EXAMPLE 10.5: MODEL PREDICTIONS

In R, the predictions made by regression model s1.lm5 for season 2020–2021 can be computed using the 'predict' function as follows:

```
# Season 2020-21
s1.pred <- predict(s1.lm5, data = build.dat)
print(s1.pred)
```

```
##
      21       22       23       24       25       26       27       28
58.89066 56.45117 44.09801 42.18244 68.39211 41.21618 56.06825 27.39557
      29       30       31       32       33       34       35       36
59.05529 59.58434 68.89329 86.55912 80.99885 45.40849 21.39681 48.44288
      37       38       39       40
57.08394 32.39671 57.56519 44.92069
```

Having obtained the predicted 'Points' values, we can compare these with the observed values for season 2020–2021 using a scatter plot as follows:

```
# Create vectors containing the new observed and predicted results
obs.s1 <- build.dat$Points
pred.s1 <- s1.pred
new.s1 <- cbind.data.frame(obs.s1,pred.s1)

# Produce scatter plot of observed and predicted results for season 2020-21
plot(pred.s1, obs.s1, pch=20, col="black", xlim=c(20,90),
   ylim=c(0,100), ylab="Predicted points", xlab="Observed points")
```

This produces the scatter plot shown in Figure 10.5.

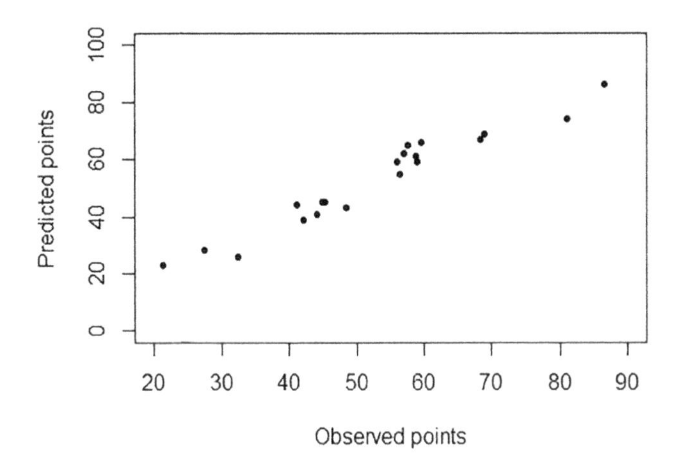

FIGURE 10.5
Scatter plot produced using the refined model (s1.lm5) and the season 2020–2021 English Premier League data.

From Figure 10.5, it can be seen that the refined model predicts the observed 'Points' values very closely. But just how close is this? Well, to answer this, we need to compute some goodness-of-fit metrics, which we can do using the following code:

```
# Compute model-fit metrics
s1.r2 <- cor(obs.s1,pred.s1)^2 # R^2
print(s1.r2)
```

```
##
[1] 0.9468251
```

```
s1.mae <- mean(abs(obs.s1 - pred.s1)) # Mean absolute error (mae)
print(s1.mae)
```

```
##
[1] 2.898508
```

This tells us that $R^2 = 0.947$ and that the MAE is just 2.9 points.

Now we turn our attention to season 2021–2022, to which we will apply the refined model s1.lm5 using the following code:

```
# Now we test the predictive ability of the linear regression model using the
# data for season 2021-22
test.dat <- s2[,c(1,3:11)]

s2.pred <- predict(s1.lm5, data = test.dat, type="response")
print(s2.pred)
```

```
##
      21       22       23       24       25       26       27       28
58.89066 56.45117 44.09801 42.18244 68.39211 41.21618 56.06825 27.39557
      29       30       31       32       33       34       35       36
59.05529 59.58434 68.89329 86.55912 80.99885 45.40849 21.39681 48.44288
      37       38       39       40
57.08394 32.39671 57.56519 44.92069
```

Now we compare the predicted 'Points' values with the observed values and produce a scatter plot for season 2021–2022.

```
# Create vectors containing the new observed and predicted results
obs.s2 <- test.dat$Points
pred.s2 <- s2.pred

# Produce scatter plot of observed and predicted results for season 2021-22
plot(pred.s2, obs.s2, pch=20, col="black", xlim=c(20,90),
    ylim=c(0,100), ylab="Predicted points", xlab="Observed points")
```

This produces the scatter plot shown in Figure 10.6.

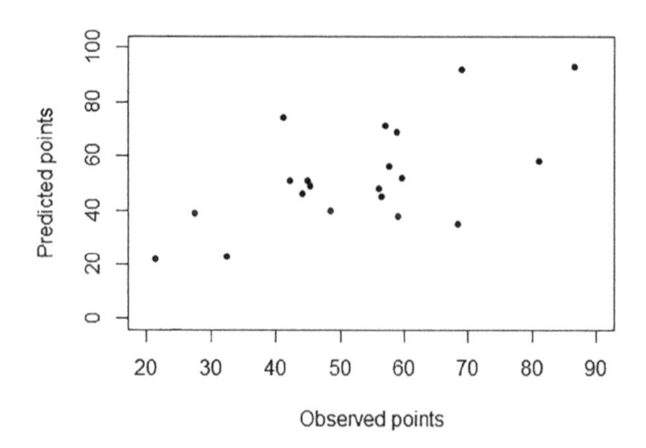

FIGURE 10.6
Scatter plot produced using the refined model (s1.lm5) and the season 2021–2022 English Premier League data.

From Figure 10.6, it can be seen that the refined model predicts the observed 'Points' values for the season 2021–2022 less well than for the season 2020–2021, as indicated by the fact that the plot is more scattered.

Finally, we compute the goodness-of-fit metrics.

```
# Compute model-fit metrics
s2.r2 <- cor(obs.s2,pred.s2)^2 # R^2
print(s2.r2)
```

```
##
[1] 0.3882609
```

```
s2.mae <- mean(abs(obs.s2 - pred.s2)) # Mean absolute error (mae)
print(s2.mae)
```

```
##
[1] 12.14548
```

This tells us that R^2=0.388 and that the MAE is 12.15 points, which is not that impressive. This casts doubt on the ability of the refined model (s1.lm5) to predict the points total achieved by the clubs in the EPL for all seasons. It looks like model s1.lm5 might have been over-fitted to the 2020–2021 data and that a more sophisticated prediction model may be required if it is to be applicable to every season.

10.6 Regression Diagnostics

Until now, we have avoided any discussion of the rules that govern the application of linear regression. This has been deliberate because it is better to get an overall feel for how regression analysis works in R before looking at the 'fine print'. But the fine print cannot be avoided forever, and so it is necessary for us now to look briefly at the assumptions (rules) that govern the application of linear regression. Many people get confused with these rules, but in reality, they are reasonably straightforward, and what is more, R has a whole set of easy-to-use diagnostic tools that we can employ to assist us. Perhaps the best way to explore the various assumptions that underpin linear regression analysis is to use the ready-made diagnostic functions in R. To this end, Example 10.6 shows us how to apply the relevant diagnostic algorithms in R to the regression models discussed in this chapter. However, before we look at Example 10.6, it is perhaps worth familiarising ourselves with the various assumptions that underpin linear regression analysis, which can be summarised as follows:

- **Linearity:** There should be a linear relationship between the dependent response variable and the independent predictor variables. This means that the least-squares best fit of the response variable is a linear combination of the regression coefficients and the predictor variables.
- **Normality:** The residual errors should be normally distributed with a mean of zero. Normality is often tested using a Q–Q plot, which is a scatter plot of the standardised residuals against the values that we would expect under the assumption of normality.
- **Homoscedasticity (constant variance):** The residual error term should be approximately the same across all values of the independent variables. Heteroscedasticity (the violation of homoscedasticity) occurs when the size of the error term differs across the values of the independent variables. Heteroscedasticity is evidenced by a 'fanning effect' that occurs between residual error and predicted values.
- **Independence of errors:** The observations should be independent of one another, which means that the residual errors should be uncorrelated with each other. However, if they are correlated, then this is called autocorrelation. Independence is usually tested using the Durbin–Watson (DW) statistic, which produces values from 0 to 4, where values from 0 to 2 indicate a positive autocorrelation and those from 2 to 4 show a negative autocorrelation. The midpoint (i.e., a value of 2) indicates that no autocorrelation exists. As a general rule of thumb, if the DW test statistic is in the region $1.5 < DW < 2.5$, then it means that there is little autocorrelation in the data and that the residual errors can be deemed independent.

In addition, as we have already seen in Example 10.4, the data should not exhibit multicollinearity, which occurs when the independent predictor variables are highly correlated. This is because multicollinearity tends to cause the model-building process to become unstable, making it difficult to draw firm inferences from any models that are constructed [1]. The most popular tool for evaluating multicollinearity is the VIF method, which is discussed in Example 10.4.

EXAMPLE 10.6: REGRESSION DIAGNOSTICS

In this example, we will evaluate the refined model (s1.lm5) using some of the standard diagnostic tools found in R.

Homoscedasticity and normality can be tested using the Breusch–Pagan and Shapiro–Wilk normality tests, respectively. If either of these tests yields a significant result (i.e., $p<0.05$), then it is assumed that these rules have been violated.

In R, the Breusch–Pagan test can be executed using the 'bptest' function in the 'lmtest' package as follows:

```
library(lmtest)
bptest(s1.lm5)
```

```
##
  studentized Breusch-Pagan test

data:  s1.lm5
BP = 4.6785, df = 5, p-value = 0.4564
```

Here the Breusch–Pagan test reveals $p=0.456$, indicating that the homoscedasticity assumption has not been violated and that the model is OK in this respect.

Now, we can run the Shapiro–Wilk test on the residuals to check whether or not normality has been violated.

```
residuals <- obs.s1 - pred.s1 # Computes the residuals for s1.lm5 using 2020-21 data.
shapiro.test(residuals)
```

```
##
  Shapiro-Wilk normality test

data:  residuals
W = 0.97429, p-value = 0.8417
```

The Shapiro–Wilk test reveals $p=0.842$, indicating that normality of the residuals has not been violated. Technically, this is saying that we don't have any evidence to demonstrate that the normality assumption has been violated, which is not the same thing as saying that we can assume normality. Nonetheless, the Shapiro–Wilk test is still helpful because it will quickly identify situations where the distribution of the residuals is heavily skewed. Having said this, the Q–Q plot is probably a more robust diagnostic test for normality.

We can also test for the independence of the residuals using the Durbin–Watson test (dwtest) function in the 'lmtest' package, as follows:

```
library(lmtest)
dwtest(s1.lm5)
```

```
##
  Durbin-Watson test

data:  s1.lm5
DW = 2.5798, p-value = 0.9029
alternative hypothesis: true autocorrelation is greater than 0
```

This gives a Durbin–Watson test value of 2.58, which is just above the 1.5–2.5 range but is not significant. Hence, we accept the null hypothesis that the residuals are not autocorrelated. This therefore indicates that the residuals are likely to be independent.

Finally, we can use the built-in diagnostic plots in R, which can be easily obtained using the following code:

```
par(mfrow=c(2,2))          # Visualize four graphs at once
plot(s1.lm5)
par(mfrow=c(1,1))          # Reset the graphics defaults
```

This produces the diagnostic plots in Figure 10.7.

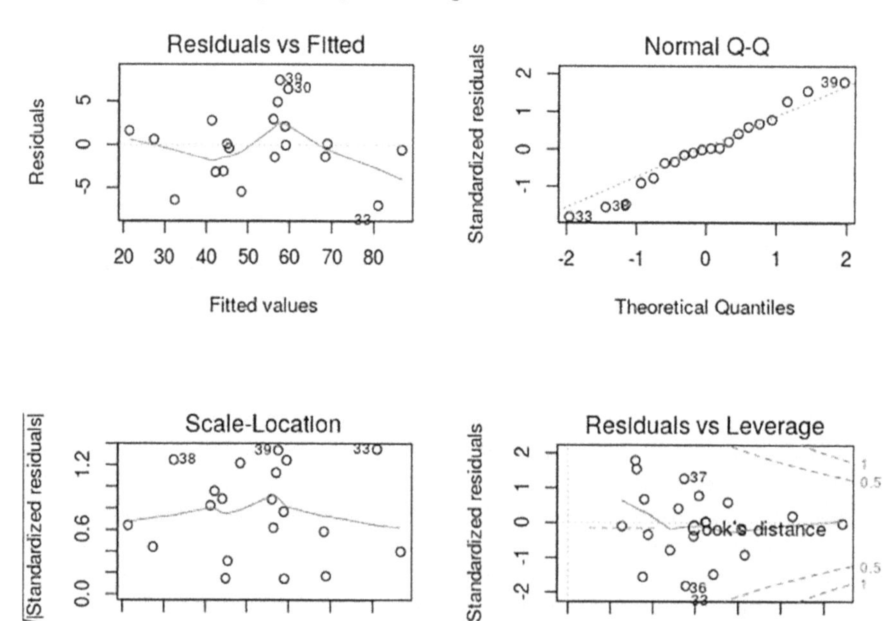

FIGURE 10.7
Diagnostic plots for the refined linear regression model (s1.lm5) using English Premier League season 2020–2021 data: (*upper left*) is a standard residuals plot; (*bottom left*) is a scale-location plot; (*upper right*) is a Q–Q plot; and (*bottom right*) is a leverage plot.

The first of the plots (upper left) in Figure 10.7 is a standard residuals plot for the fitted values, which is used to evaluate homoscedasticity. From this, it can be seen that the residuals are distributed reasonably evenly around zero and that there is no fanning effect, indicating that the assumption of homoscedasticity has not been violated, as confirmed by the Breusch–Pagan test result. The scale-location plot (bottom left) is a more sensitive way to search for heteroscedasticity. If the constant variance assumption is met, then the data points on the scale-location graph should be a random band along a horizontal line, which is pretty much what we see in Figure 10.7.

The Q–Q plot (upper right) in Figure 10.7 is the normal quantile plot of the residuals. If the residual errors are normally distributed, then the data points on the plot will follow the 45° line. Here we observe some deviation from normality in the tails, but it is not bad enough to violate the assumption of normality, as indicated by the result of the Shapiro–Wilk test.

Notice that data points which are considered to be outliers are identified by numbers on the diagnostic plots. An outlier is defined as an observation that has a large residual error, which means that the observed value for that data point is very different from that predicted by the regression model.

The last plot (bottom right) shows data points that are possible high-leverage or influential cases. A leverage point is an outlier in the predictor space that causes it to be far away from the mean of the predictions, while an influential observation is one that changes the slope of the regression line in the model. Influential data points, as their name implies, can greatly affect the goodness of fit of the regression model. Influential points can be identified using a statistic called 'Cook's distance', which is a measure of how far, on average, predicted values will move if a particular observation is dropped from the data set. In this case, we see that all the data points exhibit a Cook's distance <0.5, indicating that no influential points exist.

10.7 Final Thoughts

Although this chapter is only an introduction to the subject, having worked through it you should now have a reasonable understanding of how linear regression models work and how they can be applied using R. Hopefully, you are now in a position to try out linear regression analysis with your own data or with data downloaded from the Internet. The good news is that many of the relationships found in soccer-related performance data are linear and thus well suited to analysis using multiple linear regression. The bad news, however, is that such data sets often exhibit considerable multicollinearity [1], which means that great care needs to be taken when interpreting such data. Having said this, there are many other advanced regression techniques, such as lasso regression or principal component regression, that can be used when the data exhibits multicollinearity. These techniques are related in one way or another to linear regression, and so if multicollinearity is a problem, then it might be wise to experiment with some of these more advanced techniques – there are plenty of tutorials on the Internet that can help you.

With regard to the data set that we have analysed in this chapter, we see that the linear regression models (built using the 2020–2021 data) behaved reasonably well as explanatory models, especially when used in conjunction with the relative weights function [6]. Using this coupled approach, we were able to identify the independent variables 'SoT, 'AerislLost', and 'PassComp' as being particularly influential indicators of the number of points that a team will achieve during a season. This does appear to concur with the findings of several researchers who observed that ball possession [9–11], shots on target [10,11], and aerial duels [11] were all important indicators of success on the football pitch. From this, we can conclude that the greater the number of successful passes made by a team during a match, the more shooting opportunities will arise, and if those shots are on target, then more goals will be scored. Conversely, if teams lose fewer aerial duels, the opposition is less likely to score goals against them.

With regard to prediction, the examples presented in this chapter reveal a useful cautionary lesson. While the refined model (s1.lm5) performed very well with the EPL data for season 2020–2021, producing a very good fit (R^2=0.947), when the same model was tried out using the 2021–2022 data, it performed much less well (R^2=0.388). This illustrates the importance of validating models to ensure they have general applicability, which in this context means ensuring that the constructed model produces good predictions for all seasons and not just season 2020–2021.

References

1. Weaving D, Jones B, Ireton M, Whitehead S, Till K, Beggs CB: Overcoming the problem of multicollinearity in sports performance data: A novel application of partial least squares correlation analysis. *PLoS One* 2019, 14(2):e0211776.
2. Graham MH: Confronting multicollinearity in ecological multiple regression. *Ecology* 2003, 84(11):2809–2815.
3. Shmueli G: To explain or to predict? *Statistical Science* 2010, 25(3):289–310.
4. Johnson JW: A heuristic method for estimating the relative weight of predictor variables in multiple regression. *Multivariate Behavioral Research* 2000, 35(1):1–19.
5. LeBreton JM, Tonidandel S: Multivariate relative importance: Extending relative weight analysis to multivariate criterion spaces. *Journal of Applied Psychology* 2008, 93(2):329.
6. Kabacoff RI: *R in Action: Data Analysis and Graphics with R.* Shelter Island, NY: Manning Publications Co.; 2015.

7. Beggs CB, Shepherd SJ, Cecconi P, Lagana MM: Predicting the aqueductal cerebrospinal fluid pulse: A statistical approach. *Applied Sciences* 2019, 9(10):2131.
8. Zamboni P, Malagoni AM, Menegatti E, Ragazzi R, Tavoni V, Tessari M, Beggs CB: Central venous pressure estimation from ultrasound assessment of the jugular venous pulse. *PLoS One* 2020, 15(10):e0240057.
9. Farias VM, Fernandes WB, Bergmann GG, dos Santos Pinheiro E: Relationship between ball possession and match outcome in UEFA Champions League. *Motricidade* 2020, 16(4):1–7.
10. Shafizadeh M, Taylor M, Peñas CL: Performance consistency of international soccer teams in Euro 2012: A time series analysis. *Journal of Human Kinetics* 2013, 38:213.
11. Evangelos B, Aristotelis G, Ioannis G, Stergios K, Foteini A: Winners and losers in top level soccer. How do they differ? *Journal of Physical Education and Sport* 2014, 14(3):398.

11

Successful Data Analytics

Many students and inexperienced analysts make the mistake of thinking that data analytics is all about learning sophisticated statistical and machine learning techniques. They naively think that the more analytical techniques they know, the better analyst they will become. However, this is only part of the story, because in order to become a successful data analyst, you need to have a feel for data and also have a clear sense of what it is that you are trying to achieve.

In the same way that top strikers are comfortable on the ball and have an innate ability to 'go for goal', top data analysts are very comfortable with data and have an uncanny knack of getting to the heart of the issue, no matter how complex or messy the data is. This is because they know what they are trying to achieve and which tools (analytical techniques) are necessary for the job. As a rookie, it is all too easy to blunder around using inappropriate analytical techniques, whereas more experienced analysts will know exactly what is and what is not important. They can then select the appropriate strategy and tools to analyse the data and, where necessary, take shortcuts, thus avoiding much wasted time wandering up blind alleys, which is extremely important when you are up against tight deadlines.

If you have worked through the examples in the previous chapters of this book, you should by now have a good feel for R and know how it can be used to analyse football data. Hopefully, you will also have a reasonable understanding of the various analytical techniques that have been discussed. This should give you the confidence to experiment with new techniques using your own data. Like a child learning a language, the best way to learn new analytical techniques is to copy and adapt, and there are plenty of excellent tutorials on the Internet that can assist you. As you learn and gain in confidence, you will acquire quite an arsenal of techniques with which to analyse soccer data. But this alone is not enough to make you a good data analyst. You need something else – a sense of direction – and this is what we will address in this last chapter of the book.

So what do we mean by a sense of direction? Well, essentially, it is knowing what you want to do and how you are going to achieve it. Unlike the game of soccer, where the objective is simple (i.e., score more goals than the opposition), sports data does not come with a set of Association Football rules. Instead, it is often messy and difficult to understand, and that is even before we decide what to do with it. Consider, for example, the complex Global Positioning Satellite (GPS) data that is often collected from players during

DOI: 10.1201/9781003328568-11

matches [1]. It is relatively easy to acquire using automatic sensors, but what do we do with the data once it has been collected? If the GPS data is logged at a rate of ten times per second (i.e., at 10 Hz), then over 90 minutes, each player should in theory produce a vector containing 54,000 data points (actually, it is more complex than this because each data point may have several attributes). This is a lot of data to analyse, and processing it raises many questions. Should we analyse the data for each player individually or collectively as a whole team? Are we interested in how far each player runs during the match, or should we be focusing on how the players' running speed declines as they become fatigued? There are lots of other similar questions that could be posed, and hopefully you get the idea. The data alone does not tell us what to do or, indeed, in which direction to travel – hence the need for a sense of direction. In sport, data is generally easy to collect, but very quickly it can balloon out of all proportion, with the result that we can soon end up drowning in it.

11.1 Identifying the Primary Purpose of the Analysis

So how do we prevent ourselves from being overwhelmed by the data and its complexity? Well, the simple answer is to think ahead and plan out what we want to achieve before we start collecting and analysing data. To assist with this, it is often helpful to first identify the nature of the question or problem that we are trying to address. Broadly speaking, most tasks encountered in sport analytics can be classified as being one of the following:

- **Reporting results and descriptive statistics:** Conveying results and descriptive statistics in a concise and appropriate manner to others (many of whom may not be data literate) so that strategic decisions can be made.
- **Monitoring and assessing performance:** Day-to-day monitoring and assessment of the performance of players, coaches (managers), and teams in order to keep things on track and to ensure that targets are being met.
- **Performance optimisation:** Developing models and strategies to optimise player and team performance.
- **Identifying threats and opportunities:** This can include things like developing models to identify target players for purchase or identifying potential weaknesses in squads due to older key players nearing the end of their careers, with replacement cover not yet in place.
- **Making predictions:** Building models to predict expected match outcomes, league position, injury recovery time, etc.

- **Betting and setting odds:** Calculating match odds, etc., and developing betting strategies.
- **Exploratory analysis:** Exploring and interrogating data to identify relationships and natural groups (clusters) that otherwise might remain hidden. In so doing, the aim is to find something new and unexpected in the data.
- **Hypothesis testing:** This is generally used when undertaking formal scientific studies in which there is a hypothesis to be tested. In such studies, it is often necessary to report p-values as an indicator of statistical significance.

Of course, very few analysts will undertake work in all of these categories, with most restricting their activities to just a few. While analysts working for broadcasters or media companies may spend much of their time producing reports and sophisticated graphics to convey stats and information to the general public, those working in professional soccer might focus on monitoring players and developing strategies to optimise performance on the pitch. By comparison, those involved in the sports betting industry, be they gamblers or analysts working for bookmakers, are much more likely to be interested in setting odds, and predicting match outcomes, etc. Finally, sports scientists and those working in academia are likely to be more interested in hypothesis testing and exploratory analysis.

By identifying the category to which the work primarily belongs, it helps us focus on the key issues and narrows down the analysis options. For example, if we are involved in a scientific study at a university, we might test the hypothesis that football players who take a certain protein supplement will run faster than those who don't. In which case, we will want to perform some sort of statistical test to determine whether or not a significant difference (i.e., $p<0.05$) exists between the two groups. Knowing this should alert us to the need to control for bias, and so we might decide to randomly recruit players onto the two study arms: the intervention arm (those who take the protein supplement) and the control arm, comprising those who don't. In short, knowing the type of study being undertaken helps dictate which analysis techniques should be used. By comparison, if we are trying to develop a model for predicting the outcome of matches, all that we are interested in is how well the model works. Does it correctly predict match outcomes, and if so, how often? In which case, we are probably not interested in p-values. If the model works well most of the time, then we use it; otherwise, we rip it up and start again until we have developed a predictive model that is both consistent and reliable. In which case, we will probably need two data sets: one to build the model and a second to validate and assess its likely predictive performance in real life.

Being an introductory text, this book does not cover all the tasks outlined in the list above, some of which require the use of advanced machine learning techniques. However, this does not negate the central point, namely, that defining the nature of the task being undertaken at an early stage greatly helps with subsequent decision-making regarding the analysis of the data.

11.2 Plan before You Analyse

As with everything in life, it pays to have a good plan before you start collecting and analysing data. For example, when acquiring GPS data from players during training sessions, it is important to first decide: (i) what information you want to collect; (ii) why you want to collect it; (iii) how you are going to analyse it; and (iv) how you are going to report and communicate the results to the coaches and manager. So, rather than just recording the total distance run by each player during each training session, we might want to break things down into the distance travelled in several speed zones [2], which will give us more information about the fitness of the respective players at different intensity levels. However, in order to do this successfully, we first need to plan out how we are going to analyse the data, what we are going to do with the results, and how we are going to communicate them. If we make things too complicated, then not only will the analysis be very time-consuming, but it will also make communication of the results difficult. Remember, the decision-makers (coaches, managers, etc.) reading the analyst's report are often busy people who may not be numerically literate. Therefore, it is important to tailor the results to the expertise of the audience in such a way that the primary conclusions are clear. In other words, keep things simple so that the central message is not lost.

If our aim is to develop an effective betting strategy, as discussed in Chapter 7, then all that really matters is whether or not our strategy consistently makes a profit. After all, there is not much point in using a betting model that is going to lose money in the long run. So how can we assess whether or not any model that we have developed can be relied on to make a consistent profit? Well, the classic machine learning approach is to validate the model using a 'hold-out' data set to see how well it works [3,4], in a similar fashion to that described in Chapter 6. In short, this means separating historical match data into two data sets: a training (or exploratory) data set (perhaps consisting of seven seasons), which can be used to build the betting model, and a blinded testing (or validation) data set (perhaps consisting of three seasons), which is then used to validate the performance of the model, as illustrated in Figure 11.1. If the betting model performs well when using the testing data set, then it is likely that it will perform well in real life, and is therefore fit for use. However, if it performs badly, then it's back to the drawing board!

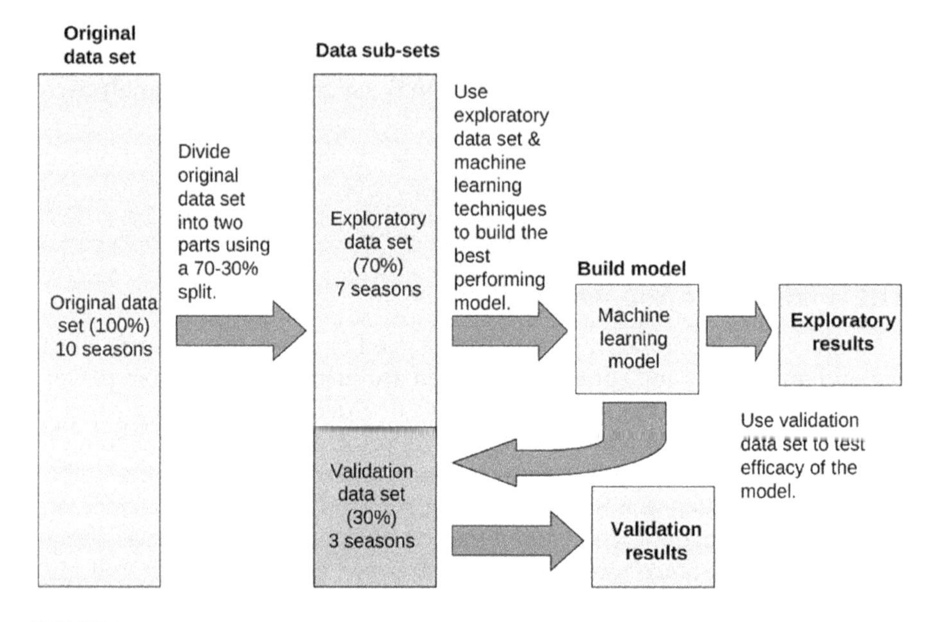

FIGURE 11.1
Processes employed when performing a 'hold-out' validation strategy with a machine learning model.

When constructing predictive models, it is also useful to identify all the variables that we might wish to incorporate beforehand. So in the case of a betting model, for example, we might want to incorporate the total shots ratio and the number of successful passes made for the respective teams, together with the bookmaker's published match odds. Therefore, we need to identify where we can acquire this information and also give some thought to how these disparate variables can be incorporated into the proposed model. With respect to this, writing down a suggested methodology and listing the variables required can be extremely helpful, as it will clarify your thoughts and highlight any potential problems.

11.3 Know Your Data

Inexperienced analysts frequently struggle when presented with new data because they don't have a feel for what the numbers represent. Often, they do not know what the variables represent or how the data has been collected. Consequently, rudimentary mistakes can be made regarding the validity and independence of the data. For example, sports science students often make the mistake of treating Body Mass Index (BMI) as an independent variable rather than as a derived variable, which is its correct status. BMI is actually a function of both height and body mass (see Equation 11.1) and therefore should not be treated as an independent variable. This means that it is incorrect to use BMI as a predictor variable in a linear regression model if body mass and height are also used as predictors. So it is important when analysing new data to know what the variables represent, especially if some (e.g., BMI, total shots ratio) are derived from other measured variables.

$$BMI = \frac{w}{h^2} \tag{11.1}$$

where w is the body mass in kg and h is the height in metres.

Often, data sets found in football exhibit large amounts of multicollinearity, with many variables highly correlated (e.g., $r > 0.7$) with each other. For example, as discussed in Chapter 10, pass completion is very strongly correlated with both shots attempted ($r > 0.8$) and aerial battles lost ($r \approx -0.8$). When such high correlations are observed, it generally means that the variables are not truly independent of each other, with several variables essentially conveying similar information. When lots of variables are strongly correlated, we say that considerable redundancy exists in the data because of the multicollinearity. Multicollinearity in the data can create major problems when performing linear regression because it causes the process to become unstable, with the result that multiple competing models may be produced [2]. This makes it difficult to draw any firm inferences (conclusions) from the data because the competing models may all say something different, despite all being mathematically correct. Furthermore, when backward exclusion (the most popular exclusion method) is performed during the regression analysis process, it can often result in one or more highly correlated variables being removed from the model. Unfortunately, inexperienced analysts don't always appreciate this, with the result that they wrongly assume that because a variable has been excluded, it must be unimportant, which is most definitely not always the case [2]. Given this, it is highly recommended that when presented with a new data set, you take a few minutes to perform a Pearson correlation analysis (as described in Example 10.2), as this will quickly establish the extent to which correlations are present in the data. Then decisions can be made about whether or not any of the variables should be considered redundant.

Another mistake that students and inexperienced analysts often make is to confuse categorical variables with continuous variables. At first sight, this might appear impossible – after all, continuous variables are made up purely of numbers (e.g. 2.3, 2, 3.4, 4, 5, 2.8, …), which can be either integers or decimal point numbers, whereas categorical variables comprise fixed classes such as 'home win', 'draw', and 'away win', or 'yes' and 'no'. So how can anyone get these two confused? Well, the confusion arises because analysts often give numerical values to the various classes (or categories) in categorical variables, which to the untrained eye can give them the appearance of a continuous variable. This can result in the ludicrous situation of the analyst trying to calculate the mean value of a categorical variable (see Example 11.1). So for example, consider a 'match location' variable that comprises two categories, 'home' and 'away', which we have numerically coded as 1 for 'home' and 2 for 'away'. If the data set contains 70 home matches and 30 away matches, then the calculated mean value for the 'match location' variable would be 1.3, which is, of course, completely meaningless. What does 1.3 mean in this context? And anyway, how can we have an average for home and away matches combined? The only thing that we can do with categorical data is count the numbers in each class and then produce a frequency distribution. The error arises purely because the categorical variable has been mistaken for a continuous variable.

In order to avoid confusing categorical and continuous variables, we can use the 'str' function in R, which displays the structure of the data. By displaying the structure of the data, it is possible to quickly determine whether variables are continuous or categorical, as illustrated in Example 11.1. If they are shown as continuous variables or character strings when they should be categorical variables, then this can be easily rectified using the 'as.factor' command in R.

EXAMPLE 11.1: TREATING CATEGORICAL VARIABLES CORRECTLY

In this short example, we will illustrate some of the pitfalls that can easily occur if we are not familiar with the data and confuse categorical variables with continuous variables.

Here we have a small data set of anthropomorphic measurements relating to ten footballers, six of whom are male and four who are female. In this data set, we include two variables to describe the gender of the players: 'Sex', which is a character string with two classes, "Male" and "Female", and 'SexID' which is a numerical vector containing 1 for males and 2 for females. Both have been included to illustrate the problems that can occur if these types of variables are not treated correctly. In this data set, we have also included BMI, which is derived from height and weight using the formula in Equation 11.1.

First we clear the workspace.

```
rm(list = ls())  # This clears all variables from the work space
```

Now we can create the data, which we do as follows:

```
Player <- c("Peter","Jane","John","Paul","Anne","Sarah","Lucy","Tom","Sean","David")
Sex <- c("Male","Female","Male","Male","Female","Female","Female","Male","Male","Male")
SexID <- c(1, 2, 1, 1, 2, 2, 2, 1, 1, 1) # Numerical classifier of gender
Age <- c(19, 21, 22, 24, 19, 21, 27, 25, 20, 18)
Height <- c(1.85, 1.64, 1.76, 1.83, 1.62, 1.57, 1.69, 1.80, 1.75, 1.81)
Weight <- c(80.4, 67.1, 75.4, 81.2, 65.2, 63.7, 66.3, 77.5, 73.4, 81.2)
BMI <- round((Weight/(Height)^2),2) # Computes BMI and rounds to 2 decimal places

dat <- cbind.data.frame(Player, Sex, SexID, Age, Height, Weight, BMI)
print(dat)
```

This produces:

```
##
   Player    Sex SexID Age Height Weight   BMI
1   Peter   Male     1  19   1.85   80.4 23.49
2    Jane Female     2  21   1.64   67.1 24.95
3    John   Male     1  22   1.76   75.4 24.34
4    Paul   Male     1  24   1.83   81.2 24.25
5    Anne Female     2  19   1.62   65.2 24.84
6   Sarah Female     2  21   1.57   63.7 25.84
7    Lucy Female     2  27   1.69   66.3 23.21
8     Tom   Male     1  25   1.80   77.5 23.92
9    Sean   Male     1  20   1.75   73.4 23.97
10  David   Male     1  18   1.81   81.2 24.79
```

Having created the data, we can now produce some descriptive statistics using the 'summary' function in R.

```
summary(dat)
```

Which produces:

```
##
   Player               Sex                 SexID          Age
 Length:10          Length:10          Min.   :1.0   Min.   :18.00
 Class :character   Class :character   1st Qu.:1.0   1st Qu.:19.25
 Mode  :character   Mode  :character   Median :1.0   Median :21.00
                                       Mean   :1.4   Mean   :21.60
                                       3rd Qu.:2.0   3rd Qu.:23.50
                                       Max.   :2.0   Max.   :27.00
     Height          Weight            BMI
 Min.   :1.570   Min.   :63.70   Min.   :23.21
 1st Qu.:1.653   1st Qu.:66.50   1st Qu.:23.93
 Median :1.755   Median :74.40   Median :24.30
 Mean   :1.732   Mean   :73.14   Mean   :24.36
 3rd Qu.:1.808   3rd Qu.:79.67   3rd Qu.:24.83
 Max.   :1.850   Max.   :81.20   Max.   :25.84
```

At first sight, this all looks OK. However, on closer inspection, we can see that something is wrong. R has assumed that 'SexID' is a continuous variable and has computed the mean as 1.4, which is of course completely meaningless because we cannot have an average of males and females. Also, when we look at the 'Sex' variable, we can see that the summary is pretty useless. All it tells us is that there are ten elements and that they are classified as characters.

So let's explore the structure of the data using the 'str' function.

```
str(dat)
```

This produces:

```
##
'data.frame':    10 obs. of  7 variables:
 $ Player: chr  "Peter" "Jane" "John" "Paul" ...
 $ Sex   : chr  "Male" "Female" "Male" "Male" ...
 $ SexID : num  1 2 1 1 2 2 2 1 1 1
 $ Age   : num  19 21 22 24 19 21 27 25 20 18
 $ Height: num  1.85 1.64 1.76 1.83 1.62 1.57 1.69 1.8 1.75 1.81
 $ Weight: num  80.4 67.1 75.4 81.2 65.2 63.7 66.3 77.5 73.4 81.2
 $ BMI   : num  23.5 24.9 24.3 24.2 24.8 ...
```

From this, we can see that R is treating 'SexID' as a numerical vector and is thus wrongly computing the mean and median values. By comparison, the variable 'Sex' is being treated as a character string, with the result that no frequency statistics are calculated. To overcome these problems, we need to convert both variables to factors, which we can do as follows:

```
# Create some factors
dat$Sex <- as.factor(Sex)
dat$SexID <- as.factor(SexID)
```

Now we can re-examine the data using the 'str' function.

```
str(dat)
```

```
##
'data.frame':    10 obs. of  7 variables:
 $ Player: chr  "Peter" "Jane" "John" "Paul" ...
 $ Sex   : Factor w/ 2 levels "Female","Male": 2 1 2 2 1 1 1 2 2 2
 $ SexID : Factor w/ 2 levels "1","2": 1 2 1 1 2 2 2 1 1 1
 $ Age   : num  19 21 22 24 19 21 27 25 20 18
 $ Height: num  1.85 1.64 1.76 1.83 1.62 1.57 1.69 1.8 1.75 1.81
 $ Weight: num  80.4 67.1 75.4 81.2 65.2 63.7 66.3 77.5 73.4 81.2
 $ BMI   : num  23.5 24.9 24.3 24.2 24.8 ...
```

We see from this that the variables 'Sex' and 'SexID' are now being treated as factors in R.

Finally, we can use the 'summary' function to produce the revised and now correct descriptive statistics, which contain frequency counts for both 'Sex' and 'SexID'.

```
summary(dat)
```

Which produces:

```
##
    Player              Sex       SexID       Age              Height
 Length:10          Female:4    1:6    Min.   :18.00   Min.   :1.570
 Class :character   Male  :6    2:4    1st Qu.:19.25   1st Qu.:1.653
 Mode  :character                      Median :21.00   Median :1.755
                                       Mean   :21.60   Mean   :1.732
                                       3rd Qu.:23.50   3rd Qu.:1.808
                                       Max.   :27.00   Max.   :1.850
     Weight            BMI
 Min.   :63.70   Min.   :23.21
 1st Qu.:66.50   1st Qu.:23.93
 Median :74.40   Median :24.30
 Mean   :73.14   Mean   :24.36
 3rd Qu.:79.67   3rd Qu.:24.83
 Max.   :81.20   Max.   :25.84
```

Notice here that BMI is a continuous variable, and while it is perfectly OK to include it when computing descriptive statistics, it is important to remember that BMI is a derived variable and therefore not independent.

When data is collected from various sources or by different individuals, mistakes can often be made, either during the measurement process or when inputting the data onto a spreadsheet. This can result in faulty analysis being undertaken if the mistakes are not spotted. One quick and easy way to help spot potential errors is to plot the data out, as illustrated in Figure 11.2, which shows the 100-metre sprint times recorded for 40 teenage athletes by four different assistant coaches. When we look at this plot, we instantly see that there is something wrong with the measurements taken by Assistant 2, whose average recorded time was just 9.60 seconds, way below that recorded by the other three assistants. Clearly, something went wrong during the data collection process for this individual, as evidenced by: (i) the consistently low values recorded, indicating a systemic problem; and (ii) the fact that the men's world record for the 100-metre race, held by Usain Bolt, is 9.58 seconds. As such, this indicates that the data collected by Assistant 2 is unreliable and therefore should be excluded from any analysis undertaken.

In addition to the errors associated with Assistant 2, it is noticeable from Figure 11.2 that one of the data points for Assistant 4 looks particularly suspect.

Apparently, if the data is to be believed, one teenager managed to complete a 100-metre sprint in just 6.32 seconds! Well, clearly, if Usain Bolt takes 9.58 seconds to run this distance, then something must be wrong with this reading. However, the fact that this one data point is so out of keeping with the other measurements collected by Assistant 4 strongly suggests that this mistake is probably a data input error. Perhaps someone else inputted Assistant 4's data onto a spreadsheet and inserted the wrong number – decimal point errors can often occur this way. Anyway, by plotting the data out, it is easy to spot inputting errors, as they generally stand out and therefore can be discounted or rectified.

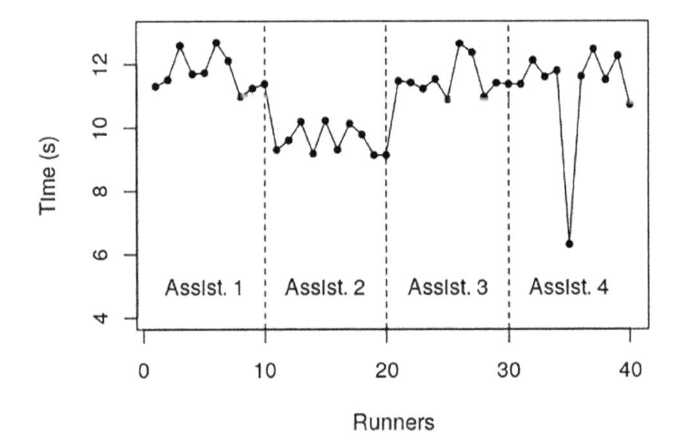

FIGURE 11.2
100-metre sprint times for 40 teenage athletes, measured by four different assistant coaches.

11.4 Knowing What Is Important (and What Is Not Important)

When trying to predict the outcome of matches or the end-of-season (EoS) league position, analysts often look for metrics that can reliably be used as predictors of team performance. However, useful and reliable metrics are often few and far between – otherwise, it would be easy to predict the results of football matches in advance. Furthermore, much of the information contained in soccer data sets is of little help when making predictions. This is because some variables are more important than others when it comes to making predictions. Therefore, it pays to know what is important and what is not important!

So how do we decide what is important and what is not? Well, there are various methods that can be utilised [2,5], but here we will concentrate on one well-known method, random forest analysis, which can be used to quantify the relative importance of the variables in a data set, as illustrated in Example 11.2. We discussed random forests in Chapter 6, so we won't go into too much detail about them here, but suffice to say they can be used to build regression models as well as classification models. In Example 11.2, we use a random forest regression model to identify the on-the-pitch performance metrics that best predict the EoS total points awarded. In so doing, we quantify the relative importance of the various predictor variables, which is something that can be extremely insightful and helpful to coaches and managers seeking to improve team performance.

In machine learning, the process of reducing the number of input variables when developing a predictive model is called 'feature selection' and is one of the cornerstones of that discipline. In Example 11.2, in order to illustrate the general approach, we shall use a random forest to perform feature selection. However, there are many other techniques (e.g., Example 10.4) that can be employed to determine the relative importance of predictive variables, although these are beyond the scope of this book.

EXAMPLE 11.2: DETERMINING VARIABLE IMPORTANCE USING A RANDOM FOREST MODEL

In this example, we revisit the team performance data set (EPL_regression_data_2020_2021.csv) that was used in Chapter 10 and which can be found in the GitHub repository at https://github.com/cbbeggs/SoccerAnalytics. This data set presents match performance data for the teams in the English Premier League (EPL) for seasons 2020–2021 and 2021–2022. The aim of this example is to demonstrate how a simple random forest regression model can be used to quantify the relative importance of the predictor variables in the data set. As in Example 10.5, the response variable is the number of points that the various teams achieve at the EoS, and the predictors are the other ten variables in the data set.

First, we clear any existing data from the RStudio workspace.

```
rm(list = ls()) # This clears all variables from the work space
```

Then we input the data in the form of a CSV file, which we have stored here in a directory called 'Datasets'.

```
# Load data (NB. Data acquired from
# https://fbref.com/en/comps/9/2021-2022/2021-2022-
# Premier-League-Stats)
mydata <- read.csv("C:/Datasets/EPL_regression_data_2020_2021.csv", sep=",")
rfdata <- mydata[,4:14]
```

This produces a working data set called 'rfdata', the first six rows of which can be displayed as follows:

```
head(rfdata)
```

```
##
  Points Shots SoT ShotDist PassComp Dribbles Tackles Crosses
1     69   580 185     17.3    16110      295     573     433
2     45   461 159     17.1    12521      275     686     452
3     46   441 139     15.6    11551      233     693     433
4     51   483 140     16.8    16342      334     722     485
5     35   406 117     16.4     9508      255     629     449
6     74   585 199     16.9    20878      360     660     496
  Intercepts AerialWon AerialLost
1        476       553        662
2        587       599        639
3        589       830        824
4        558       650        629
5        617       984        912
6        557       623        547
```

Now, we can build the base random forest regression model using the 'randomForest' library package, which we construct using all ten predictor variables.

```
# Install 'randomForest' package
install.packages("randomForest") # This installs the 'randomForest' package.
# NB. This command only needs to be executed once to install the package.
# Thereafter, the 'randomForest' library can be called using the 'library' command.

library(randomForest)
set.seed(123) # Set a seed so that the results are repeatable.
rf.mod1 = randomForest(Points ~., data = rfdata)
print(rf.mod1)
```

This displays the following, which tells us that the model explains 76.98% of the variance in the Points data.

```
##
Call:
 randomForest(formula=Points~., data=rfdata)
               Type of random forest: regression
                     Number of trees: 500
No. of variables tried at each split: 3

          Mean of squared residuals: 72.09508
                    % Var explained: 76.98
```

Having built the base random forest model, we can now investigate the relative importance of the various predictor variables using the following code:

```
importance(rf.mod1)
```

Which produces:

```
##
             IncNodePurity
Shots            1692.1697
SoT              3498.3960
ShotDist          436.5131
PassComp         2562.5970
Dribbles          292.7123
Tackles           632.4058
Crosses           448.3872
Intercepts        481.0269
AerialWon         434.8878
AerialLost       1393.7468
```

This list is rather difficult to utilise, so a more useful form is a scree plot, which can be produced using:

```
varImpPlot(rf.mod1) # Scree plot
```

This produces Figure 11.3, from which it can be seen that an elbow in the scree plot occurs at 'Tackles', with the variables 'SoT', 'PassComp', 'Shots', and 'AerialLost' identified as being much more important than the other variables.

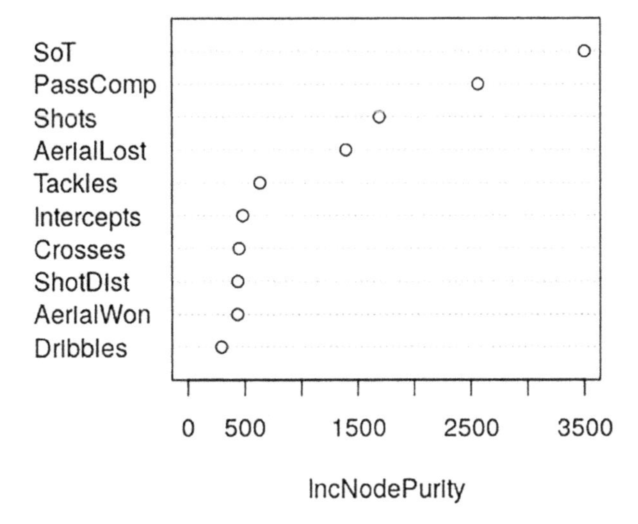

FIGURE 11.3
Variable relative importance scree plot for the base random forest model (rf.mod1).

Now, we can build a refined random forest model using just the variables 'SoT', 'PassComp', 'Shots', and 'AerialLost' as predictors.

```
# Refined model
set.seed(123) # Set a seed so that the results are repeatable.
rf.mod2 = randomForest(Points ~ SoT + PassComp + Shots + AerialLost, data = rfdata)
print(rf.mod2)
```

```
##
Call:
 randomForest(formula=Points~SoT+PassComp+ Shots+AerialLost,
data=rfdata)
               Type of random forest: regression
                     Number of trees: 500
No. of variables tried at each split: 1

          Mean of squared residuals: 71.23045
                    % Var explained: 77.26
```

From this, we can see that, far from impairing the predictive performance, removing the redundant variables from the model actually slightly improves it. This implies that the other excluded variables were not influential predictors of the EoS points total.

Now, we can make some predictions using the two random forest models. First, we make a prediction using the base model, 'rf.mod1', which we plot in Figure 11.4.

```
# Prediction using full model
mod1.pred <- predict(rf.mod1, data = rfdata)

# Create vectors containing the new observed and predicted results
obs.mod1 <- rfdata$Points
pred.mod1 <- mod1.pred
mod1.dat <- cbind.data.frame(obs.mod1,pred.mod1)

# Produce scatter plot of observed and predicted results for the 2 seasons
library(ggplot2) # Load package 'ggplot2'
ggplot(mod1.dat, aes(x=pred.mod1, y=obs.mod1)) +
  geom_point(shape=1) +   # Use hollow circles
  geom_smooth(method=lm, se = FALSE) +  # Add linear regression line
  xlab("Predicted points") +
  ylab("Observed points")
```

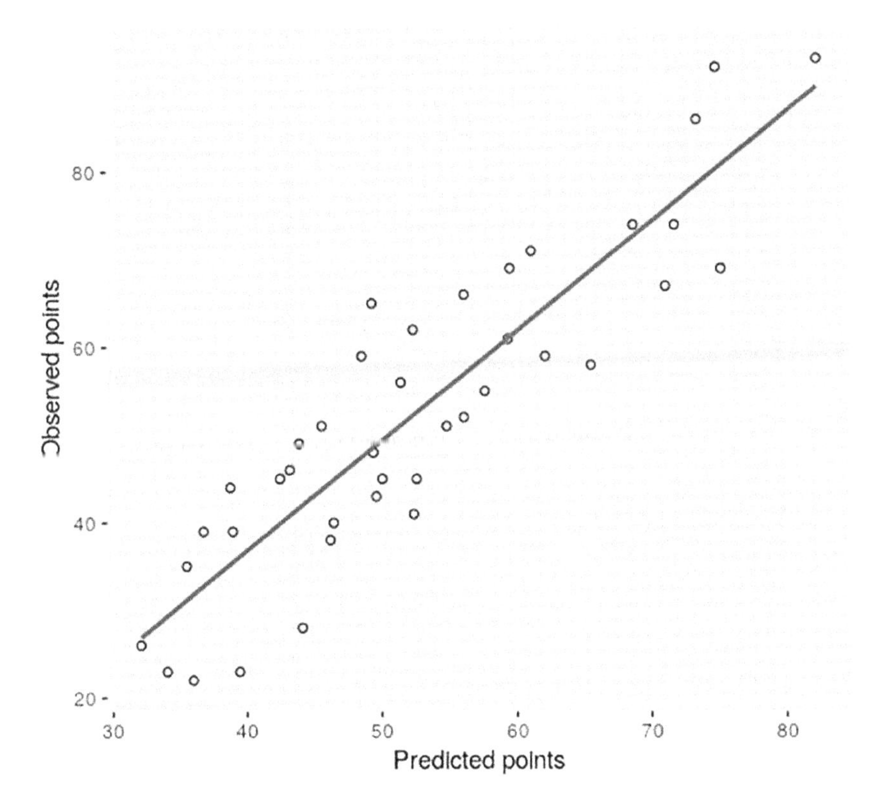

FIGURE 11.4
Scatter plot of observed and predicted points for the base random forest model, 'rf.mod1', with all ten predictor variables included. Also shown is the least-squares best-fit line.

The mean absolute error (MAE) for this model can be computed as follows:

```
# Compute mean absolute error
mod1.mae <- mean(abs(obs.mod1 - pred.mod1)) # Mean absolute error
print(mod1.mae)
```

```
##
[1] 7.132962
```

Although the MAE is only 7.13 points, it should be noted from the least-squares best-fit line in Figure 11.4 that on average the 'rf.mod1' model tends to overestimate the EoS points total of the bottom teams and underestimate the points earned by the top teams, whereas it tends to perform better with the mid-table teams.

Now, we predict the EoS points using the refined model, 'rf.mod2', and plot the results in Figure 11.5.

```
# Prediction using refined model
mod2.pred <- predict(rf.mod2, data = rfdata)

# Create vectors containing the new observed and predicted results
obs.mod2 <- rfdata$Points
pred.mod2 <- mod2.pred
mod2.dat <- cbind.data.frame(obs.mod2,pred.mod2)

# Produce scatter plot of observed and predicted results for the 2 seasons
library(ggplot2) # Load package 'ggplot2'
ggplot(mod2.dat, aes(x=pred.mod2, y=obs.mod2)) +
  geom_point(shape=1) +    # Use hollow circles
  geom_smooth(method=lm, se = FALSE) +  # Add linear regression line
  xlab("Predicted points") +
  ylab("Observed points")
```

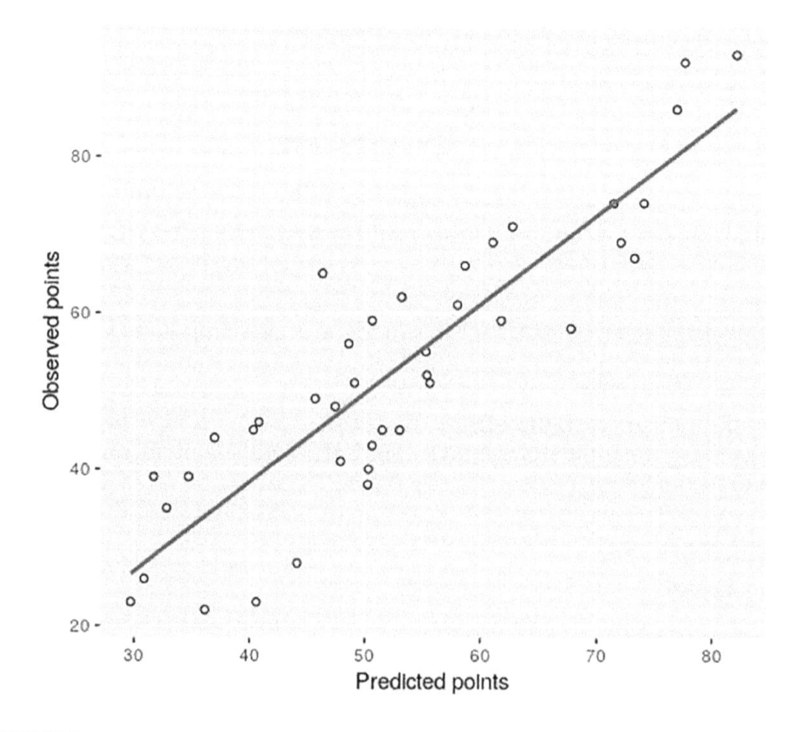

FIGURE 11.5
Scatter plot of observed and predicted points for the refined random forest model, 'rf.mod2', with 'SoT', 'PassComp', 'Shots', and 'AerialLost' as predictors. Also shown is the least-squares best-fit line.

The MAE for the refined model can be computed as follows:

```
# Compute mean absolute error
mod2.mae <- mean(abs(obs.mod2 - pred.mod2)) # Mean absolute error
print(mod2.mae)
```

```
##
[1] 7.115546
```

From this, we see that most of the measured variables were redundant and that only 'SoT', 'PassComp', 'Shots', and 'AerialLost' were needed to predict the EoS points total. Furthermore, we find that the refined model managed to predict the EoS points total pretty well, with an accuracy of ±7.1 points. As such, this example illustrates very well that when it comes to prediction, it is often the case that a minority of the measured variables are more important than all of the other variables. Of course, the challenge for the analyst is to identify which variables in any given data set are important and which can safely be ignored.

11.5 p-Values: Don't Let Them Become a Straitjacket

Anyone who has worked in science or academia will know that *p*-values are important. Scientists have little hope of getting their work published if they cannot demonstrate that their experimental results exhibit *p*-values of less than $p=0.05$ (see Key Concept Box 2.3 for more details on *p*-values). Scientific journal articles are peppered with *p*-values and statements about statistical significance. If $p<0.05$, then the results are significant, and we can believe what the scientists say. But if $p>0.05$, then it means that the scientific hypothesis being tested can't be trusted, and so we don't believe it. It's simple; we can trust *p*-values because scientists use them, so they must be important. The $p<0.05$ threshold is the gold standard, so it should be obeyed absolutely. Well, perhaps not! So this is what we are going to explore now, because many people get confused about *p*-values and often interpret them incorrectly [6]. Indeed, even though they use them all the time, sadly, many researchers don't fully understand *p*-values.

Data analysis in the real world is not the same as that undertaken by students, many of whom have an incomplete understanding of statistics and who generally think that *p*-values are beyond reproach. Consequently, many people (including some experienced scientists) make the mistake of relying too much on p-values, especially when it comes to decision-making. For example, in Chapter 1 (Section 1.7), we used a Chi-square test to compare the performance of two Manchester United managers: Alex Ferguson, who had an excellent win percentage of 59.7%, and Ralf Rangnick, who had a win percentage of just 37.9%. To our surprise, we found that the difference between the performances of the two managers was not statistically significant ($p=0.062$). So how can this be, given that Rangnick's performance record was so much worse than Ferguson's? Well, in layman's terms, what a p-value of 0.062 is telling us is that statistically, there is still a slim chance that the observed difference between Ferguson's and Rangnick's performance might have been due to natural variation that could have occurred just by chance. Therefore, we cannot rule out the possibility that the choice of manager is independent of team performance and that the observed difference might be purely coincidental. In statistical terms, we say we cannot reject the null hypothesis. However, the overwhelming likelihood is that Ferguson's superior performance compared with Rangnick's was due to a real effect that was not statistically detected due to the small sample size, resulting in what is called a Type II error (see discussion below). So the correct interpretation is that although we cannot reject the null hypothesis, there is a strong likelihood that a Type II error may have occurred. Unfortunately, many people don't realise this, and when they see $p>0.05$, they simply interpret this as meaning that there is no real difference between the two sets of observations, which in the case of Ferguson and Rangnick is completely erroneous. So we should always be wary when communicating p-values, especially to non-numerate audiences, because of the risk of misinterpretation.

Although students are often taught that p-values are important, they are frequently surprised to hear that a sizable number of professional scientists and statisticians challenge the supremacy of p-values [7–9], suggesting that they have become a bit of a monster – a straitjacket that their inventor, Ronald Fisher, never intended them to be [7]. Back in the 1920s, when he introduced the concept of p-values, they were intended simply to be an informal way of judging whether or not results and evidence looked promising, and worthy of a second look. But they soon got swept into a movement to make decision-making as objective as possible, and so we ended up with the $p<0.05$ threshold [7] – despite the fact that there are some inherent problems associated with the concept of the p-value.

Probably the most obvious problem associated with p-values is that they are as much a measure of sample size as anything else [10]. If we consider the formula for a t-test (Equation 11.2), we see that if the sample sizes N_1 and N_2 are small, then the t-statistic is also likely to be small, making it unlikely that the test will achieve a result that is statistically significant (i.e., $p<0.05$), unless, of course, the difference between the mean values of the two samples (i.e., the absolute effect size [10]) is very large. Conversely, if N_1 and N_2 are large, then it is highly likely that a significant result will be achieved, even if the absolute effect size is very small. Indeed, when dealing with big data sets containing tens of thousands of observations, such as those found in public health, it is often very hard not to achieve a statistically significant result, unless, of course, the two sample means are exactly the same, which is highly unlikely [10].

$$t = \frac{\bar{x}_1 - \bar{x}_2}{\sqrt{\frac{S_1^2}{N_1} + \frac{S_2^2}{N_2}}} \tag{11.2}$$

where $\bar{x}_1$ is the mean of the first sample, $\bar{x}_2$ is the mean of the second sample, S_1 is the standard deviation of the first sample, S_2 is the standard deviation of the second sample, and N_1 and N_2 are the numbers in each sample.

The impact of sample size on significance is illustrated in Example 11.3, where the heights of men in two fictional countries are compared to see if the observed difference is statistically significant or not. In Example 11.3, when the sample size is 30 for each country, we see that the *p*-value is less than 0.05 only 20.9% of the time, despite the fact that the men in country B were on average 2 cm higher than those in country A. This means that for this sample size, 79.1% of the time the *t*-test produced what statisticians call a Type II error (see Key Concept Box 11.1) – that is, the *t*-test results imply that there was no significant difference between the two populations, when in fact there is. This huge error rate exists simply because the sample size is too small and the test was statistically underpowered.

If we increase the sample size, as illustrated in Table 11.1, then we see that the percentage of *t*-tests that yield a significant result (i.e., $p < 0.05$) also increases. But even with a sample size of 100, only 52.1% of tests yield a significant result, with 47.9% resulting in a Type II error, which is not very good. Indeed, in order to achieve >80% of tests being significant (i.e., a statistical power of >0.8), we have to increase the sample size to 200 (i.e., 200 subjects from each country, making 400 subjects in total).

TABLE 11.1

Impact of Increasing Sample Size on Effect Size and Computed *p*-Value for a Fixed Difference in Population Means

	Sample Size = 10	**Sample Size = 30**	**Sample Size = 100**	**Sample Size = 200**
Difference in population means (cm)	2.00	2.00	2.00	2.00
Mean difference in sample means (cm)	2.03	2.07	1.993	2.04
Mean Cohen's *D*	0.44	0.34	0.29	0.29
Mean *p*-value	0.443	0.324	0.139	0.042
Significant results (%)	9.6	20.9	52.1	83.3
Type II errors (%)	90.4	79.1	47.9	16.7

Note: The sample size refers to the number of subjects in each arm of the study.

From Table 11.1 it can be clearly seen that if the sample size is too small, then the chances of obtaining a significant result are slim, which quite frankly is very unsatisfactory, given that we actually know the mean heights of the male population in both countries are 178 and 180 cm, respectively. Unless the sample size is drastically increased, most of the time Type II errors will occur, leading to the erroneous conclusion that there is no difference in the heights of men from the two countries. The implications of this are profound for any analysts working in sport because: (i) the observed effects in sport (e.g., differences in player performance, team performance, etc.) are often pretty small; and (ii) typical sample sizes in sport are generally small, which will inevitably mean that Type II errors are highly likely to occur. For example, say a coach at a soccer club wants to know if a new training regime will increase the sprint speed of his players. Well, the squad size may only be 25 players in total, and the likely reductions in sprint time, if any, are going to be in the region of fractions of a second. Therefore, any test is highly unlikely to yield a statistically significant result, even if the new training regime has actually improved player sprint times.

So, what should an analyst working in sports do if they find themselves in the predicament described above? Well, perhaps a better approach is to view p-values as Ronald Fisher always intended them to be, a useful way of judging whether or not results look promising and worthy of a second look [7]. If, for example, we perform a t-test using a sample size of just 15 and obtain a p-value of 0.08, then, provided that the sample is not biased, it is highly likely that the result will reach significance if the sample size is increased. Therefore, we should view $p=0.08$ as a promising result worthy of a second look because the test was statistically underpowered in the first place and the chances are that we might have a Type II error. Another thing that we can do is calculate a metric of standardised effect size, such as the widely used Cohen's d (see Key Concept Box 11.1), as illustrated in Example 11.3. Standardised effect size metrics such as Cohen's d can be extremely informative because they are actually the main finding of any quantitative study [10], while the p-value is only an indicator of how confident we are that the effect actually exists. Therefore, if Cohen's d is greater than 0.2, which is deemed a small but not trivial effect (see Key Concept Box 11.1), then it may be indicative that the effect is real, even if $p > 0.05$, as we observe in Table 11.1.

Although computing the effect size can be helpful, it is important to remember that Cohen's d tends to be overinflated when sample sizes are small, as illustrated in Table 11.1, where it is 0.44 when $n=10$. However, once samples are >50 subjects, the value of Cohen's d becomes fairly consistent and does not change much with increased sample size. This over-inflation of Cohen's d is of particular relevance in sport, where study cohorts are often relatively small. Consequently, it means that in sport, there may be a danger of overestimating the potential impact of interventions, such as a new training regime or the use of a new hydration drink, etc. So caution should always be exercised when interpreting the results of studies involving small numbers of participants, particularly if the observed effect is relatively small, as is often the case in sport where margins between athletes and teams are generally pretty fine.

EXAMPLE 11.3: P-VALUES AND SAMPLE SIZE

The impact of sample size on significance is illustrated in this example, where the heights of men in two fictional countries, A and B, are compared to see if the observed difference is statistically significant or not. Importantly, we actually specify the mean heights of the male population in both countries as being 178 and 180 cm, respectively. So we know that there is a real absolute effect of 2 cm before we perform our statistical test, which in this case is a 2-tailed independent *t*-test. For good measure, we also compute the Cohen's *d* effect size

In this example, we will perform the simulation using a sample size of 30 subjects, randomly selected from each country. However, any sample size could be used, as illustrated by the results in Table 11.1.

First, we clear any existing data from the RStudio workspace.

```
rm(list = ls())   # This clears all variables from the work space
```

Next, we specify the means and standard deviations for the populations of both countries. From which, we generate two artificial populations, each with a normal distribution and containing a million individuals, using the 'rnorm' function.

```
# Country A
meanA <- 178 # Height in cm.
sdA <- 7 # Standard deviation in cm

# Country B
meanB <- 180 # Height in cm.
sdB <- 7 # Standard deviation in cm

npop <- 1000000 # Population size

# Create two vectors for each population (each npop long).
A <- rnorm(npop, meanA, sdA)
B <- rnorm(npop, meanB, sdB)

# Create table of descriptive statistics
CountryA <- c(mean(A),sd(A))
CountryB <- c(mean(B),sd(B))

results <- cbind.data.frame(CountryA,CountryB)
row.names(results) <- c("Mean","SD")
print(round(results,1))
```

```
##
      CountryA CountryB
Mean       178      180
SD           7        7
```

Now we take a random sample of 30 individuals from each population (i.e., 30 from country A and 30 from country B) and perform a *t*-test, a process that we repeat a thousand times. Here we will use the 'CohensD' function in the package 'lsr' to compute the Cohen's *d* values.

```
# Specify sample size
nsamp <- 30

# Specify number of simulations
nsim <- 1000
pval <- matrix(0,nsim,4)

# Install 'lsr' package
install.packages("lsr")  # This installs the 'lsr' package.
# NB. This command only needs to be executed once to install the package.
# Thereafter, the 'lsr' library can be called using the 'library' command.

library(lsr) # This library is required to compute Cohen's d.
set.seed(123) # Sets seed for repeatability
for(i in 1:nsim){
  sampA <- sample(A,nsamp)
  sampB <- sample(B,nsamp)
  dh <- mean(sampB)-mean(sampA) # Difference in sample means (the effect)
  es <-cohensD(sampA, sampB) # Computes Cohen's d effect size
  t <- t.test(sampA,sampB, pair=FALSE) # Independent t-test
  pval[i,1] <- round(dh,2)
  pval[i,2] <- round(t$p.value,3)
  if(pval[i,2]<0.05){
    pval[i,3] <-1
  }
  pval[i,4] <- round(es,2)
}

# Display first six rows of pval
colnames(pval) <- c("DiffMeans","p-value","Significant","CohansD")
head(pval)
```

This produces the following result, of which we display the first six rows.

```
##
     DiffMeans p-value Significant CohansD
[1,]     -1.32   0.440           0    0.20
[2,]      3.29   0.088           0    0.45
[3,]     -0.27   0.859           0    0.05
[4,]      2.95   0.178           0    0.35
[5,]      2.56   0.155           0    0.37
[6,]      2.55   0.171           0    0.36
```

Having done this, we can compute the mean absolute effect, *p*-value, and Cohen's *d* results.

```
# Display results
round(mean(pval[,1]),3) # Mean difference between the sample means
```

```
##
[1] 2.074
```

```
round(mean(pval[,2]),3) # Mean computed p-value
```

```
##
[1] 0.324
```

```
round(mean(pval[,4]),3) # Mean effect size (Cohen's d)
```

```
##
[1] 0.338
```

```
# Compute percentage of significant results
sig.perc <- (sum(pval[,3])/nsim)*100
print(sig.perc)
```

```
##
[1] 20.9
```

Finally, we can display the simulation results in the form of a histogram, as shown in Figure 11.6.

```
# Display histogram
p_values <- pval[,2]
hist(p_values,100)
abline(v=0.05, lty=2, lwd=2)
```

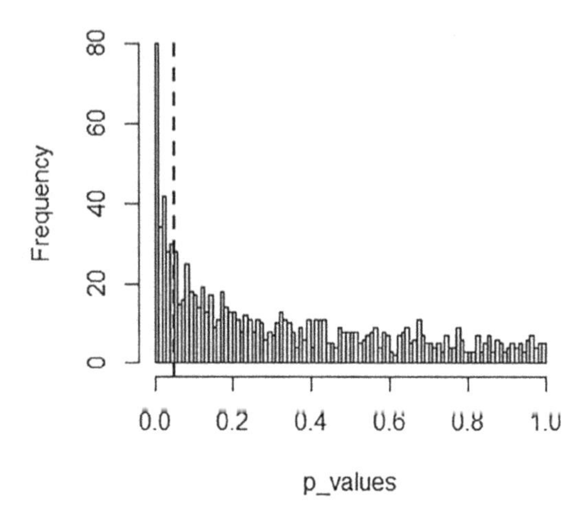

FIGURE 11.6
Histogram of p-value results from a 1000 simulations with a sample size of 30, and a difference between the population mean heights of 2 cm. Vertical dashed line indicates $p=0.05$.

From Figure 11.6, we can conclude that with a sample size of 30, we are only likely to achieve a significant result about 21% of the time, despite the fact that we know there is a real difference between the population mean heights of 2 cm. In other words, with this sample size, 79% of the time we end up with a Type II error.

KEY CONCEPT BOX 11.1: EFFECT SIZE AND STATISTICAL POWER

When performing hypothesis testing, many people become confused about the difference between effect size and p-values. They wrongly think that the two measure the same thing, with large effect sizes always equating to smaller p-values and vice versa. Yet this is incorrect because effect size and p-values are measuring two separate and completely different things. Effect size is simply a measure of the magnitude of the observed effect, which may be big or small, whereas the p-value is a measure of how confident we are that the observed effect (which is derived from sampling) actually reflects the true difference between the two populations being studied.

So, for example, say we want to know whether women in country A are taller than those in country B. To establish this, we could sample ten women from country A and ten from country B and measure their heights. If we found that the average heights of the women in the samples from countries A and B were 164 and 162 cm, respectively, we might draw the conclusion that the females in country A are taller than those in country B. But of course, this might not be true. Although we found a 2 cm difference (i.e., an absolute effect of 2 cm) between the two samples, it might be that our samples were not representative of the populations of women in each country and that there might actually be no difference in average height between A and B. After all, we only sampled ten subjects from each country, so we cannot be confident that the observed effect of 2 cm is actually representative of the two populations as a whole. However, say we then went on to sample a million women from each country and still got the same absolute effect size. Well, then we would be much more confident that the observed effect was real, and it is this confidence that p-values represent. The p-value is simply an indicator of confidence; being it is the probability that the observed difference between two groups is due to chance. So if the p-value is small, say $p < 0.05$, it means that the probability of the observed effect occurring by chance (due to natural variability) is less than 5%, which by convention means that the effect is more likely to be real and something in which we can have confidence. Indeed, when $p < 0.05$, we say that the statistical test has reached significance, and we reject the null hypothesis that the observed effect occurred just by chance.

From Equation 11.2, the formula for a t-test, we can see that p-values are strongly affected by sample size. If the sample sizes are small, then the t-statistic is also likely to be small, making it unlikely that the test will reach statistical significance (i.e., $p < 0.05$) unless, of course, the difference between the mean values of the two samples (i.e., the absolute effect size) is very large. Conversely, if sample sizes are very large, it is highly likely that a significant result will be achieved, even if the absolute effect size is very small. This means that it is often the case that the result of a statistical test will fail to reach a significant p-value, not because the effect isn't real but rather because the sample size is simply not large enough. In other words, the test fails to detect a statistical difference that really exists, which in effect gives us a false-negative result. In statistics, we call this type of error a Type II error, and in reality, this type of error is extremely common, especially in sport, where both sample sizes and observed effects are generally small.

The other type of error that occurs during hypothesis testing is a Type I error, which happens when we wrongly reject the null hypothesis (i.e., we wrongly say the observed effect is real when in fact it is just down to natural variance in the data). Traditionally, in statistics, avoiding Type I errors has been the major concern of analysts when performing hypothesis testing, despite this being at the expense of an increase in Type II errors. However, there is good reason for this, particularly in medical science, where minimising Type I errors is of paramount importance because doctors need to be absolutely confident that the drugs they prescribe will actually work.

As we can see from the above discussion, there is a trade-off between Type I and Type II errors. Basically, if an effect is real and there is actually a difference between the two populations being sampled (assuming a two-sample independent *t*-test), then as the sample size increases, the number of Type II errors will decrease (as observed in Table 11.1). However, changing the sample size has no effect on the probability of a Type I error. So increasing the sample size will not reduce the chance of a Type I error occurring. Statisticians therefore need to strike a balance between Type I and Type II errors, which they do by performing a statistical power calculation, which typically permits a 20% Type II error rate (i.e., equivalent to a statistical power of 0.8). This is used to calculate how many subjects should be recruited for any given study in order to ensure that the Type II error rate remains low.

In order to perform a statistical power calculation, it is necessary to have an indication of the likely effect size, which can often be obtained from a review of the published scientific literature. With respect to this, the standardised effect size metric, Cohen's *d*, is often used. Cohen's *d* is the ratio of the difference in the means and the pooled standard deviation of the two samples, as shown in Equation 11.3.

$$d = \frac{\bar{x}_1 - \bar{x}_2}{s_p} \tag{11.3}$$

where $\bar{x}_1$ is the mean of the first sample, $\bar{x}_2$ is the mean of the second sample, and S_p is the pooled standard deviation of the two samples.

For two independent samples, the pooled standard deviation can be calculated using Equation 11.4, where n_1 and n_2 are the numbers in the respective samples.

$$s_p = \sqrt{\frac{\sum_{i=1}^{n_1} (x_1 - \bar{x}_1)^2 + \sum_{i=1}^{n_2} (x_2 - \bar{x}_1)^2}{n_1 + n_2 - 2}} \tag{11.4}$$

Once Cohen's *d* is known, it is possible to compute the sample sizes required to ensure a statistical power of 0.8, which for a 2-tailed hypothesis test (assuming $p<0.05$) with the study cohort divided into two groups of equal size, produces the results shown in Table 11.2. From this, we see that the effect size has a huge impact on the sample size required to achieve a power of 0.8. When Cohen's *d* is 0.2, almost 200 subjects are required in each arm of the study, whereas when Cohen's *d* is 0.6, only 23 subjects are needed in each group, which is a dramatic decrease.

TABLE 11.2

Required Sample Size for Various Effect Sizes, Assuming a Statistical Power of 0.8, and a 2-Tailed Hypothesis Test When the Study Cohort Is Divided into Two Groups of Equal Size

Cohen's *d*	0.2	0.4	0.6	0.8	1.0	1.2	1.4	1.6	1.8	2.0
Total cohort size	393	99	45	26	17	12	9	7	6	5
Number in each group (arm)	197	50	23	13	9	6	5	4	3	3

For Cohen's *d*, effect sizes are often classified as small, medium, large, etc., as shown in Table 11.3. Given that in professional sport, teams and athletes are performing at the highest level, it means that the benefits and gains achieved by interventions such as changes in training regime or diet are likely only to be marginal. Consequently, observed effect sizes in this context tend to be in the range 'very small' to 'small', which often makes a statistical power of 0.8 very difficult to achieve in practice.

TABLE 11.3

Effect Size Classifications Generally Used with Cohen's *d*

Effect Size	**Cohen's *d***
Very small	$0.01 < d < 0.2$
Small	$0.2 < d < 0.5$
Medium	$0.5 < d < 0.8$
Large	$0.8 < d < 1.2$
Very large	$1.2 < d$

11.6 Why Do We Need *p*-Values Anyway?

There are many roles involving professional analysts where nobody cares much about *p*-values. For example, professional gamblers generally don't care whether or not their results are statistically significant. All they care about is making money. If their betting strategy consistently makes a profit, then they are doing things right – if not, then a new strategy is required. Similarly, in professional football, if a data analyst is helping to improve team performance, then he or she is doing a good job. However, if no improvement in performance can be demonstrated, then the club might be looking for a new analyst. So why do some data analysts rate *p*-values so highly while others don't? Well, basically, it comes down to whether the primary role of the analysis is to predict things or, alternatively, explain things. Analysts explaining observed phenomena find p-values helpful, whereas those making predictions tend not to use them as much.

Gamblers, bookmakers, and soccer club analysts are all primarily interested in predicting outcomes and making judgement calls, whereas sport scientists are generally more interested in proving things and explaining why certain phenomena occur. For example, professional gamblers develop betting strategies that they predict will yield a profit, and then they make decisions about how much money should be wagered on each bet. Likewise, when deciding whether or not to purchase a new striker, data analysts working for soccer clubs will make predictions about the number of goals the target player is likely to score. In both cases, the decisions taken are based on the predictions made, and it is relatively easy to say whether or not the predictions are accurate. Did the new betting strategy make a profit close to that predicted, and did the new player score the predicted number of goals when he or she joined the club?

So how can we assess the accuracy of our predictions? There are several techniques that can be utilised. If the prediction yields a continuous value (i.e., either an integer or a decimal point number), then we can use the mean absolute error (MAE) to quantify the accuracy of the predictions made, as illustrated in Example 11.2. So for instance, if for season 2021–2022, we made the points total predictions shown in Table 11.4 (column 2) for the various EPL teams, we could then compare these with the actual points awarded (column 3) and compute the difference (column 4). However, as we can see, some of the difference values are negative, and therefore computing the mean of column 4 would produce a misleading result. Therefore, we take the absolute values of the differences (column 5) and compute the mean of these instead, which is the MAE. From this, we can see that MAE is 5 points, which is not too bad.

TABLE 11.4

Predicted and Actual Points Totals for Teams in the English Premier League for Season 2021–2022, with Mean Absolute Error (MAE) Shown

Team	Predicted Points	Actual Points	Difference	Absolute Difference
Manchester City	96	93	3	3
Liverpool	94	92	2	2
Chelsea	80	74	6	6
Tottenham Hotspur	67	71	–4	4
Arsenal	71	69	2	2
Manchester United	73	58	15	15
West Ham United	54	56	–2	2
Leicester City	51	52	–1	1
Brighton & Hove Albion	45	51	–6	6
Wolverhampton Wanderers	47	51	–4	4
Newcastle United	42	49	–7	7
Crystal Palace	48	48	0	0
Brentford	33	46	–13	13
Aston Villa	45	45	0	0
Southampton	46	40	6	6
Everton	52	39	13	13
Leeds United	41	38	3	3
Burnley	45	35	10	10
Watford	25	23	2	2
Norwich City	21	22	–1	1
MAE	–	–	–	**5**

While the MAE is a useful metric for assessing predictions involving integers and continuous variables, an alternative approach needs to be taken when assessing predictions involving classification results. Examples of classification predictions might be:

- Predicting the outcome of a soccer match. There are only three options: a home win, an away win, or a draw. So the prediction for any given match must be one of these three classes.
- Predicting whether teams will be relegated or not. Here, there are only two classes: relegated and not relegated. The prediction must be one or the other.
- Predicting whether or not teenage academy players at a club will make it to the first team squad. Here again, there are only two classes: first team and not first team, and the prediction must be one or the other.

At the simplest level, we could just count up the number of classification predictions that we got correct and express this as a percentage. So for example, if for season 2021–2022 we had predicted that Brentford, Brighton, and Norwich would be relegated from the EPL, we would have correctly predicted the outcome of 16 of the 20 teams in the league, which is an impressive 80% accuracy. This sounds pretty good, but unfortunately, we failed to spot two of the three relegated teams, Burnley and Watford, which is not so impressive. So how do we get around this problem? Well, we could create four outcome classes, as follows:

- **True positive:** This is when we correctly predict the positive outcome, which in this case is when we correctly predict that a team will be relegated.
- **False positive:** This is when we wrongly predict the positive outcome, which in this case is when we incorrectly predict that a team will be relegated.
- **True negative:** This is when we correctly predict the negative outcome, which in this case is when we correctly predict that a team will not be relegated.
- **False negative:** This is when we wrongly predict the negative outcome, which in this case is when we incorrectly predict that a team will not be relegated.

So now our prediction can be classified as follows: one true positive (TP); two false positives (FP); fifteen true negatives (TN); and two false negatives (FN). Having done this, we can compute the *sensitivity* and *specificity* scores as follows:

$$\text{Sensitivity} = \frac{\text{TP}}{\text{TP} + \text{FN}} \tag{11.5}$$

$$\text{Specificity} = \frac{\text{TN}}{\text{TN} + \text{FP}} \tag{11.6}$$

The *sensitivity* (or true positive rate; TPR) and *specificity* (or true negative rate; TNR) scores mathematically describe the accuracy of a test, which reports the presence or absence of a given condition. In the case of our prediction as to who will be relegated, the sensitivity is just 0.333 (33.3%) and the specificity is 0.882 (88.2%). So while we were very good at saying who would not be relegated, we were not very good at predicting who would be relegated, which in this case is what really matters.

While in many contexts, minimising the number of false positives is important, this is not always the case. Indeed, there are certain circumstances where minimising the number of false negatives takes precedence because the consequences of false negatives can be catastrophic. For example, when screening for cancer, a false negative is a much more serious error to make compared with a false positive. If a doctor wrongly diagnoses that a patient has cancer when in fact they do not, the patient might get very worried, but they will not in fact die of cancer. However, if the patient is wrongly told that they do not have cancer when in fact they do, they might feel relieved, but the outcome is still going to be very bad for them, especially if they are not treated. For this reason, when screening for cancer, we want to keep the false-negative rate as low as possible.

Similarly, in sport, there are situations where false negatives may take precedence over false positives. For example, because youth academies are expensive to run, it is common practice to release young players whom the clubs think are not good enough to progress to the senior professional level [3]. In this case, they cull players who they predict are not going to make it to the senior squad. However, while this might save money in the short term, if the cull is too severe, then there is the risk that they might inadvertently release two or three academy teenagers who, given time, might develop into Premier League players. If this happens, then the exercise becomes a false economy because these players might have ended up in the first team squad or, alternatively, have been sold for considerable sums of money. In which case, the false-negative rate becomes an important factor that must be considered alongside the false positive rate.

When making predictions, it is important to remember that there is always the possibility that we might be getting things right just by luck and not because our model is any good. So how do we ensure that an observed outcome is not purely down to chance – after all, it might just be luck that our new betting strategy succeeded? Well, yes, that may happen once or twice, but if the betting strategy consistently yields the predicted profit year after year, or the new striker scores the predicted number of goals for the next six seasons, then it is highly unlikely that this is down purely to chance. So when making predictions, it is vital to continuously monitor their accuracy. That way, it is possible to make adjustments to prediction models where necessary to ensure that high-quality decision-making is maintained. It is also helpful as evidence when seeking to demonstrate the efficacy of the overall approach taken. Remember, within organisations, there may be senior managers who doubt the need to take a data science approach to decision-making, and so it is always useful to have hard evidence at hand to support your position.

11.7 Correlation and Causality

One of the holy grails of data science is trying to demonstrate causality. If we can prove that something causes an effect, be it good or bad, then we can decisively act to improve things. For example, once it was shown that cigarette smoking caused lung cancer, it was possible to introduce public health measures to reduce smoking and improve the situation. But until the causal link between the two could be conclusively demonstrated, there was always the doubt that some other factor might be causing the cancer rather than smoking itself. So establishing causality is an issue of great interest to most data analysts, and those working in sport are no exception, because if we can establish what causes teams to win and players to recover faster from injury, etc., then it will lead to a competitive advantage that can be exploited.

With respect to establishing causality, one of the most common mistakes that students and inexperienced analysts often make is to assume that correlation implies causation. However, this is not necessarily the case, because, for example, two correlated events A and B might both be caused by a third event C that has not been measured. Alternatively, there may be no connection between A and B, and it might be that the observed association is just a coincidence. Accordingly, this has led to the much-quoted mantra that *"correlation does not imply causation"*. While this is undoubtedly true, the idiom is held a little too firmly by some, with devotees refusing to accept any possible causal link unless absolute proof of causation can be demonstrated incontrovertibly. However, history tells us that this approach can be disastrous. For example, the association between tobacco use, cancer, and cardiovascular disease was regularly reported from 1898 onwards, with observational evidence published as early as the 1930s that smokers live shorter lives [11]. Yet this work was criticised for not showing causality – something that is extremely difficult to demonstrate. Consequently, it was not until 1964, when the Surgeon General's Report on Smoking and Health was published in the USA [12], that bans on certain cigarette advertising were introduced, along with the requirement for warning labels on tobacco products. However, during the period between correlation being first observed and causation being generally accepted, millions of smokers died needlessly. So although correlation does not always imply causation, sometimes it does, and therefore, when strong correlations are observed, they are always worth exploring because they might yield insights into the causal mechanisms shaping the data.

One of the reasons why causal links are sometimes overlooked is because the person analysing the data is from a mathematical or statistical background and does not have the subject-specific knowledge required to understand what the data means. Consequently, they may dismiss important correlations using the mantra that *"correlation does not imply causation"*, when someone with subject knowledge (e.g., a doctor, a biologist, an engineer) would instantly spot a causal link. Consider, for example, the observation that windmills rotate faster when wind speeds are higher. To a statistician from another planet with no understanding of how wind behaves on Earth, the correlation between the two phenomena could be interpreted as: wind is caused by windmills turning, or vice versa. Indeed, to them, it might even be that a third unknown phenomenon is causing both the wind and the windmills to rotate independently. However, an engineer from Earth who designs wind turbines would instantly know that it is the wind that is causing the windmill blades to rotate. As such, this highlights the importance of subject knowledge when analysing data because specific *a priori* knowledge can help identify causal links that otherwise might be overlooked. For example, in soccer, goals that are scored arise from shots on target, which in turn arise from shots that are attempted (as illustrated in Figure 11.7); clearly, therefore, all three variables are strongly correlated with each other. However, without subject knowledge of football, a naïve analyst might ignore the causal link between shots on target and goals scored and wrongly treat these two variables as being completely independent of each other.

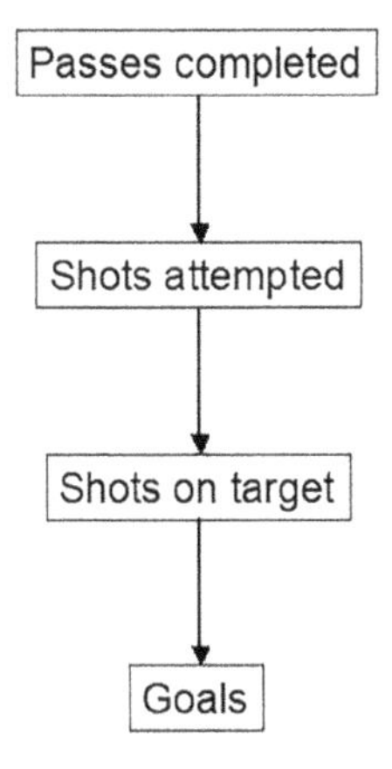

FIGURE 11.7
Graphical representation of the relationships between several variables in a typical soccer data set. The arrows represent causal relationships (e.g., more shots on target cause more goals to be scored).

Football data often exhibits considerable multicollinearity, with many variables highly correlated. This can create problems when trying to establish cause and effect. Therefore, in order to better understand the relationships that exist in such data, it is often helpful to produce diagrams such as that shown in Figure 11.7. From this, it can be seen that two clear causal relationships exist: (i) shots on target cause goals to be scored; and (ii) shots attempted cause some shots on target to occur. Therefore, by inference, shots attempted cause goals to be scored, which is obvious when we think about it. But what about less obvious relationships such as pass completion and goals scored? Well, of course, some goals are scored from corners, free kicks, interceptions, etc., but most arise in open play due to moves that involve passes between multiple players. Therefore, we would expect a strong positive correlation between pass completion and shots attempted, which is exactly what we observed in Example 10.2, where for season 2021–2022 in the EPL, the correlation between these two variables was $r=0.835$. When we think of the system in this way, the causal link becomes obvious. Clearly, just as windmills do not cause the wind to blow, the number of shots attempted or goals scored does not drive the pass completion rate; rather, it is the other way around. Without possession, it is not possible to create many shooting opportunities, and thus, it is not possible to score many goals. However, despite this causal link, it should be remembered that on occasion, the team that completes the most passes successfully does not necessarily score the most goals because sometimes counter-attacking teams can win matches against the run of play.

By sketching out the relationships between the variables in the data (as illustrated in Figure 11.7), it is possible to determine the topological ordering of events and identify possible causal relationships. In data science, the formal name for this type of diagram is the *Directed Acyclic Graph* (DAG) [13]. A DAG is a graph that provides a visual representation of causal relationships between a set of variables. These causal relationships can either be known to be true or, perhaps more commonly, may only be assumed to be true. Each arrow in a DAG represents a known or assumed causal relationship (e.g., smoking cigarettes has a causal effect on the risk of lung cancer). By representing the variables in a DAG, it is possible to distinguish those that have a causal relationship from those that act as confounders (i.e., things that confuse by acting on multiple variables), which helps to bring clarity when constructing statistical models.

While a full discussion of the construction and application of DAGs is well beyond the scope of this book, it is important to note that DAGs can be used to help identify causation [14,15]. Therefore, simply by sketching out the relationships between variables and the order in which events occur (e.g., shots occur before goals), a kind of approximate DAG can be produced, which can then be used to gain insights into complex data sets and identify possible causal mechanisms that might be at work.

11.8 Concluding Remarks

So we have come to the end of this chapter and the end of the book. Hopefully you have enjoyed reading it, and that now you have a much better understanding of R and how to use it compared with when you first picked up this book. Hopefully also, you now know how to analyse football data. But before you leave, it is perhaps worth re-emphasising that becoming a successful data analyst involves much more than just learning lots of sophisticated statistical and machine learning techniques; rather, it involves developing softer skills like getting to know the data and understanding what it means; having a clear sense of what you are trying to achieve; knowing what is important and what is not important; and knowing how to interpret and communicate results in an appropriate manner so that people can understand what you are saying. Sadly, many students and would-be analysts are not aware of this. They may be great at coding and using mathematical techniques, but their soft skills don't necessarily match their technical abilities. Consequently, it is all too easy for analysts to be overlooked and their opinions ignored, particularly if they are not good communicators and their boss is not numerate.

Remember, data analysts are often employed because their bosses are not numerically literate. What the boss needs is to receive information in a way that they can easily understand – which is where good communication skills come to the fore. So, if you recognise that you have a tendency to hide away and bury your head in coding for hours on end (something the author enjoys, as do many others), perhaps it is worth spending some time working on the softer skills outlined in this chapter, as they may well be of benefit to you in your career.

One problem in the world of work is that, as they grow older, many people stop learning and become 'fixed' in their outlook and opinions. Worse still, they can develop a silo mentality and become suspicious of 'newfangled' ideas and techniques imported from other disciplines. They know, what they know and they are happy with that! From a data science point of view, this means that analytical techniques that are widely used in some disciplines may be almost unheard of in other disciplines. In sport, for example, many machine learning and data science techniques are rarely encountered, despite the fact that they are highly relevant and widely used in other industries such as the financial sector. So, if you want to avoid congealing as your career develops, it is important to keep learning and to expose yourself to new ideas and techniques. With regard to this, it is often helpful to read relevant scientific articles, as these will be thought-provoking, and will keep you abreast of advances in your field. By taking a thoughtful and exploratory approach to data analysis rather than running on tramlines and using standard techniques just because others use them, you will develop and become better at what you do. Of course, this will probably involve hard work, but the effort will be worth it because, in the end, you will gain insights that others do not possess and your horizons will be raised.

So now it's over to you. Good luck with your data analysis, and thanks for reading this book!

References

1. Dalton-Barron N, Whitehead S, Roe G, Cummins C, Beggs C, Jones B: Time to embrace the complexity when analysing GPS data? A systematic review of contextual factors on match running in rugby league. *Journal of Sports Sciences* 2020, 38(10):1161–1180.
2. Weaving D, Jones B, Ireton M, Whitehead S, Till K, Beggs CB: Overcoming the problem of multicollinearity in sports performance data: A novel application of partial least squares correlation analysis. *PLoS One* 2019, 14(2):e0211776.
3. Till K, Jones BL, Cobley S, Morley D, O'Hara J, Chapman C, Cooke C, Beggs CB: Identifying talent in youth sport: A novel methodology using higher-dimensional analysis. *PLoS One 2016,* 11(5):e0155047. .
4. Yadav S, Shukla S: Analysis of k-fold cross-validation over hold-out validation on colossal datasets for quality classification. In: *2016 IEEE 6th International Conference on Advanced Computing (IACC): 2016.* IEEE, Bhimavaram ; 2016: 78–83.
5. Beggs CB, Shepherd SJ, Cecconi P, Lagana MM: Predicting the aqueductal cerebrospinal fluid pulse: A statistical approach. *Applied Sciences* 2019, 9(10):2131.
6. Greenland S, Senn SJ, Rothman KJ, Carlin JB, Poole C, Goodman SN, Altman DG: Statistical tests, P values, confidence intervals, and power: A guide to misinterpretations. *European Journal of Epidemiology* 2016, 31(4):337–350.
7. Nuzzo R: Statistical errors. *Nature* 2014, 506(7487):150.
8. Wasserstein RL, Lazar NA: The ASA Statement on P-Values: Context, Process, and Purpose, *The American Statistician* vol. 70: Taylor & Francis; 2016: 129–133. DOI: 10.1080/00031305.2016.1154108
9. Lang JM, Rothman KJ, Cann CI: That confounded P-value. *Epidemiology* 1998, 9:7–8.
10. Sullivan GM, Feinn R: Using effect size: Or why the P value is not enough. *Journal of Graduate Medical Education* 2012, 4(3):279–282.
11. Doll R: Uncovering the effects of smoking: Historical perspective. *Statistical Methods in Medical Research* 1998, 7(2):87–117.
12. Surgeon General of the United States. Smoking and health: Report of the Advisory Committee to the Surgeon General of the United States. Public Health Service Publication No. 1103. US Department of Health, Education, and Welfare; 1964.
13. Digitale JC, Martin JN, Glymour MM: Tutorial on directed acyclic graphs. *Journal of Clinical Epidemiology* 2022, 142:264–267.
14. Arnold KF, Berrie L, Tennant PWG, Gilthorpe MS: A causal inference perspective on the analysis of compositional data. *International Journal of Epidemiology* 2020, 49(4):1307–1313.
15. Williams TC, Bach CC, Matthiesen NB, Henriksen TB, Gagliardi L: Directed acyclic graphs: A tool for causal studies in paediatrics. *Pediatric Research* 2018, 84(4):487–493.

Index

Note: **Bold** page numbers refer to tables; *italic* page numbers refer to figures.

For Product Safety Concerns and Information please contact our EU representative GPSR@taylorandfrancis.com
Taylor & Francis Verlag GmbH, Kaufingerstraße 24, 80331 München, Germany

www.ingramcontent.com/pod-product-compliance
Ingram Content Group UK Ltd.
Pitfield, Milton Keynes, MK11 3LW, UK
UKHW021444080625
459435UK00011B/367

* 9 7 8 1 0 3 2 3 5 7 5 8 4 *